Energy efficiency refurbishments

Edition **DETAIL** Green Books

Energy efficiency refurbishments

Principles
Details
Examples

Clemens Richarz
Christina Schulz

Imprint

Authors:
Clemens Richarz, Prof. Dipl.-Ing. Architect
Christina Schulz, Dipl.-Ing. Architect

Co-authors:
Prof. Dr. Volker Quaschning, Berlin
Prof. Werner Schenk, Rosenheim
Prof. Dr. Joachim Stoll, Munich
Dipl.-Ing. (FH) Medin Verem, Gröbenzell
Prof. Dipl.-Ing. Friedemann Zeitler, Penzberg

Project management:
Jakob Schoof, Dipl.-Ing.

Editorial work:
Christina Schulz, Dipl.-Ing. Architect
Jakob Schoof, Dipl.-Ing.
Jana Rackwitz, Dipl.-Ing.

Illustrations:
Ralph Donhauser, Dipl.-Ing. (FH)

Cover design:
Cornelia Hellstern, Dipl.-Ing. (FH)

Translation:
Sharon Heidenreich

Proofreading:
Roderick O'Donovan

DTP & layout:
Roswitha Siegler

Reproduction:
Martin Härtl OHG, Munich

Print:
Kösel GmbH & Co. KG, Altusried-Krugzell
1st edition 2013

Institut für internationale
Architektur-Dokumentation GmbH & Co. KG
Hackerbrücke 6, D-80335 München
Telephone: +49/89/38 16 20-0
Telefax: +49/89/39 86 70
www.detail.de

A specialist book from Redaktion DETAIL

ISBN: 978-3-920034-90-4

MIX
Paper from responsible sources
FSC® C013736
www.fsc.org

The FSC-certified paper used for this book is manufactured from fibres proved to originate from environmentally and socially compatible sources

Contents

Approximately 80 % of the world's resources is being consumed by 20 % of the world's population. When considering the growth forecasts for the newly industrialised countries, it is apparent why the reduction of energy consumption has become the focus of our attention. The problem concerning the finite nature of energy reserves with the consequent uncontrollable upward spiral of prices and the fight to get a share of the dwindling resources represents only one side of the coin. The other side is the risk to our environment caused by the build-up of carbon dioxide (CO_2) in the Earth's atmosphere which is released through the combustion of fossil fuels. In highly industrialised countries, the building sector, in particular, accounts for about 50 % of the total CO_2 emissions through the construction and operation of buildings. In Germany, for example, 20 % of the emissions stem from the production and transportation of building materials and 30 % from conditioning the buildings (heating, ventilation, cooling and lighting).
A conservative approach to the use of resources, one which is based on long life cycles and therefore positive in terms of sustainability, will increasingly influence all aspects of construction. In this context, making existing building stock energy efficient through refurbishment is of utmost importance. The long-term viability of these buildings requires that upgrades incorporate energy efficiency and energy saving features.

Reduction of CO_2 emissions through refurbishment

The preface to the German Thermal Insulation Ordinance 1995 (Wärmeschutzverordnung), even at the time of its introduction, stated the following: "The real CO_2 reductions must be achieved in the existing building stock. The cross-ministry working group for CO_2 reduction assumes that a potential saving of about 100 million tonnes is available in existing buildings, which requires an investment of approximately DM 350 – 400 billion (€180–200 billion)."
It is gradually becoming clear that reduction in the volume of CO_2 emissions caused by the conditioning of buildings can only be achieved by performing the relevant upgrades to the vast majority of existing buildings.

Sustainable development of building stock

The importance of refurbishment that is sustainable as distinct from solely energy efficient is best illustrated by taking a look at residential buildings. Owing to their poor layouts, many apartments can only be let at huge discounts, regardless of whether their heating energy consumption already meets more recent standards or not. So long as serious deficiencies, such as lack of contact with the outside world due to the inadequate size or even absence of balconies and windows, outdated sanitary facilities and, generally, insufficient floor space are not eliminated, these apartments will remain unattractive and difficult to let. Experience has shown that this in turn leads to undesirable segregation processes in the demographic structure of an apartment block, the housing complex or even the whole district with a related potential for social conflict. Energy efficiency upgrades to such apartments, ignoring the need to improve the overall living standards, do not fulfil the aims of a sustainable, long-term and holistic approach.
Besides improving the energy standards, sustainable refurbishment should involve changing the layouts to such an extent that they meet the needs of current and future tenants. Such work includes, for example, bathroom modifications to accommodate the needs of the elderly or alterations to the interior layouts with the aim of achieving greater flexibility of use

(increase or decrease of room sizes, various user groups, etc.). Naturally, the same problems apply to non-residential buildings. In general, the conversion or further development of existing buildings should not be restricted to energy aspects, but should also address functional and architectural issues. The involvement of architects is therefore indispensable when it comes to sustainable refurbishments.

Content of this book

Since inadequate energy efficiency standards are frequently the trigger for more extensive refurbishment work, it is essential that architects possess basic knowledge in energy-related matters. Without know-how in this field, it will become increasingly difficult to secure appropriate commissions. Hence a danger exists that architects might be excluded from the entire refurbishment process.

If architects however display the necessary skills, their competencies will be in demand from the inception of projects. During the course of the planning work, they will then be able illustrate the potential of a building as a whole and help ensure that energy performance measures are incorporated into a viable, sustainable concept.

This book is designed to encourage the inclusion of energy-related aspects in the planning process for refurbishment work. In order to establish a direct reference to statutory provisions, all energy efficiency issues are considered according to the objectives of the German Energy Saving Ordinance, the Energieeinsparverordnung 2009 (EnEV). The energy efficiency upgrade of a building is therefore investigated for both winter and summer conditions. The refurbishments presented are not limited to residential buildings and include various types of buildings each with their own particular characteristics. Furthermore, the plant technology is considered as an integral component of the overall energy efficiency refurbishment concept. The approach to finding solutions is in accordance with the methods used in the EnEV and the associated standards, since these, irrespective of their legal relevance, allow for a comprehensive understanding of all the relevant factors affecting the energy balance.

The nature and extent of refurbishment measures to existing buildings are frequently influenced by external circumstances and regulations which impact on the project management. These include, for example, fire protection issues (suitability of components upon renewal), building legislation (altered distance to neighbouring buildings as a result of extensions and insulation measures), tax law (repayment options for maintenance and construction costs, preservation orders for monuments), building costs (subsidy programmes, rent increases) and, above all, tenancy law (tenants' obligation to accept changes, apportionment of expenses). All issues mentioned above must be solved in detail and individually for the property concerned during the course of the planning work and in collaboration with the appropriate experts. The outcome will highlight whether and to what extent the issues concerning project management will affect the refurbishment process itself. As is the case for all building projects, it is these boundary conditions which define the feasibility of a scheme and they should therefore be clarified in advance.

We hope that this book will help readers to gain an insight into this topic and stimulate them to consider sustainability-related issues when refurbishing existing building stock.

Energy demand

The increasing demand for energy – not least because of the heating, cooling and ventilating of buildings – and the continuous growth of world population are leading to an unprecedented demand for the primary resources oil, gas and coal. At the same time, geologists expect oil production to reach a high over the next years and then to start declining (fig. 1.1 and 1.6, p. 11).
It can therefore be assumed that it will no longer be possible to meet the demand for the resources oil and gas over the course of the next fifty years. This will have economic, and naturally also political, consequences. One of the consequences of resource depletion will be a sharp increase of carbon dioxide (CO_2) emissions. This in turn is leading to a progressive change of the Earth's atmosphere with rising global warming and severe harm to the climate. Industrialisation and world population growth have already led to the CO_2 content of the air being higher today than the level in any of the preindustrial periods (fig. 1.2).
In the meantime, all predictions concerning CO_2 emissions expect the primary energy demand and the related level of CO_2 emissions to almost triple from the year 2000 to 2050, if the current development persists.
Since the various nations show considerable differences in their energy demand and CO_2 emissions per capita, one of the main questions of concern is how the resources that appear to be growing increasingly scarce should be distributed (fig. 1.5, p. 11). The question concerning energy consumption is therefore also a political issue, which can, at the end of the day, only be answered by politicians in a global context. Attempts to find answers – unfortunately so far to no avail – have been made at the climate conferences held by the United Nations.

Global aims

There are different predictions concerning the development of resource consumption up to the year 2050. These different approaches can generally be presented as three scenarios with different objectives. In each case, the annual carbon dioxide emissions are used as a point of reference (fig. 1.3):

- Scenario 1 is based on the assumption that there will be continuous growth mainly due to the efforts being made by the emerging countries to catch up, which is leading to an incredible increase in resource consumption. Even though resource consumption in the industrialised countries will almost remain stable, the other countries will raise their energy consumption, reaching the same level as that of the industrialised countries owing to better living standards. The CO_2 emissions in scenario 1 are at approximately 90 billion tonnes in 2050.
- Scenario 2 is based on the assumption that the industrialised nations will halve their CO_2 emissions by 2050 in accordance with agreements made and the other countries will only reduce or increase their level until it meets that of the industrialised countries. This scenario is based on a limited world population growth – regardless of how this might be achieved – stopping at around seven billion people worldwide. The CO_2 emissions in scenario 2 are at approximately 35 billion tonnes in 2050, so at approximately today's level.
- In principle, scenario 3 corresponds with the aims of scenario 2, however, the overall reduction targets are higher. The industrialised countries will lower their CO_2 emissions by 80% and the other countries will only increase or reduce their levels until reaching the lower level of the industrialised countries. This target is also linked to a controlled world population growth. A world population of seven billion people is also regarded as a possible limit in this case. Under these circumstances, the CO_2 emissions would be approximately 18 billion tonnes in 2050.

A smooth and structured transformation into an era which allows human life without the use of finite resources to generate energy is only realistic if the limit described in scenario 3 is met (worldwide reduction to 18 billion tonnes of CO_2 per year by 2050).

Situation in Germany
In terms of decreasing CO_2 emissions, the attempts of governmental institutions in Germany are related to the targets referred to in scenario 3. They are to be met exclusively by improving energy efficiency, increasing the use of renewable energies and extending the running times of nuclear power plants. Questions concerning the living standards and the life style of the society, i.e. questions concerning the reasons for the high demand of energy, are not considered. It therefore appears extremely unlikely that these targets will actually be met.
Since Germany is responsible for only 2% of the global CO_2 emissions, the country's attempts will have only marginal influence on the global carbon footprint. The purpose of the measures undertaken in Germany and other smaller industrialised countries is aimed far more at developing holistic models for life styles that are able to manage with a maximum of two tonnes of CO_2 emissions per capita and year.
A significant aspect of this strategy is the development of know-how concerning the improvements to buildings that are beneficial in terms of energy consumption.

Renewable energy supply
The overall aim of all efforts must be to secure the generation of power without the use of finite resources. The problem concerning security of supply, which generally occurs in the case of renewable power production due to the time lag between supply and demand, must be given top priority. A combination of several measures, which are currently being developed and improved, appears to be most promising. This solution includes:

- intelligent power grids
- a network of various power generators (sun during the day, wind at night)
- a network of various storage facilities (e.g. compressed air storage, storage of heat generated by solar radiation during the day and use of the heat at night to produce steam, hydrogen production through electrolysis and the production of methane involving a reaction with CO_2)

A global emission deal including a fine for those who exceed a certain CO_2 limit, which would then be used to finance specific renewable energies, would offer an additional incentive to accelerate the transformation to renewable energies on a global scale. A levy of approximately €20 for each tonne of CO_2 exceeding the determined limit is currently under discussion.
If a target of two tonnes per capita were agreed on up until 2050, Germany, from 2010 onwards, would have to invest approximately €360 billion levied by fines in renewable energies during this period. In relation to the target year 2050, this would account for an average of €9 billion per year. This amount would increase the Federal Budget 2011 (€307 billion) by approximately 3 %.
A study performed by the authors Mark Jacobson (Stanford University) and Mark DeLucchi (University of California) dem-

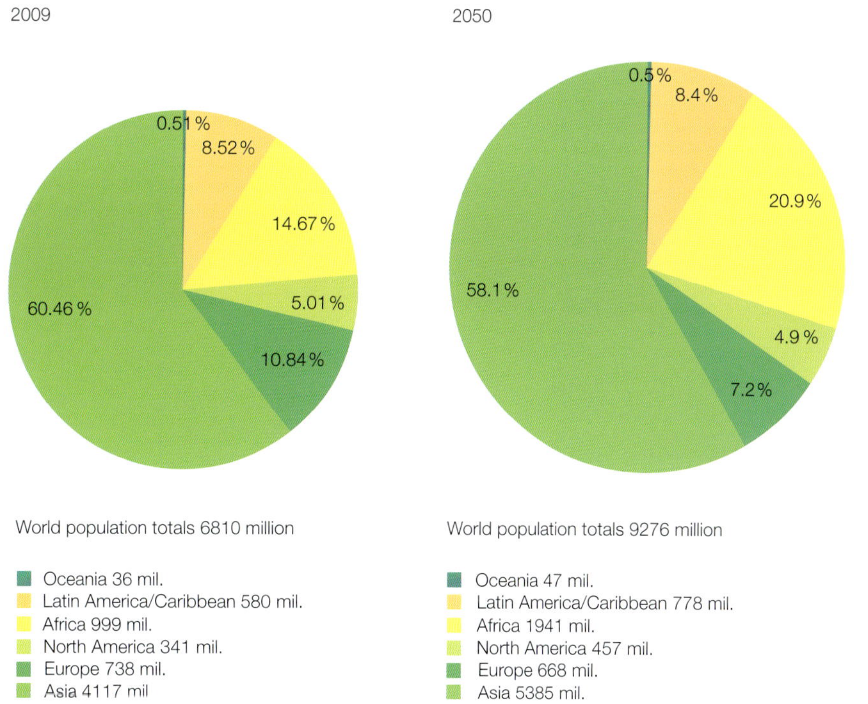

1.1

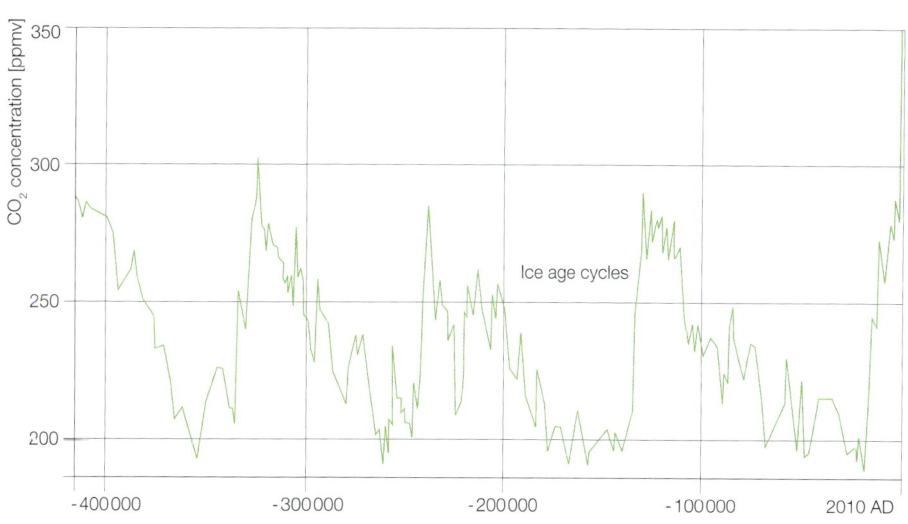

1.2

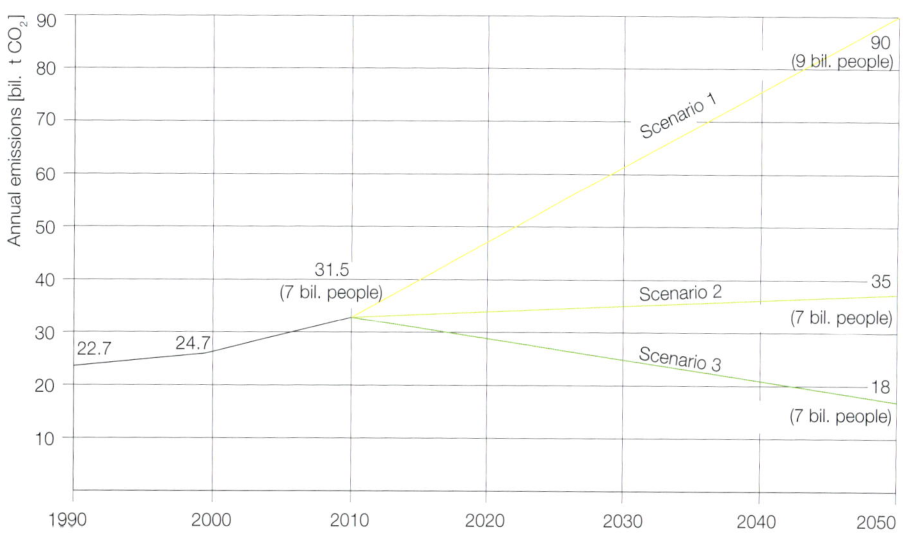

1.3

1.1 Predictions of the development of the world population up until 2050
1.2 Historical development of CO_2 concentration in the air. The recurring fluctuations caused by natural processes are frequently applied to relativise the problems of climate change. There is clear evidence of the rise caused by mankind over the last 200 years.
Forecasts predict that the CO_2 content in the air will increase to 600 ppm if further development remains unchanged. The significance of this assessment becomes clear when one considers that the so-called Pettenkofer limit for interior space is limited to 1000 ppm. Rooms should be ventilated once the concentration rises above this value, as otherwise comfortable conditions can no longer be maintained.
1.3 Development of global CO_2 emissions up until 2050 in accordance with the different scenarios

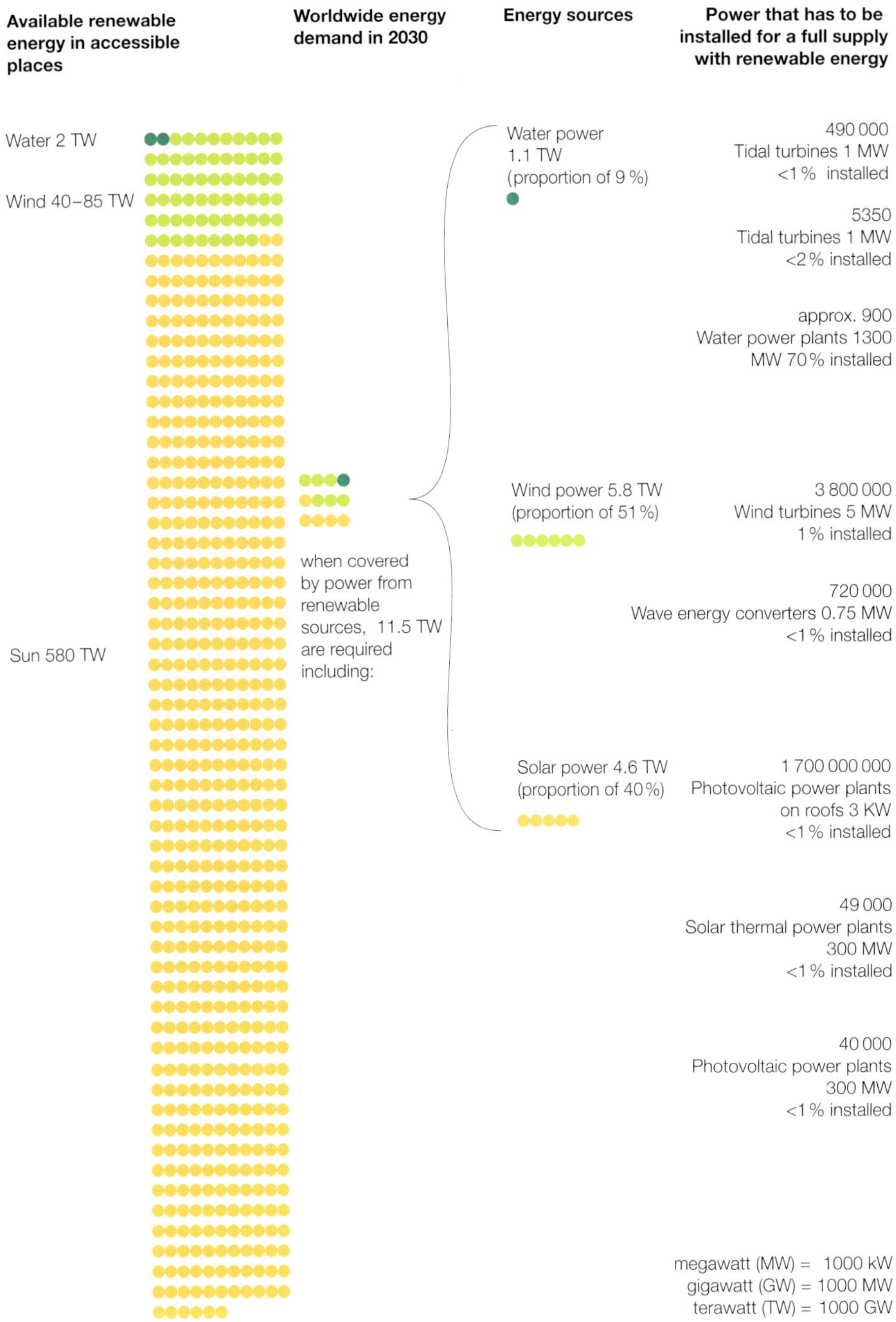

1.4

onstrates how energy could be supplied at short notice by 2030 using already available technology and exclusively wind, water and solar power (fig. 1.4) [1]. Whereas large-scale solar power plants generate electricity close to the equator, wind farms are used in the northern hemisphere. In order to cover the full power demand, the solar and wind plants are supplemented by water power plants. High performance power grids are able to balance peak demand periods and energy gaps by distributing the power generated by renewable sources on a large scale during alternating periods and in a demand-oriented way. In practise, however, the development of the necessary globally coordinated transmission networks frequently has not been very successful due to national interests or due to resistance stemming from locally organised citizens' movements. Jacobson and DeLucchi have estimated costs of 100 trillion dollars for the transformation of the global energy system. Based on similar assumptions, a number of studies have forecasted a doubling of electricity prices in Germany in comparison to today's level if the power is to be supplied exclusively by renewable energy sources.

In summary, the following can be said: The future form of energy will be that generated from renewable sources, possibly also energy stored in its transformed form as methane gas.
This type of energy will lead to increased costs. For economic reasons alone, it is therefore necessary to reduce energy consumption before a high demand is covered by renewable but expensive energy. In the case of buildings, this means that a reduced energy demand achieved through structural measures still has the highest priority when it comes to assessing the energy efficiency of buildings – even in an era of renewable power generation.

1.4 Scenario for a totally renewable energy supply in 2030. In order to provide a continuous supply, the facilities that are to be installed generate more power than is actually required.
1.5 Country-specific levels of CO_2 emissions in million tonnes. The table shows that all measures undertaken to reduce CO_2 emissions have been in vain.
The industrialised countries have been able to almost retain or even slightly reduce their level of CO_2 emissions since 1992, however, due to the technological catch-up and the population growth in other countries, the overall demand for primary energy and therefore also the emission of carbon dioxide continues to rise. Without serious, globally effective and binding agreements, there is no end to this spiral movement.
1.6 Historical development of oil extraction for conventional and unconventional oil as well as condensate (NGL) including a forecast for the possible development up until 2050. The forecast is based on the development of extraction over the past 25 years, which was determined by global free trade with uniform prices. It is therefore fairly optimistic.
The real price development generally depends on whether a uniform world market will prevail or whether individual countries (e.g. China) will secure access to specified production rates based on bilateral agreements.
1.7 Relation between useful energy, final energy, primary energy and CO_2 emissions according to EnEV and GEMIS

Energy terms

According to the accounting methods determined by the German energy saving directives (EnEV, GEMIS [2]), the energy demand of buildings can be described from four different perspectives. The relevant terms are explained in the following paragraphs (fig. 1.7).

Useful energy demand

In order to use a building in the climate conditions prevailing in Germany, energy

is necessary. It is required to heat the building (heating energy demand), to illuminate the building (lighting energy demand) and possibly also cool the building (cooling energy demand).
The amount of energy required by a building is dependent on:
- the climate zone in which the building is located
- its use
- the thermal comfort conditions anticipated by the occupants (interior temperature, lighting, etc.)
- the useful floor area
- the quantity of building envelope, i.e. the structural elements separating the heated interior space from the unheated exterior space (see compactness, p. 24)
- the quality of the building envelope. In terms of energy, the quality of the envelope is determined predominantly by the thickness of the insulation as well as by the orientation, the size and type of windows (sun protection, type of glazing).
- furthermore, whether and to what extent storage mass is available in the interior of the building

Final energy demand
In order to determine how much energy is actually required by a building, the amount of energy necessary to run the generator of cold or heat itself – for instance the boiler – has to be added to the useful energy demand. A certain amount of plant loss is associated with every type of power generator, further increasing the total energy demand.
Plant loss occurs in association with the heat generation (boiler, chimney), the heat storage (buffer tank, hot water storage), the heat distribution (distribution pipes in the basement) and the heat output (thermostatic valves). As regards the generation of cold, plant loss occurs in the same way. Electricity powered installations generally do not show any plant loss. The useful energy demand is therefore equivalent to the final energy demand (e.g. as regards lighting or electric heating).
Auxiliary energy required to operate technical equipment (pumps, fans, etc.) also contributes to the final energy demand.

Primary energy demand
When performing the energy balance for a building, it is necessary to compare and add together different types of energy. The simple addition of the amounts of energy generated by gas, oil, electricity

Development of CO_2 emissions according to country [mil t]

	1992	1996	2000	2004	2008	population 2008 in mil.	t CO_2 per capita
USA	5489	5963	6369	6528	6575	304	21.6
Former Soviet Union	3447	2430	2250	2488	2533	142	17.8
China	2573	3216	2740	4881	6389	1300	4.9
Japan	1231	1328	1344	1391	1393	128	10.9
Germany	960	956	903	901	861	80	10.8
South America	750	889	968	990	1159	550	2.1
Africa	674	786	828	932	1020	970	1.0
France	432	421	434	438	421	61	6.9
Middle East [1]	781	962	1085	1362	1573	1550	1.0
World	22565	23903	24677	28424	30892	6751	4.8

[1] Iran, Afghanistan, Pakistan, India, Sri Lanka, Bangladesh, Nepal, Bhutan

1.5

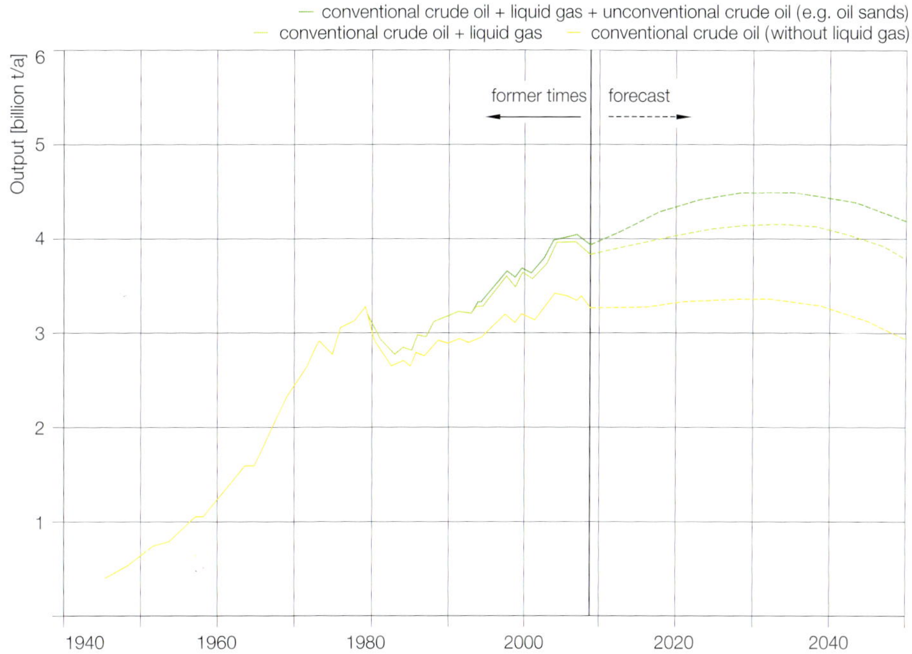

1.6

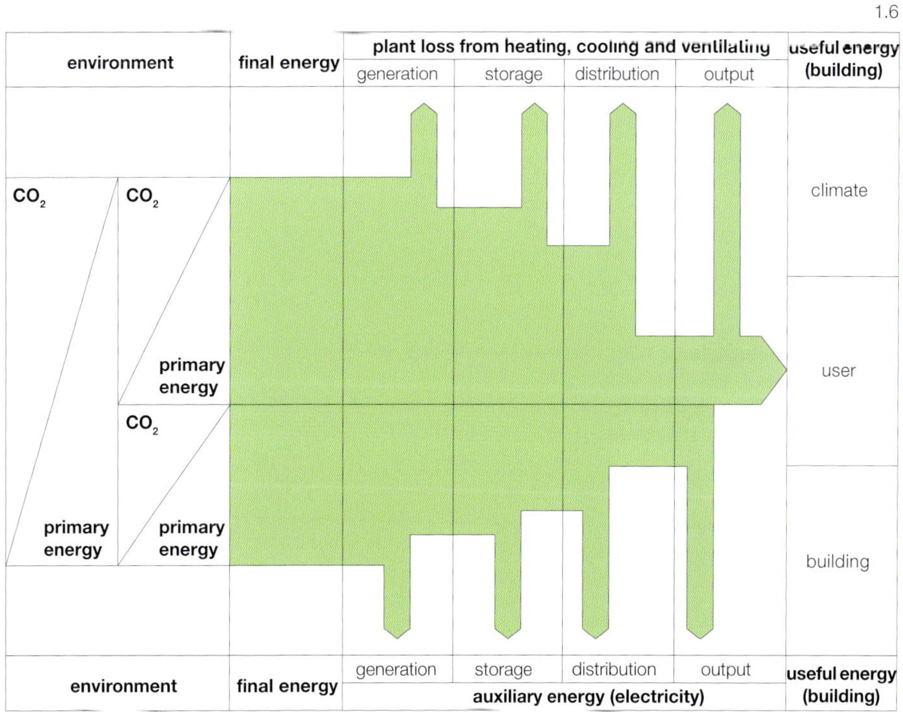

1.7

CO_2 emissions caused by the generation of electricity

	Germany emphasis: coal		France emphasis: nuclear power		Norway emphasis: water power		China emphasis: coal	
Energy source	amount of electricity [TWh]	proportion [%]	amount of electricity [TWh]	proportion [%]	amount of electricity [TWh]	proportion [%]	amount of electricity [TWh]	proportion [%]
Coal	278.5	43.6	19.8	3.4	2.2	1.5	2785.74	78
Nuclear energy	148.8	23.3	440.3	72.7	0	0	82.14	2.3
Natural gas	83.0	13.0						
Wind power	40.2	6.3	0	0	0	0	13.00	0.36
Mineral oil products	10.5	1.6						
Water power	19.6	3.1	63.4	11.0	140.5	98.5	628.58	17.6
Other energy sources	58.1	9.1	50.9	8.9	0	0	62.14	1.74
Total	639.1	100	574.4	96	142.7	100	3571.46	100
CO_2 emissions [g/kWh]	**604**		**61**		**0.1**		**1000**	

1.8

Cumulative energy input and greenhouse gas emissions of various energy sources and processes

		Cumulative energy input [kWh_{prim}/kWh_{end}]			Greenhouse gases CO_2 equivalent [g/kWh_{end}]	Primary energy factors	
Type of energy	**process** [1]	total	non-renewable proportion	renewable proportion [2]		total	non-renewable
Fuel [3]	fuel oil EL [4]	1.11	1.11	0	302	1.1	1.1
	natural gas H [5]	1.12	1.12	0	244	1.1	1.1
	liquid gas	1.11	1.11	0	263	1.1	1.1
	bituminous coal	1.08	1.07	0	438	1.1	1.1
	brown coal	1.21	1.21	0	451	1.2	1.2
	wood chip	1.07	0.06	1.01	35	1.2	0.2
	wood pellets	1.16	0.14	1.03	41	1.2	0.2
Electricity	electricity mix	2.96	2.61	0.34	633	3.0	2.6
District heat [6]	district heat 70 % CHP	0.77	0.76	0.01	219	0.7	0.7
	district heat 35 % CHP	1.15	1.14	0.01	313	n/a	n/a
	district heat 0 % CHP	1.52	1.51	0.01	407	1.3	1.3
Local heat [6]	local heat 70 % CHP	0.71	0.70	0.01	-79	0.7	0.7
	local heat 35 % CHP	1.08	1.07	0.01	119	n/a	n/a
	local heat 0 % CHP	1.46	1.44	0.01	318	1.3	1.3

[1] upstream chain for the final energy up until the transition point in the building incl. material expenses for heat generator, excl. auxiliary energy within the building
[2] the renewable proportion includes secondary resources, e.g. residual timber and waste
[3] reference value: lower heating value H_L
[4] EL = extra light
[5] H = high, natural gas with high methane level
[6] electricity credit for coal power
local heat supply with natural gas BTTP (= proportion of CHP) + natural gas peak load boiler
district heat with bituminous coal condensation power plant (= proportion of CHP) + heating oil peak load boiler
(CHP = combined heat and power)

1.9

1.8 CO_2 emissions caused by power generation using a variety of energy sources in 2008. The table shows that the globally most effective way to reduce CO_2 emissions (also from an economical point of view) would be to replace the coal power plants in China by other power plants with a proportion of renewable energy.

1.9 Cumulative energy input and greenhouse gas emissions of various energy sources and processes (calculated with GEMIS Version 4.5) and primary energy factors according to DIN V 18599-1 (table A.1)

or other kinds of energy media in kilowatt-hours (kWh) is possible from a physical point of view, however, this method gives no indication of the ecological relevance of the power generated by the different media.
Due to the production (generation, extraction) and the distribution, each form of power generation is dependent on the deployment of primary (finite) resources (oil, gas, coal). This is expressed in the primary energy factor f_P. The factor is used to correct the final energy amounts and leads to the "primary energy demand".
Owing to its complex generation, electricity is assigned a primary energy factor of 2.6 in Germany. Approximately fifty percent of all electricity is generated in coal power stations, and almost two thirds of the energy contained in the coal is lost as waste gas. The factor expresses how much primary energy is required to generate a kilowatt hour of electricity (fig. 1.8).
Gas and oil on the other hand require only 10 % of the energy primarily contained in the raw material for their production and transportation. For this reason, they are assigned the factor 1.1. Wood as the most common form of biomass has a primary energy factor of 0.2. Wood burning consumes hardly any primary energy since biomass is a regrowable raw material. The low amount of primary energy is merely required for the extraction of the wood (saw), processing it (pellets) and transportation. However, the use of biomass on a large scale does not help towards solving the CO_2 problem since wood emits a large amount of CO_2 when burnt (approximately 390 g CO_2/kWh). The only reason this does not have to be accounted for is because wood absorbs almost the same amount of CO_2 from the atmosphere during its growth. Wood in the form of biomass is also a type of CO_2 store. Eliminating it therefore contributes towards intensifying environmental problems.

CO_2 emissions

In comparison to primary energy, CO_2 emissions are a better indicator of energy demand, since the individual energy media are assessed according to their climate-damaging effect. This is not the case when using the primary energy factor.
The term used in the accounting method "CO_2 equivalent emissions" also incorporates the amount of CO_2 emitted during the energy generation process. It further-

more includes the amount of other gases that are weighted according to their global warming potential and also contribute towards climate warming.
The differences between primary energy and CO_2 emission are best explained by using the example of coal combustion:
In order to generate 10,000 kWh of useful energy through the combustion of mineral coal, 10,800 kWh of primary energy are required since 8 % of the energy generated by the coal is actually needed for the extraction of the coal ($f_{P\ Coal}$ = 1.08).
When considering the environmental load, the CO_2 emissions amount to 4380 kg. If the same quantity of useful energy were generated using gas ($f_{P\ Gas}$ = 1.1), the primary energy demand would be slightly higher, measuring 11,000 kWh, whereas the CO_2 emissions would be almost halved, amounting to only 2440 kg. This comparison between energy media in terms of their environmental load cannot be made using the primary energy demand, but is only possible using the indicator "CO_2 emissions" (fig. 1.9). However, the environmental load caused, for example, by using nuclear power cannot be assessed by the CO_2 indicator.

Relationship between useful, final and primary energy and CO_2 emissions
The useful energy demand, i.e. the energy required to heat, cool and ventilate a building, should be reduced as far as possible by the design concept and the implementation of construction-related measures.
The remaining useful energy demand is then supplied to the building by using technical equipment. Plant loss, which is dependent on the type of technology chosen (generation and supply of heat, cold and hot water) contributes further towards the useful energy demand. The aim of all conceptual considerations should therefore be to minimise plant loss through a coordinated arrangement of the distribution systems. This can be achieved, for example, by arranging floor plans in such a way that the supply makes do with very few riser pipes. The useful energy demand and the energy demand to cover plant loss together with the auxiliary energy add up to the final energy demand. This is the amount of energy which eventually must be supplied to the building and paid for. At least a proportion of the final energy should be covered by using renewable energy.
The primary energy demand gives evidence of the effect the final energy demand has on the environment through the consumption of finite resources. However, the CO_2 emissions, that result from the selected energy media, are a much clearer indicator of the harm caused to the atmosphere through the consumption of final energy (fig. 1.7, p. 11).

Energy efficiency refurbishments
In the case of refurbishments, the overall aim is, of course, to first of all reduce the building's useful energy demand through construction-related measures. These include in particular:
- insulating the opaque envelope
- upgrading transparent components by using better frame constructions and glazing
- improving summer heat protection by adding sun protection devices, or by uncovering or increasing the storage mass (an interesting consideration for existing building stock is the possibility of retrofitting storage mass in form of latent heat stores contained in composite gypsum boards)
- improving the use of daylight
- creating possibilities for natural ventilation in order to reduce the run time of air handling units or air conditioning plant.

In existing building stock, the employment of insulation measures is often restricted as the complicated connections involved could lead to severe damage to the building structure. Exterior insulation cannot be used if it destroys design features that are important to the urban context. In these cases of existing building stock, the greatest potential lies in upgrading the technical building services.

Building services
The useful energy demand, which has already been reduced as far as possible through construction-related measures, can be further reduced by using suitable plant technology. By including renewable energy sources it is even possible to meet the primary energy standards of new builds.
The replacement of old boilers by efficient heating appliances can lower energy consumption by almost 30 %.
Further improvements can be obtained by incorporating heat pumps for heating purposes. However, in circumstances where the building itself has not been refurbished, such as in old uninsulated buildings, bivalent systems should be used combining a heat pump with a traditional boiler since this is the only way to meet the necessary high forward flow temperatures.
In residential buildings, energy demand can be reduced further by approximately 500 kWh per person a year by using thermal solar energy to support domestic hot water production.
When old lighting plant is upgraded or replaced in non-residential buildings, the electric power demand per square meter can be lowered by approximately 10 kWh per year.
Cooling systems can frequently be removed or reduced significantly in existing buildings that have been upgraded in terms of construction-related measures. So long as the cooling requirements are not too great, it is possible to extract heat from the building during the summer months, store it in a ground store and reuse it with the help of a heat pump to heat the building in winter.
In particular in old buildings, it makes sense to install air-handling units where windows have been exchanged to avoid moisture buildup and improve the air quality. Controlled interior space ventilation with heat recovery requires auxiliary energy to drive the fans amounting to approximately 3 kWh/m² per year, however at the same time the heating energy requirement is reduced by around 20 kWh/m² during the same period.
In the medium term, it will no longer be possible to supply power using finite energy sources. Buildings will rather be heated, cooled and ventilated using power generated from renewable resources. However, since most technical installations have a fairly short life expectancy, this outlook does not apply to the upgrading of plant until after 2030.
The type and scale of all services in existing buildings, as is the case in new builds too, is always dependent on the structural concept and the building's use. For this reason, it is not simply possible to focus solely on improving building services since this does not adequately deal with the complexity of the actual task.

Thermal comfort

The term thermal comfort refers to the conditions in an enclosed space and expresses whether and to what extent these cater to the physiological and psychological needs of the occupants.
Up until recently the term thermal comfort was determined by a small number of physical factors including:

- air temperature (20–26 °C)
- relative humidity (30–60 %)
- air movement/velocity (< 0.3 m/sec)
- temperature of the enclosing surfaces (which should be no more than 3 K below the air temperature)

Simple statements like those above are used by numerous standards and other guidelines, both technical and legal. DIN 4108-2, which defines the limits for the indoor temperature in table 6, is a good example (Fig. 1.11). In summer, these temperatures may not be exceeded during more than 10 % of the occupancy periods.
Recently, due partly to application of computer-assisted calculation programs, more complex methods of determining standards of thermal comfort have become established. Thermal comfort cannot simply be ascertained by using physical factors, it is also necessary to define room conditions in a climatic, cultural and social context. The social context in particular, along with factors such as age, gender and physical condition, are becoming more and more important. The two most useful tools to describe thermal comfort, the Mollier diagram and the DIN EN ISO 7730, are described in the following paragraphs.

Mollier diagram

The starting point for every physical evaluation is the condition of the air in the enclosed space and awareness of how it responds when heated, cooled, humidified or dehumidified. In 1923, the physicist Richard Mollier depicted these factors and the way in which they influence

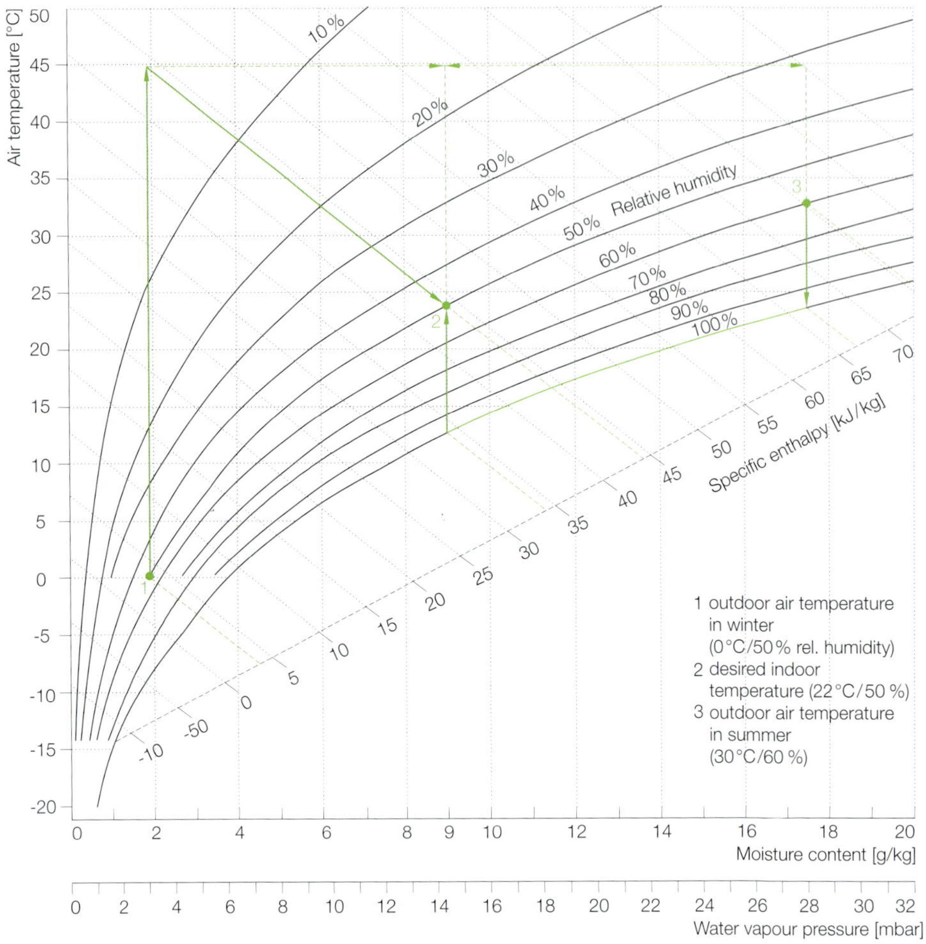

1.10

Temperature limits for interior space according to DIN 4108-2

Summer-climate region	regional feature	indoor temperature limit [°C]	maximum of the mean monthly outdoor temperature [θ °C]
A	cool summer (e.g. Kiel)	25	θ ≤ 16.5
B	moderate (e.g. Munich)	26	16.5 < θ < 18
C	hot summer (e.g. Freiburg)	27	θ ≥ 18

1.11

Thermal comfort criteria according to DIN EN ISO 7730

Type of building/ room	activity [W/m²]	category	room temperature [°C]		maximum mean air velocity [1] [m/s]	
			summer (cooling period)	winter (heating period)	summer (cooling period)	winter (heating period)
Single office open-plan office conference room auditorium cafeteria/ restaurant classroom	70	A	24.5 ± 1.0	22.0 ± 1.0	0.12	0.10
		B	24.5 ± 1.5	22.0 ± 2.0	0.19	0.16
		C	24.5 ± 2.5	22.0 ± 3.0	0.24	0.21 [2]
Kindergarten	81	A	23.5 ± 1.0	20.0 ± 1.0	0.11	0.10 [2]
		B	23.5 ± 2.0	20.0 ± 2.5	0.18	0.15 [2]
		C	23.5 ± 2.5	22.0 ± 3.5	0.23	0.19 [2]
Department store	93	A	23.0 ± 1.0	19.0 ± 1.5	0.16	0.13 [2]
		B	23.0 ± 2.0	19.0 ± 3.0	0.20	0.15 [2]
		C	23.0 ± 3.0	19.0 ± 4.0	0.23	0.18 [2]

[1] The maximum mean air velocity is based on a turbulence level of 40 % and an air temperature which is equivalent to the operative temperature. A relative humidity of 60 % and 40 % are applied in summer and winter, respectively. The lowest temperature level is used both in summer and winter to determine the maximum mean air velocity.
[2] below a threshold value of 20 °C

1.12

1.10 The Mollier diagram according to the physicist Richard Mollier with an appropriate example: in order to transform the air from state 1 to state 2, 40 kJ/kg or 13.6 kWh/m³ of energy must be invested and the moisture content of the air must be increased by 7 g/kg or 0.008 l/m³. To change the condition of the air from state 3 to state 2, 30 kJ/kg or 10.2 kWh/m³ of heat and 8 g/kg or 0.01 l/m³ of moisture must be extracted from the air.
1.11 Indoor temperature limits for the various summer climate regions in Germany according to DIN 4108-2:2003-07
1.12 Design criteria for thermal comfort in different room types according to DIN EN ISO 7730

one another in the so called Mollier diagram (fig. 1.10). With the aid of the diagram, it is possible to interrelate the air temperature, the moisture content of the air, the air's relative humidity and its heat content and therefore clearly determine the condition of the air. The diagram can also be used to identify strategies to change the air quality with plant technology, herewith improving thermal comfort within the interior space.

DIN EN ISO 7730

DIN EN ISO 7730 "Ergonomics of the thermal environment – analytical determination and interpretation of thermal comfort using the calculation of the PMV and PPD indices and the local thermal comfort criteria" expands the four basic physical factors of thermal comfort described at the beginning of this chapter by adding further parameters. For example, the room temperature considered comfortable is also dependent on the activities performed by the occupants as well as the insulating effect of their clothing. The type of activity is either expressed in W/m² or as an abstract MET value (metabolic unit to measure the energy cost of an activity), the type of clothing by the clothing factor clo or by using the thermal resistance value of clothing measured as m²K/W. Office work with light clothing can, for example, either be expressed by the units 70 Wm² or 1 met and the clothing factor 0.5 clo. These values are then included in the calculation of the PMV and PPD (see below).
Furthermore, DIN EN ISO 7730 no longer considers the parameters room air temperature, relative humidity, temperature of the enclosing surfaces and air velocity separately, but evaluates them in relation to one another (fig. 1.12).

PMV and PPD

Algorithms based on empirical data have been developed to make predictions on the occupants' satisfaction with the thermal environment in different conditions. The result of these computer-assisted calculations is the specification of the two following indices:

- PMV (Predicted Mean Vote)
  The PMV index, the presumably mean opinion of occupants in a defined thermal environment, makes a statement about the average sensation of the climate in a specified room. The climate condition prevailing is expressed on a scale from 0 to 6 for too warm and on a scale from 0 to -6 for too cool.
- PPD (Predicted Percentage of Dissatisfied)
  The PPD index, the presumably mean opinion of occupants in a defined thermal environment, makes a statement about the percentage of occupants who feel uncomfortable in the climate conditions prevailing in the specified room. The scale ranges from 0% to 100% of dissatisfied persons.

Naturally, both values, PPD and PMV, correlate, which is expressed graphically in the standard DIN 7730 (fig. 1.13, p. 16).
In order to maintain a room climate still considered comfortable, the standard clearly states values that must not be exceeded. Three comfort levels are established, category A, B and C, which are each characterised by a number of dissatisfied persons. Category A applies when less than 6% are dissatisfied, Category B when less than 10% and Category C when less than 15% are dissatisfied (fig. 1.16, p. 16).

Independent of these complex relations generated and evaluated by using EDP, the standard makes some important statements about further aspects concerning thermal comfort. These are briefly explained in the following sections.

Vertical air temperature difference

A large difference in the air temperature as measured at head and foot level can cause the occupants to feel uncomfortable. This can occur in particular in ground floor apartments with insufficient insulation to the ground or basement levels, or in rooms with underfloor heating, which is operated with very high forward flow temperatures due to a large heating load. Figure 1.14 (p. 16) illustrates that already 6% of all occupants are dissatisfied in rooms with a temperature difference of only three Kelvin.

Asymmetric radiation temperature

Asymmetric radiation is a term used to describe the different temperatures of the surfaces enclosing a room. Asymmetric radiation is frequently a cause of discomfort. Figure 1.15 (p. 16) shows that ceiling surfaces that are warmer than the room air quickly lead to a feeling of discomfort, whereas cooler ceiling surfaces do not cause any problems.
Occupants are less sensitive to wall surfaces that are either cooler or warmer than the room air temperature. Wall surfaces with surface temperatures of 40°C

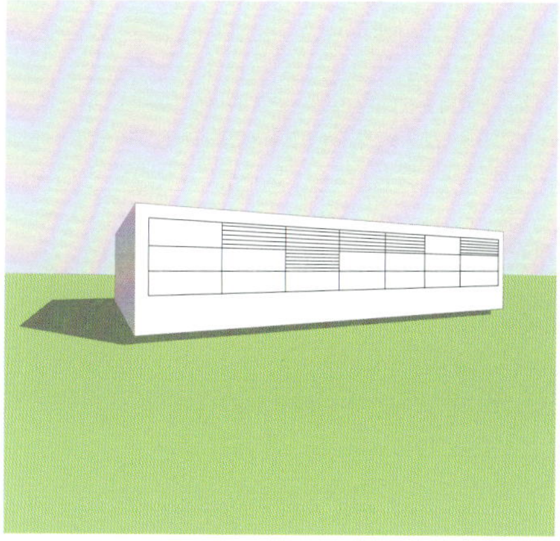

Example: thermal comfort

The presented building serves as a reference property throughout the book. It is used to explain specific issues in an exemplary way.
This chapter examines the effect different interior climate conditions have on the occupants in the pavilion's office space.

Pavilion data

net floor area	196 m²
volume	588 m³
heated volume	811 m³
total envelope surface area	682 m²
• wall surface	179 m²
• window surface	43 m² (5 m² west/east/north, 38 m² south)
• roof surface	225 m²
• covered area	225 m²

The following factors are the basis of this study:

• sedentary activity	1.2 met (=70 W/m²)
• light clothing	0.43 clo (=0.068 m²K/W)

Situation 1

air temperature	22°C
mean radiation temperature	22°C
air velocity	0.1 m/sec
relative humidity	60%
PMV	- 0.75
PPD	17%

Comment

Due to the very high humidity, the interior climate tends to be considered too cool and therefore unpleasant.

Situation 2

air temperature	23.5°C
mean radiation temperature	25.5°C
air velocity	0.1 m/sec
relative humidity	60%
PMV	- 0.01
PPD	5%

Comment

Due to a slight increase in room temperature and a significant increase in the mean radiation temperature of the enclosing surfaces, the interior climate is perceived as comfortable.

Situation 3

air temperature	27.0°C
mean radiation temperature	27.0°C
air velocity	0.3 m/sec
relative humidity	60%
PMV	0.44
PPD	9%

Comment

The effect the increased air velocity has on thermal comfort is compensated for by increasing the room temperature and the mean radiation temperature significantly.

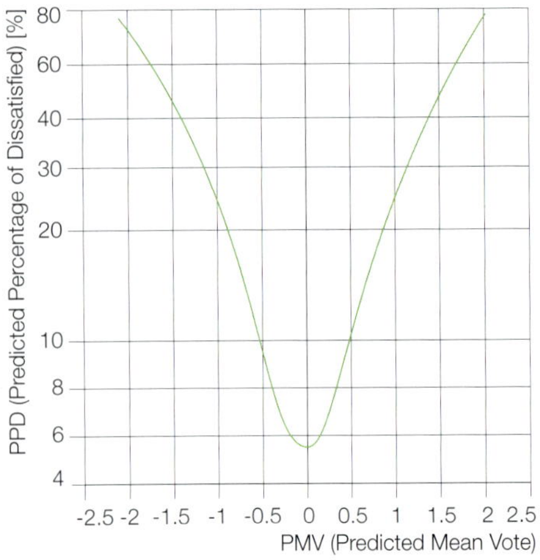

1.13

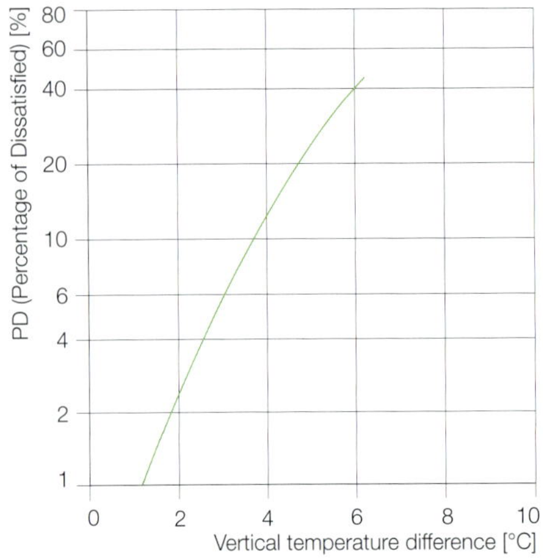

1.14

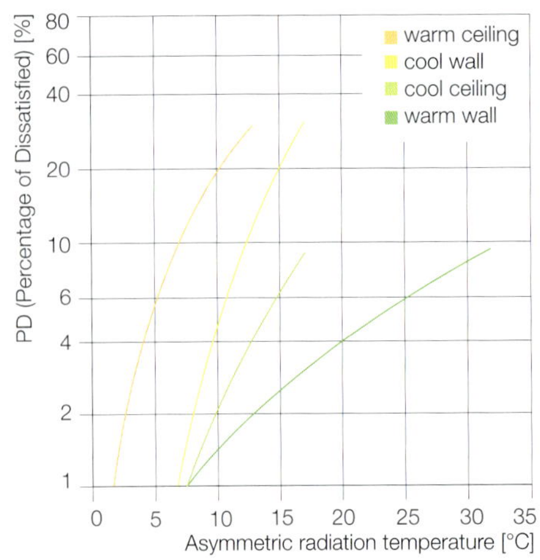

1.15

1.13 Predictions concerning the occupants' sensation of thermal comfort: relationship between PMV (Predicted Mean Vote) and PPD (Predicted Percentage of Dissatisfied) in accordance with DIN EN ISO 7730
1.14 Local discomfort due to vertical temperature differences in accordance with DIN EN ISO 7730
1.15 Local thermal discomfort caused by asymmetric radiation temperatures
1.16 According to DIN EN ISO 7730, there are three thermal comfort categories (A, B and C) to distinguish requirements concerning thermal comfort in interior space
1.17 Critical values concerning the discomfort caused by formaldehyde in interior space according to IfAU (Cooperative Institute for Research in Environmental Sciences, www.ifau.org)

are therefore generally not considered unpleasant, which means that wall heating as a type of surface heating, in contrast to ceiling or floor heating, is an option worth considering.

Further aspects

Dependent on the activity and the clothing, draughts can be the cause of major discomfort. However, the ways in which air movement is perceived differ considerably, depending on the room temperature and the type of clothing.

As regards thermal comfort, relative humidity tends to be of minor importance, whereas being able to adjust the room climate to meet individual needs generally contributes towards a sensation of feeling comfortable.

Energy efficiency refurbishments

It cannot be denied that the growing demand for greater thermal comfort is increasing energy consumption. A critical analysis of user requests should therefore always be part of the considerations and efforts to save energy. Numerous demands for thermal comfort could be reduced significantly without notably affecting the quality of life. Alongside energy saving aspects, energy efficiency refurbishments also aim at removing elements disrupting thermal comfort by implementing construction-related and building services measures. Concepts that are based on a holistic approach are designed to reduce the energy input required to heat, cool and ventilate the building and, at the same time, increase the comfort factors defined in the standard whilst using the interior space. Thermal comfort can be increased within the scope of an energy efficiency refurbishment by employing the following measures:

- The surface temperature of the exterior walls is increased by insulating the building envelope. This measure reduces asymmetric radiation. The same applies to new windows.
- By adding insulation to the ceiling of the basement or the surface of the ground floor, thermal comfort is improved. The vertical temperature difference between head and feet is reduced due to the higher surface temperature at floor level.
- The higher temperature of the enclosing walls, achieved by insulating the exterior walls and fitting new windows, reduces the air movement arising from temperature differences. These circumstances improve the thermal comfort conditions indoors.
- By upgrading the glazing and the sun protection devices, summer heat gain is reduced. This has a positive effect on the development of room temperature during the summer months.
- By uncovering or retrofitting storage mass, the rise in summer room temperature can be minimised.
- Improvements made to plant technology can reduce plant-related problems concerning thermal comfort. Causes of discomfort are high temperatures used in underfloor heating, high air velocities due to air-handling systems and large temperature asymmetries stemming from cooling systems.

Immediately perceptible defects in buildings are frequently the impetus for considering an energy efficiency refurbishment. The Mollier diagram can be used as a tool to understand and remove the problem of mould growth, which is known to be the cause of great discomfort, in a target-oriented and source-related way. This topic is dealt with thoroughly in the section "Interior climate" (p. 17ff.).

Comfort categories and requirements according to DIN EN ISO 7730

	Thermal condition of the occupant in total			Local discomfort PD [1] [%]		
Category	PPD [%]	PMV	DR [2] [%]	vertical temperature difference	warm or cold floor	asymmetric radiation
A	< 6	- 0.2 < PMV < + 0.2	< 10	< 3	< 10	< 5
B	< 10	- 0.5 < PMV < + 0.5	< 20	< 5	< 10	< 5
C	< 15	- 0.7 < PMV < + 0.7	< 30	< 10	< 15	< 10

[1] PD = percentage of dissatisfied concerning a single aspect
[2] DR = percentage of dissatisfied due to draughts

1.16

Indoor climate

In economically highly developed societies, people spend 90 % of their lifetime indoors. Since each person breathes in 10–20 m³ or, converted into kilograms, 12–24 kg of air per day, the air quality of interior space is of major importance and national regulations have been introduced precisely for this reason. In Germany, limits are used to determine the pollutant content of the air in habitable spaces. Specific and detailed regulations apply for rooms where, due to the activities performed, certain hazardous substances can lead to air pollution.
Due to the increasing airtightness of buildings, the natural air exchange is being reduced continuously. This can, in contrast to the situation outdoors, be the cause for an increased concentration of pollutants in indoor environments. If nothing else, it is the offgassing of building materials and furniture which leads to conditions of severe discomfort and possibly also damaging effects on health. To make matters worse, these pollutants are absorbed by textile materials and therefore remain permanently in the indoor space.
Buildings completed after 1950 are particularly affected by a high pollutant content since these frequently used artificially produced materials, such as wood preservatives or PVC-containing building materials.
Impairments to health caused by these kind of environmental pollutants are often not detected or cannot necessarily be attributed to a specific cause. The occupants' reaction to this kind of permanent pollution is usually a variety of unspecific health symptoms. Furthermore, finding evidence of the pollution is laborious, as the environmental pollutants are absorbed by adipose tissue and are therefore not or hardly detectable in blood or urine samples. Normally, the physical symptoms do not start to subside until the affected person has spent a longer period in a new location. This further complicates studies to detect the cause or origin of a disorder.

Volatile organic compounds

Volatile organic compounds (VOC) are carbon-based compounds produced through chemical reactions. They are volatile because they are released into the atmosphere quickly due to their low boiling point. The contamination of room air by VOC is caused by natural sources (plants, pets) or by using materials that offgas VOC. Organic solvents (e.g. benzene) or solvent mediators are also classified as VOCs. The use of benzene as a solvent is actually forbidden. However, benzene, which is produced during the combustion of petrol, can find its way into the interior of buildings as traffic-generated air pollution.
The replacement of volatile organic compounds, such as trichlorethylene and perchlorethylene, by substances that offgas at a slower pace, such as glycol compounds in carpet adhesives or water varnishes, has led to the misleading labelling of these products as "solvent-free" despite the fact that these solvents might perhaps pollute the room air for years even if to a lesser degree. Biological building materials also use solvents (limonene, D-3-carene), which can be the cause of health disorders.
Depending on the concentration, solvents can be hazardous to health in different ways, particularly in the case of allergy sufferers. Frequent symptoms in this context are headaches, nervous coughs, elevated liver levels and nervous disorders.
Since indoor air contains many different organic compounds and the availability of reference values for single air pollutants is fairly limited, the Federal Environment Agency in cooperation with responsible authorities has determined reference values for the total air pollution with volatile organic compounds (fig. 1.19, p. 18). The total of volatile organic compounds (TVOC) is divided into five concentration levels, which are each assessed from a hygienic perspective. If a single type of gas can be distinguished in the room air, the reference values of the Federal Environment Agency apply (fig. 1.18, p. 18). The appropriate course of action with regard to the air pollution can be determined from this assessment.

Formaldehyde

Formaldehyde (CH_2O) is a base material used to produce synthetic resin and adhesives, especially common in furniture (particle boards). Formaldehyde is also used to produce synthetic varnishes for furniture and parquet flooring. The air humidity then causes the materials to emit their volatile gas into the room air. Despite the E1 emission standard, which came into effect in 1988, formaldehyde concentrations in habitable rooms have not been reduced during the last years. One of the reasons for this is the increasing proportion of inexpensive furniture made of particle board.
From a toxicological point of view, a certain concentration of formaldehyde causes irritation of the respiratory tract, which over time can lead to bronchial asthma and an intensification of already existing allergies. The World Health Organization regards formaldehyde as a possible human carcinogen with an average danger level. For this reason, concentrations of formaldehyde are not to exceed the limits presented in figure 1.17.

Nonvolatile gases

If the boiling point of a gas is above 200 °C, it is referred to as nonvolatile. Nonvolatile gases are, for example, pentachlorophenol (PCP), polychlorinated biphenyls (PCBs), plasticizers (phthalates) and polycyclic aromatic hydrocarbons (PAHs), which are contained in wood preservatives, coating materials, adhesives and plastics. The disadvantage of these gases is that it takes a long time for them to vaporise. They are therefore responsible for causing long-term pollution to the indoor environment.
In particular, if the room air is polluted with PCP, usually due to wood preservatives, there is urgent need for action since this frequently means that highly toxic dioxins are also present. The use of products containing PCP was prohibited in Germany in 1989. However, PCP products are still being released on to the market through imports.
Contamination with PCB occurs when joint sealants that contain PCB are used or PCB starter units are integrated into conventional ballasts. Their use was prohibited in 1978.
Phthalates are added to all kinds of plastic products as softeners or plasticizers (e.g. in sealants, skirting boards, cable sheathing). The function of PAHs in rubber products and bituminous adhesives is exactly the same as that of phthalates. Bituminous adhesives were used well into the seventies to fix parquet flooring. These mentioned contaminants are suspected of being carcinogenic and of having a negative effect on the immune system.

Formaldehyde levels according to IfAU

minor pollution	up to 25 µg/m³
average pollution	25–75 µg/m³
median value	55 µg/m³
test level for emission reducing measures	above 75 µg/m³
test level for emission reducing measures in the case of children	above 50 µg/m³

1.17

Limits for the concentration of pollutants in indoor environments

Volatile gases VOC	Limit II [mg/m³]	Limit I [mg/m³]	Introduction [year]
benzaldehyde	0.2	0.02	2010
benzyl alcohol	4	0.4	2010
monocyclic monoterpenes (lead substance d-limonene)	10	1	2010
aldehydes, C4 – C11 (saturated, acyclic, aliphatic)	2	0.1	2009
carbon dioxide	2000 ppm (hygienically inacceptable)	1000 ppm (hygienically harmless)	2008
C9 – C14 alkanes/ isoalkanes (dearomatised)	2	0.2	2005
naphthalene	0.020	0.002 [1]	2004
bicyclic terpenes (lead substance pinene)	2	0.2	2003
tris (2-chlorethyl) phosphate (TCEP)	0.05	0.005	2002
diisocyanates	see fig. 1.19		2000
TVOC	see fig. 1.19		1999
mercury (as metallic vapour)	0.00035	0.000035	1999
styrene	0.3	0.03	1998
nitrogen dioxide (NO_2)	0.35 (30 min. value) 0.06 (7 day value)	–	1998
dichlormethane	2 (24 h)	0.2	1997
carbon monoxide	60 (½ h) 15 (8 h)	6 (½ h) 1.5 (8 h)	1997
pentachlorphenol (PCP)	0.001	0.0001	1997
toluene	3	0.3	1996

[1] Limit I should also prevent odour nuisance

1.18

Hygienic assessment of highly volatile hydrocarbon compounds (TVOC)

Level	TVOC concentration levels [mg/m³]	Hygienic assessment
1	≤ 0.3 mg/m³	hygienically harmless
2	> 0.3 to 1 mg/m³	hygienically harmless, unless single substances or substance groups exceed the recommended limits
3	> 1 to 3 mg/m³	hygienically suspicious
4	> 3 to 10 mg/m³	hygienically dangerous
5	> 10 mg/m³	hygienically inacceptable

1.19

Maximum acceptable annual mean values for particulate matter	
PM_{10} = 20 µg/m³ annual mean	can be exceeded for single days
$PM_{2.5}$ = 10 µg/m³ annual mean	can be exceeded for single days
Maximum acceptable 24-hour mean values for particulate matter	
PM_{10} = 50 µg/m³ 24-hour mean	no additional days for possible exceedance
$PM_{2.5}$ = 25 µg/m³ 24-hour mean	no additional days for possible exceedance

1.20

1.18 Limits recommended by the Federal Environment Agency for the concentrations of single substances in indoor air.
Limit I represents the concentration which, even in the case of life-long exposure, does not cause any adverse impacts to the health of individuals.
Limit II is to be regarded as a limit which should by no means be exceeded.

1.19 Hygienic assessment of TVOC (Total Volatile Organic Compounds) in five levels according to the Federal Environment Agency.

1.20 The guideline values recommended by the World Health Organization for the different sizes of particulate matter (PM 10 = 10 micrometers, PM 2.5 = 2.5 micrometers). PM 2.5 is considered respirable and often described as fine particulate matter.

The possibilities of detecting nonvolatile gases by using air pollution metering equipment are limited. In this case, it is more suitable to have an analysis of house dust performed by a qualified surveyor, who is registered as such in the Chamber of Craftsmen.

Particulate matter

The danger of indoor air contamination with particulate matter has long been underestimated. Damage to health can be caused by breathing in the fine particles. Fine particulate matter originates from motor vehicles (diesel combustion, tyre abrasion), from wood burning processes but also from office environments with copying machines and laser printers. Vacuum cleaners with inadequate dust filters also contribute towards the pollution of air with particulate matter.
Health hazards caused by fine particulate matter range from mucosal irritation, the aggravation of allergies to carcinogenic effects. Legal limits for the concentration of fine particulate matter differ throughout the EU. For reasons of clarity and better comprehension, the levels recommended by the World Health Organization, which are well below the EU's guideline values, are generally used as reference values (fig. 1.20).

Fibres

Due to the structure of fibres, older insulating materials made of mineral fibres, which were produced up until 1995 and used up until 2000, are classified as carcinogenic. The health hazard potential is determined by using the so-called carcinogenicity index (CI) in accordance with TRGS 905 (Technische Regeln für Gefahrstoffe – Technical Rules for Hazardous Substances). In the case of a refurbishment, these products should be removed.
The same applies to the utilisation of products containing asbestos fibres. These fibres are contained in the gaiter between pipe flanges, in old fire doors, usually containing loose asbestos insulation, and old cement fibre slabs (manufactured up until 1991) with compressed asbestos fibres. TRGS 519 lists all of the technical rules that must be observed when removing and disposing asbestos-containing substances. Asbestos must be deposited at a specially approved disposal plant labelled with the appropriate waste code in accordance with the European Waste Catalogue (EWC).

Moisture

Depending on the temperature, an increased moisture content in the air leads to a rise of relative humidity. An indoor relative humidity above 60% encourages the growth of mould, mites and other forms of infestation.

Mould infection

By using the Mollier diagram (fig. 1.10, p. 14), the problems of mould infection, which are known to substantially reduce the feeling of comfort and greatly impair health, can be examined and removed in a target-oriented and source-related way. Depending on the use of the room, an inadequate exchange of air usually leads to an increase in the relative humidity indoors. This is almost always the case in older buildings when old, draughty windows are replaced by new ones but the ventilation habits of the occupants remain the same. The extreme airtightness of the new windows reduces the former "automatic", but adequate, air exchange through unsealed joints. As a result – this initially goes unnoticed – the moisture content of the indoor air increases. This situation becomes critical when the surface temperatures of the enclosing structural components in total (due to insufficient thermal insulation) or partially (in the vicinity of thermal bridges) drop so low that the relative humidity increases to 80% or more. If these conditions persist over several days for more than six hours, the growth of mould spores, which are always present in the air, is encouraged to such an extent that hazardous concentrations can occur.

The critical moisture content of 80% is already met in the vicinity of the cooler surfaces in normal room conditions (50% relative humidity, 20°C air temperature) and with surface temperatures of 12°C (fig. 1.10, p. 14). It is not at all unusual to meet these critical temperatures on the room-enclosing surfaces in older buildings, especially in zones with structural thermal bridges (change of material in the facade) or geometrical thermal bridges (corners, reveals). These problems are illustrated in the isotherms of the simulations in the chapter "Heat sinks" (pp. 38ff.).

Even if the exterior walls are insulated sufficiently, occupants may unintentionally facilitate the growth of mould by not ventilating properly. For example, windows that, in winter, are kept in a tilt position for longer periods reduce the temperatures of the adjoining surfaces and therefore give rise to critical conditions.

Radiation

Even low level doses of radiation can be hazardous to health. However, at present, there is no reliable information concerning the severity of the hazard.

Exposure to radon, for example, is one of the main risks of ionising radiation. In this case, the ground is the source of the radiation and rooms in basement or ground floor apartments are especially affected in certain regions.

In the case of non-ionising radiation, a distinction is made between high frequency radiation, deriving from wireless networks, and medium frequency radiation, deriving from standard electrical appliances. Studies concerning the effects of non-ionising radiation on health have not yet been completed.

However, in view of the abundance and diversity of possible contaminants in indoor air, the following strategy can help to minimise pollution:

Indoor air quality is best improved by using a combination of two measures. These include firstly the reduction of pollution sources by carefully selecting building materials and furnishings and secondly ensuring a sufficient exchange of air, for example, by retrofitting some kind of mechanical ventilation system. The new DIN 1946-6 published in May 2009 also follows this approach. The standard includes limits for residential buildings regarding the minimum quantities of fresh air that must be drawn into a room (see Ventilation, pp. 76ff.).

Energy efficiency refurbishments

An indoor air pollution survey should be performed at the beginning of each refurbishment in order to detect whether the room air is contaminated. This survey can set the tone for the scope of all refurbishment measures.

Common problems are:

- products that contain asbestos in technical equipment, in fire protection products and sheathing
- wood preservatives containing PCB
- high formaldehyde concentrations, especially in the case of timber and prefab houses

The first part of an indoor air pollution analysis should always be a visual inspection. Numerous materials, such as building materials that contain asbestos and mineral wool, can often be identified at first sight. An air analysis reveals whether there is a high VOC content, which can then be differentiated further by performing more specialised tests.

Material samples should be taken and examined in laboratories to detect whether non-volatile gases or particulate matter are causing pollution.

If the cause of indoor air pollution is outdoors, it is necessary to develop concepts that allow a fresh air intake via controllable air vents. The hazardous substances can then be removed from the air by using appropriate filters.

Fine particulate matter sources indoors, such as copying machines and laser printers in offices, can be moved into separate rooms during the course of a refurbishment's reorganisation measures. The fine particles can then be removed directly through an air extraction system and should therefore no longer pose a health hazard.

In the case of refurbishments, the same care that is applied to new builds should be taken in selecting building materials that do not pollute indoor air. With a view to avoiding hygienic impairments due to high humidity levels, construction-related and possibly also building services measures must be taken to ensure sufficient air exchange.

Finally, the occupants themselves must help to prevent large quantities of pollutants being brought into the building by selecting interior fittings conscientiously; this especially applies to furniture and floor coverings.

It also makes sense to use textile materials sparingly since these absorb pollutants and can therefore contribute towards a permanent deterioration of indoor air quality.

In order to regulate the moisture content of room air, the surfaces of constructional elements should be designed in a way that enables them to absorb moisture for a short period and release it again once the number of occupants has been reduced or the function of the room changes. In this context, the use of laminate, stone or ceramic floor coverings and wallpaper or emulsion paint for the walls is not advantageous in terms of absorbing moisture.

Outdoor climate

The outdoor climate in accordance with the structural context and the occupants' demands concerning thermal comfort determines the amount of technical equipment required to provide a room climate regarded as comfortable.
The temperature of the outdoor air, the length of exposure to solar radiation and the angle of incidence plus the amount of precipitation are dependent on the latitude and longitude coordinates of the property. In particular the amount of solar radiation in conjunction with other geographical factors (ground surface characteristics, vegetation, large areas of water) characterise the outdoor climate.
Traditional building types have evolved from each climate zone or climate region as a response to the respective climatic boundary conditions. Dependent on the occupants' requirements for thermal comfort, these buildings require little or no mechanical plant for heating, cooling or ventilating the rooms.

Climate description

In order to clearly describe the characteristics of the prevailing climate in a specific region, Wladimir Peter Köppen and Rudolf Geiger developed a typological model that classifies the regional conditions according to a graduated set of criteria (climate zone, climate type, climate subtype) (fig. 1.21–1.23) [3].
The climate classification system can be used to give a methodically detailed description of global climate conditions. It is a very useful tool to prepare, during the early planning phases, "climate responsive" designs for projects in different regions around the world.

Climate zones

A Tropical rainforest or savannah climate without winter
The average temperature remains above 18 °C throughout the year.
Example: Indonesia

B Dry climate
Precipitation remains below a dry threshold which is dependent on temperature and the distribution of precipitation. The threshold (limit of r) is determined according to the total annual precipitation (r in cm) and the average annual temperature (t in °C)
- in the case of prevailing winter rain $r = 2t$
- in the case of an even distribution of precipitation $r = 2(t + 7)$
- in the case of prevailing summer rain $r = 2(t + 14)$

Example: Saudi Arabia

C Temperate climate
The temperature of the coldest month lies between -3 and + 18 °C. The total amounts of annual precipitation are above the aforementioned dry threshold.
Example: Great Britain

D Boreal forest climate
The temperature of the coldest month is below -3 °C; the temperature of the warmest month remains above +10 °C.
Example: Finland

E Snow climate
The average temperature of the warmest month is below +10 °C.
Example: Greenland

Climate types

The climate types categorise the climate zones according to their amount of precipitation. The letters that are used to differentiate the climate zones above are explained below:

F Permafrost climate
All twelve months have average temperatures below 0 °C
Example: Antarctica

S Steppe climate
A dry region in which the amount of precipitation permits regular vegetation. Precipitation remains above a dry threshold, which is dependent on temperature and the distribution of precipitation and is used to distinguish between steppe and desert. The threshold (limit of r) is calculated with r = total annual precipitation in cm and t = average annual temperature in °C
- in the case of prevailing winter rain $r = t$
- in the case of an even distribution of precipitation $r = t + 7$
- in the case of prevailing summer rain $r = t + 14$

Example: Alicante

T Tundra climate
The average temperature is above 0 °C for at least one month of the year.
Example: Spitsbergen

W Desert climate
Dry climate with very little or no precipitation. Precipitation remains below the threshold of the steppe climate.
Example: Kuwait

f Moist
All months are wet. The driest month in the A climate zone has a precipitation amount of at least 60 mm/m².
Example Af: Kuala Lumpur
Example Cf: Stuttgart

m Monsson or hybrid form between f and w
In the Am climate zones, precipitation remains below 60 mm/m² for one or several months. This shortage of rainfall is made up for by the rainfall in the other months allowing tropical rainforest to thrive. The amount of precipitation in the driest month is more than 4 % of the difference between 2500 mm/m² and the total annual rainfall. If, for example, the annual rainfall amounts to 1500 mm / m², the driest month still has precipitation amounting to 40 mm (4 % of 2500–1500).
Example: Rangoon

s Dry summer
The driest summer month has less precipitation than 40 mm/m² and less than one third of the wettest winter month's total precipitation.
Example: Rome

w Dry winter
The average precipitation of the driest winter month is less than one tenth of the wettest summer month's total precipitation.
Example Aw: Bombay
Example Cw: Lhasa
Example Dw: Beijing

Climate subtypes

Differentiations between temperatures can be made among the climate subtypes by adding a further letter.

a Hot summer
Average temperature of the hottest month is above +22 °C.
Example: Marseille

Climate zone	A tropical				B dry		C temperate				D cold				E polar	
climate type (amount of precipitation)	f	w	s	w	S	W	f	w	s		f	w			E	T
climate subtype (temperature)					h	k	a	b	c	d	a	b	c	d		

1.21

1.21 Climate zones, climate types and climate subtypes according to Köppen/Geiger. Würzburg's classification is highlighted green. It is the reference climate for the energy assessment performed according to the German Energy Saving Ordinance (Energieeinsparverordnung, EnEV).
1.22 Climate zones according to Köppen/Geiger
1.23 Average annual global radiation in kWh/m²

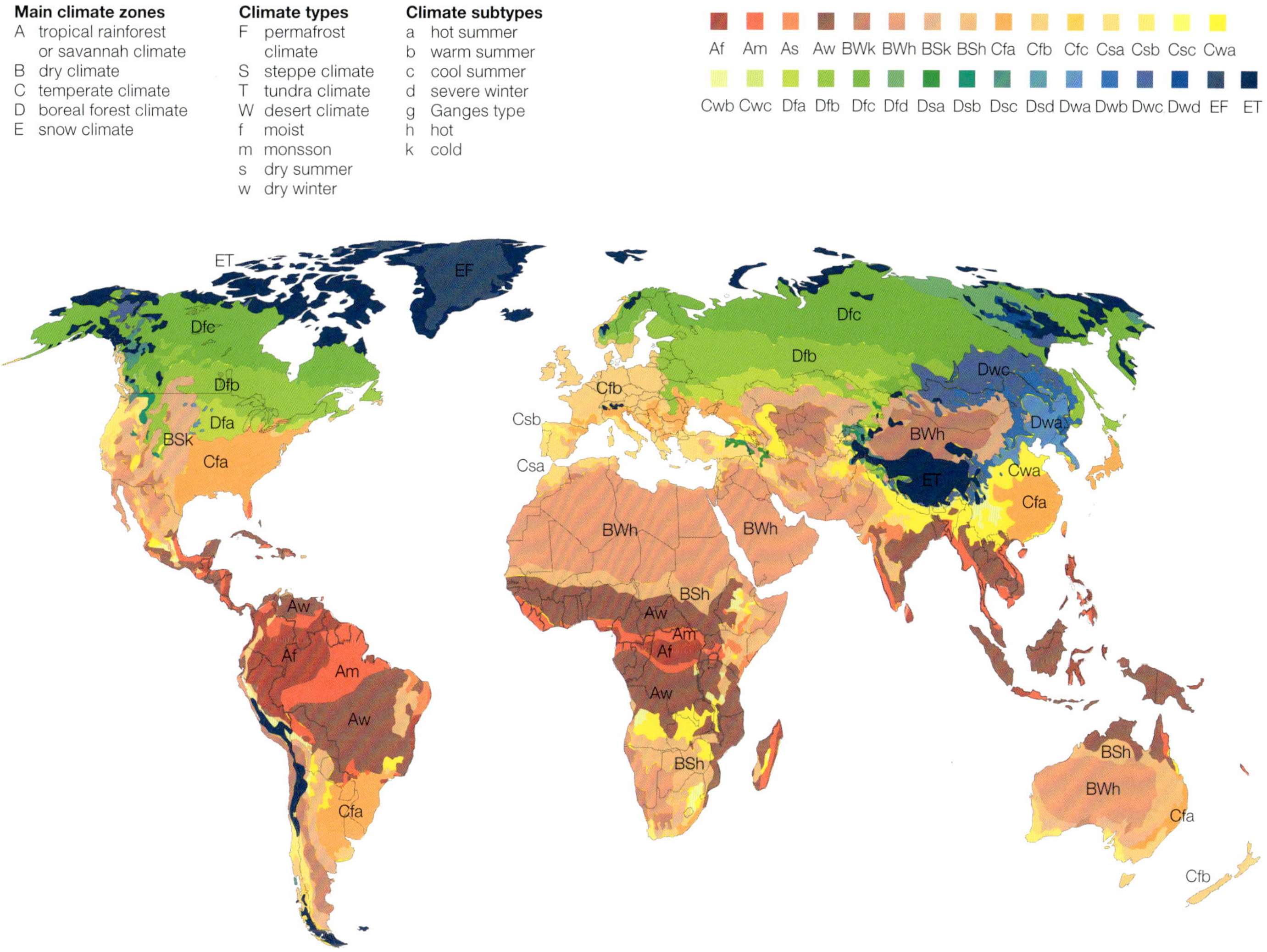
Main climate zones
A tropical rainforest or savannah climate
B dry climate
C temperate climate
D boreal forest climate
E snow climate
Climate types
F permafrost climate
S steppe climate
T tundra climate
W desert climate
f moist
m monsson
s dry summer
w dry winter
Climate subtypes
a hot summer
b warm summer
c cool summer
d severe winter
g Ganges type
h hot
k cold
Af Am As Aw BWk BWh BSk BSh Cfa Cfb Cfc Csa Csb Csc Cwa
Cwb Cwc Dfa Dfb Dfc Dfd Dsa Dsb Dsc Dsd Dwa Dwb Dwc Dwd EF ET
ET
EF
Dfc
Dfb
Dfa
BSk
Cfa
Cfb
Csb
Csa
Dfc
Dfb
Dwc
Dwa
BWh
ET
Cwa
Cfa
BWh
BWh
BSh
Aw
Am
Af
Aw
Af
Am
Aw
BSh
BSh
BWh
Cfa
Cfb

1.22

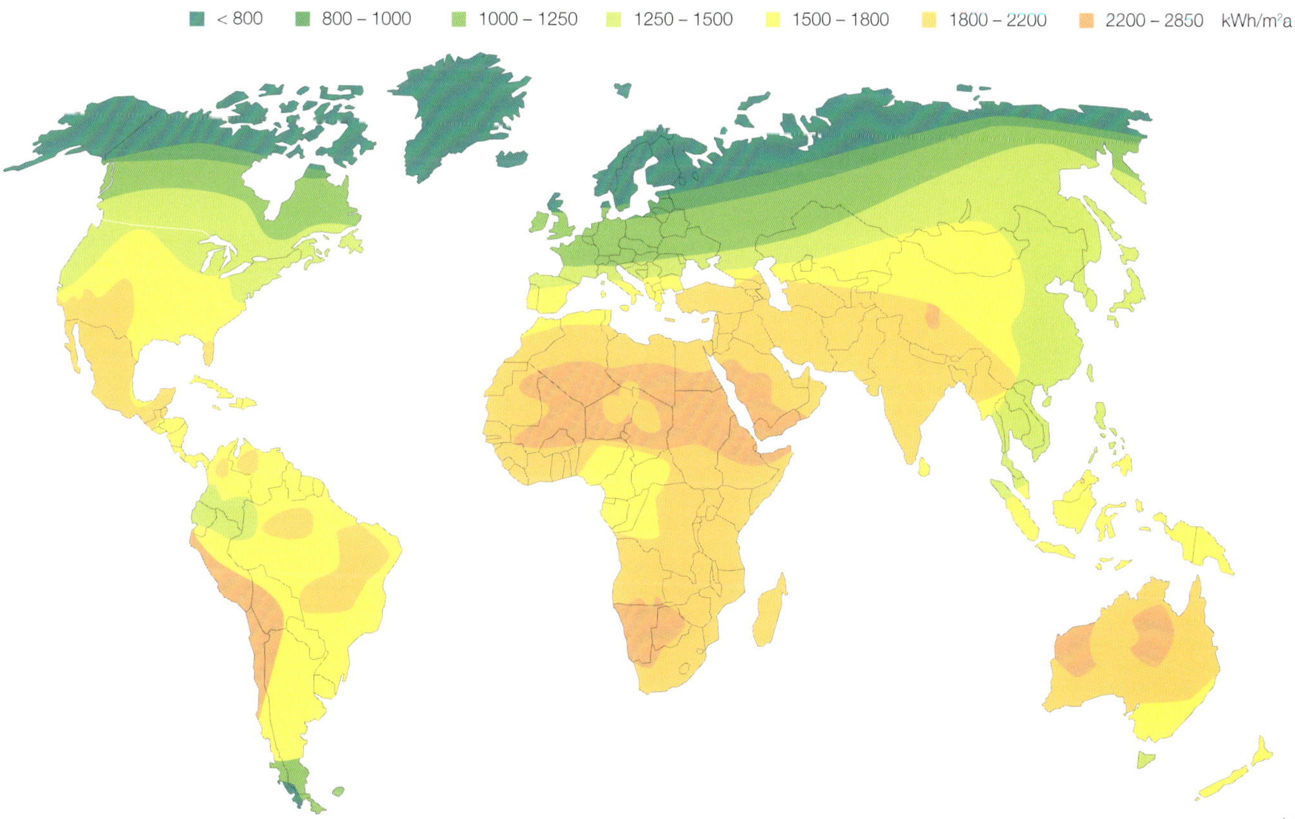
< 800
800 – 1000
1000 – 1250
1250 – 1500
1500 – 1800
1800 – 2200
2200 – 2850
kWh/m²a

1.23

Climate data for selected locations

Location: Cairo	30° 3'53'' N / 31° 14'58'' E / 68 m ASL											
	J	F	M	A	M	J	J	A	S	O	N	D	Year
Ø temperature [°C]	13.9	15.5	17.7	21.6	24.8	27.5	28	27.9	26.7	23.6	19.2	15.0	21.7
Ø daily maximum [°C]	19	20	21	24	27	29	30	31	30	28	25	21	
Ø daily minimum [°C]	10	10	12	14	17	21	23	23	22	19	16	12	
rel. humidity [%]	65	61	56	50	50	54	58	62	59	62	63	63	
precipitation [mm/m²]	6	4	4	0	0	0	0	0	0	1	4	5	29
sunshine hours [h]	9	9	8	9	11	11	10	11	8	9	7	6	
solar radiation on horizontal surface [kWh/m²]	93	96	139	167	204	224	216	195	163	129	92	85	1803
ideal inclination [°]	54	45	32	17	4	-3	1	13	28	41	51	56	26
solar radiation on ideally inclined surface [kWh/m²]	128	120	157	171	194	205	202	194	178	156	122	119	1946

Location: Würzburg	49° 47'39'' N / 9° 55'38'' E / 177 m ASL											
	J	F	M	A	M	J	J	A	S	O	N	D	Year
Ø temperature [°C]	0.3	1.3	5.3	8.8	13.7	16.6	18.7	18.4	14.2	9.1	4.2	1.7	9.4
Ø daily maximum [°C]	2	4	9	14	19	22	24	23	20	14	7	3	
Ø daily minimum [°C]	-4	-3	0	3	7	11	12	12	9	5	1	-2	
rel. humidity [%]	84	78	74	68	67	65	69	69	76	83	86	85	
precipitation [mm/m²]	42	34	42	39	53	66	57	52	42	50	45	55	577
sunshine hours [h]	2	3	4	5	7	7	8	7	5	3	2	1	
solar radiation on horizontal surface [kWh/m²]	22	41	74	113	149	151	166	134	94	55	28	17	1044
ideal inclination [°]	62	58	45	32	20	13	17	28	42	54	61	62	34
solar radiation on ideally inclined surface [kWh/m²]	34	60	92	126	152	147	166	145	115	76	42	26	1181

Location: Helsinki	60° 10'11'' N / 24° 56'17'' E / 14 m ASL											
	J	F	M	A	M	J	J	A	S	O	N	D	Year
Ø temperature [°C]	-6.9	-6.8	-2.9	2.9	9.9	14.9	16.6	15.0	10.0	5.4	0.1	-4.1	4.5
Ø daily maximum [°C]	-3	-4	0	6	14	19	22	20	15	8	3	-1	
Ø daily minimum [°C]	-9	-9	-7	-1	4	9	13	12	8	3	-1	-5	
rel. humidity [%]	88	85	77	69	60	68	69	73	78	84	87	88	
precipitation [mm/m²]	41	31	34	37	35	44	73	80	73	73	72	58	651
sunshine hours [h]	1	2	5	7	8	9	9	9	6	2	1	1	
solar radiation on horizontal surface [kWh/m²]	8	26	62	111	161	168	169	118	66	31	10	4	934
ideal inclination [°]	75	71	58	43	29	21	25	35	50	62	69	70	41
solar radiation on ideally inclined surface [kWh/m²]	8	33	65	124	167	187	176	123	73	32	11	5	1003

Location: Bangui	5° 12'0'' N / 25° 10'59'' E / 367 m ASL											
	J	F	M	A	M	J	J	A	S	O	N	D	Year
Ø temperature [°C]	24.8	26.8	27.3	26.8	26.1	25.3	24.6	24.5	25	25.1	25.1	25.7	25.6
Ø daily maximum [°C]	29	32	33	32	31	29	28	28	29	29	29	29	
Ø daily minimum [°C]	19	21	23	23	23	22	22	22	22	22	22	20	
rel. humidity [%]	68	60	71	76	79	83	85	86	85	85	84	76	
precipitation [mm/m²]	15	32	105	128	155	167	191	229	193	202	73	27	1517
sunshine hours [h]	8	8	7	6	7	5	5	5	5	5	6	8	
solar radiation on horizontal surface [kWh/m²]	206	193	205	181	180	167	164	166	168	170	181	198	2179
ideal inclination [°]	35	24	9	-9	-23	-29	-26	-15	2	18	32	37	9
solar radiation on ideally inclined surface [kWh/m²]	223	202	206	176	170	156	154	160	167	175	194	216	2201

1.24

Standard climate Germany	J	F	M	A	M	J	J	A	S	O	N	D	Year
average monthly temperature [°C]	-1.3	0.6	4.1	9.5	12.9	15.7	18.0	18.3	14.4	9.1	4.7	1.3	8.9
radiation on 90° south-facing surface [kWh/m²]	41.7	41.0	59.5	98.6	88.5	93.6	100.4	83.3	82.8	60.2	38.9	24.6	810
radiation on 90° east/west-facing surface [kWh/m²]	18.6	24.8	39.4	93.0	94.3	108	116	85.6	64.8	37.9	20.2	11.2	713
radiation on 90° north-facing surface [kWh/m²]	10.4	15.5	25.3	47.6	58.3	73.6	74.4	52.0	34.6	24.6	13	7.4	433

1.25

b Warm summer
Average temperature of the hottest month is below +22 °C; at least four months have an average temperature of at least +10 °C.
Example: Karlsruhe

c Cool summer
Average temperature of the hottest month is below +22 °C; up to three months have an average temperature of at least +10 °C.
Example: Reykjavik

d Severe winter
Average temperature of the coldest month is below -38 °C.
Example: Werchojansk

g Ganges type of annual temperature profile
The annual maximum occurs before summer solstice and the summer's rainy season.
Example: New Delhi

h Hot
Average annual temperature is above +18 °C.
Example: Karachi

k Cold
Average annual temperature is below +18 °C.
Example: Teheran

Würzburg in southern Germany, used by the Energy Saving Ordinance (Energieeinsparverordnung, EnEV) 2009 as the reference location for an average climate in Germany, is employed as an example here to illustrate the method and its practical value in planning energy-efficient buildings.

- climate zone C
- climate type f
- climate subtype b

The climate situation in Würzburg is therefore designated as Cfb, which in layman's terms means a moist temperate climate with warm summers.

Detailed description
A detailed description of the climate is made by using single values that are recorded on a monthly basis. This climate data can then be incorporated in the energy analysis (fig. 1.24 and 1.25). The evaluation of data concerning the useful amount of solar radiation is particularly important for the development of an energy supply concept using renewable energy sources (fig. 1.23, p. 21).
The energy efficiency of buildings in Germany is assessed according to DIN V 18599 based on a standard cli-

mate with an average monthly temperature and a distinction of solar radiation according to the four cardinal directions and the panel inclinations 0°, 30°, 45°, 60° and 90°. The standard does not provide data to perform a differentiated site-related assessment in Germany.

Energy efficiency refurbishments

An energy efficiency refurbishment should improve a building in such a way that its heating, cooling and ventilation is affected by incorporating the climate conditions prevailing on site. In addition to shortcomings concerning the insulation, existing buildings often have issues with summer heat protection. They frequently overheat due to a lack of or insufficient structural measures and have to be cooled by using some kind of technical plant.
Uncomfortable room conditions in summer are caused by too large window areas without sun protection or interior shading devices, insufficient possibilities to ventilate the building with cool night-time air, a lack of storage mass as well as the internal heat gains added by occupants and equipment.
The construction-related measures that should be performed in the context of a refurbishment in a central European climate (climate zone Cfb) include:

- A perfectly insulated building envelope reduces the considerable heat loss in winter.
- Depending on the function and the orientation of the building, a window surface area of 30 to 60% of the facade's total surface is appropriate to, on the one hand, provide perfect daylight conditions but, on the other hand, to limit heat gain in summer. All windows facing west, east and south must be fitted with flexible, external sun protection devices. Interior glare protection can be added if required.
- Structural heat-absorbing components in the building's interior are able to store excess heat in summer, which is dissipated to the cool outside air at night (see Heat sources, pp. 69ff.). It is for this reason that storage mass should be uncovered during the course of a refurbishment. Alternatively, it is possible to retrofit latent heat stores by installing suspended ceilings with extra storage mass.

By carrying out the appropriate alterations and additions to the construction, existing buildings can be improved as regards protection against summer heat and prepared to operate with considerably less plant. The logical consequence is a significant reduction not only in energy consumption, but also in operating costs.
The refurbishment of an office building in the centre of Stuttgart offers a good example of the removal of air conditioning plant, which was made possible by carrying out the appropriate construction-related measures (fig. 1.26).
By replacing the former fixed glazing of sun protection glass, it is now possible to respond to the different climate conditions in a much more flexible way. The most important structural component in this context is a second skin made of movable glass lamellae with sun protection blinds mounted on the inside, between the new and old facade. The controllable second skin facade creates a buffer zone which is used to reduce heat loss in winter and allow natural ventilation of the building during the day and at night in summer. It is at the same time intruder proof, independent of bad weather conditions and permits the use of the sun protection devices in any wind conditions. Furthermore, daylight penetration is improved by removing the former sun protection glazing.

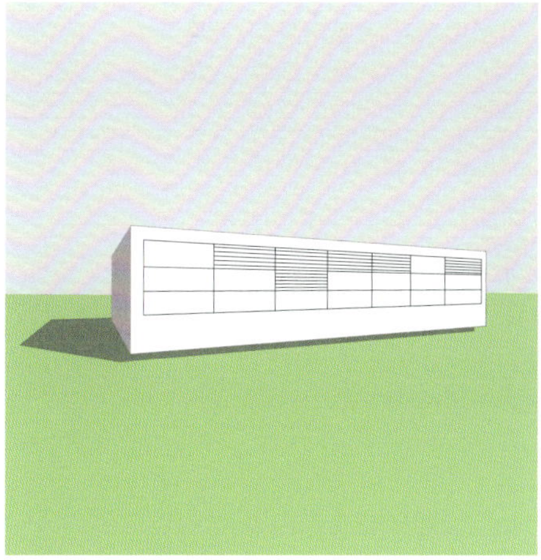

Example: site climate

Here the monthly heat sources and heat sinks, the resulting useful heat demand and useful cold demand as well as the heating period in Freiburg and Garmisch-Partenkirchen are illustrated for the reference building. The heat sinks are the heat loss due to transmission and ventilation; the heat sources are the solar yield through transparent surfaces as well as the heat gains from the building's use.

Freiburg (278 m ASL)

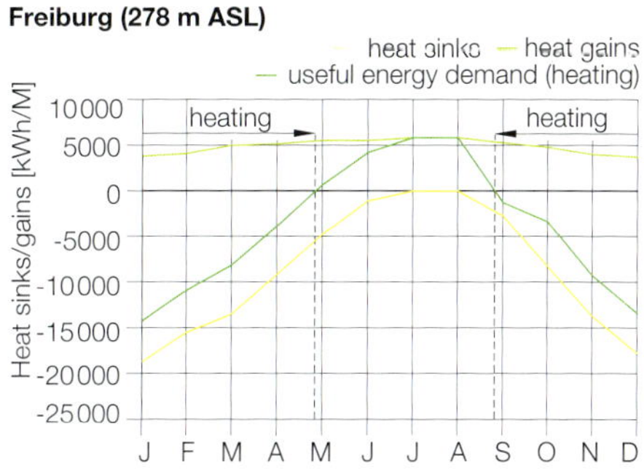

Garmisch-Partenkirchen (708 m ASL)

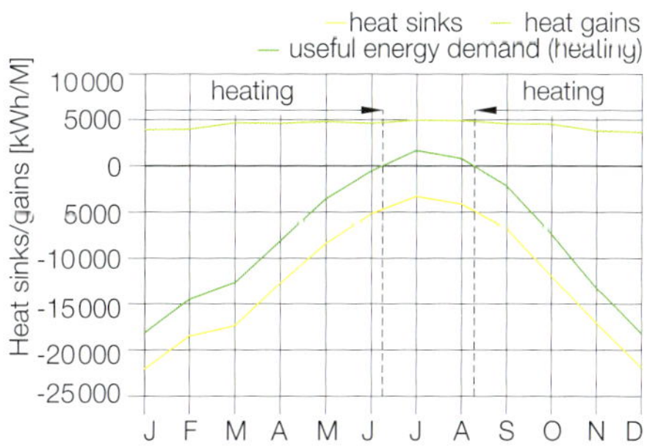

Comment

The diagrams show that, despite the same construction, the specific heat sources and heat sinks lead to a different useful energy demand due to the climate conditions.
The heating period commences as soon as the heat loss is greater than the heat gain. The excess heat in summer may require that the building be cooled.

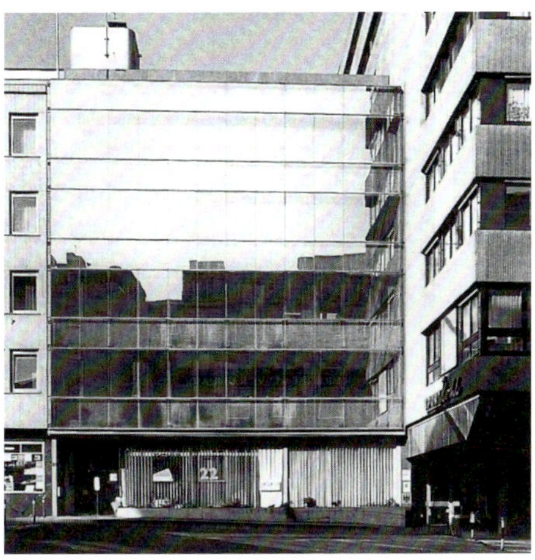

a

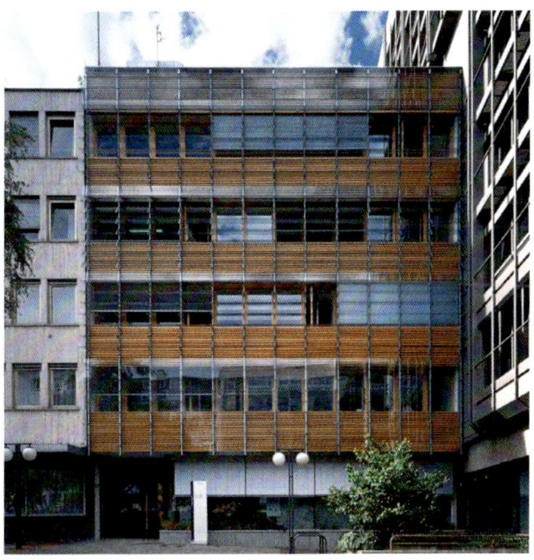

b 1.26

1.24 Applicable climate data for selected locations in different climate zones
1.25 The "standard climate" in Germany which is used to assess the energy efficiency according to DIN V 18599 (extract)
1.26 Office building in Stuttgart 1969/1994
Architect: Behnisch und Partner
a facade before the refurbishment
b double skin facade after the refurbishment

Design parameters

Important parameters concerning the energy required to operate, but also to construct the building, are already determined during the early design stages. These include in particular the compactness, arrangement of functions and the orientation of the building. The opportunities to later make alterations to these fundamental aspects are very limited. The improvements that can then be made by structural measures are marginal in comparison.

Compactness

The compactness of a building is a principal feature of a design concept's energy efficiency. The definition of compactness differentiates according to the perspective taken (fig. 1.28):

- A / V_e is the ratio of the building envelope enclosing the heated volume to the heated volume
- V_e / A_{NFA} is the ratio of the heated or cooled volume to the conditioned net floor area
- A / A_{NFA} is the ratio of the building envelope enclosing the heated or cooled volume to the conditioned net floor area
- A / A_{UFA} or A / LA is the ratio of the building envelope enclosing the heated or cooled volume to the conditioned usable floor area or conditioned living area
- V_e / A_{UFA} or V_e / LA_{cond} is the ratio of the heated or cooled volume to the conditioned usable floor area or conditioned living area.

Only the last two of these definitions can be applied if a holistic approach is to be taken, because it is only in these two cases that an optimised design of the building – with for example a reduction of circulation and other service areas – is considered from the outset. If, for example, a residential building includes a heated basement, tall spaces incorporating two or more storeys, open-plan living rooms under the eaves or extravagant entrance areas, the surface area of the building envelope and the heated volume increase, but the living area remains small. However, these additional envelope areas are included in the assessment of energy efficiency only if the conditioned usable floor area or living area are referred to in the definition of compactness. To date, the only assessment that is performed according to these more realistic terms is the assessment for passive house standards. This is also the reason why passive houses do not allow

User profile for open-plan offices according to DIN V 18599

Occupancy	Unit	Quantity
occupancy per day	hours	7am–6pm
days of occupancy per year	d/a	250
annual hours of occupancy during the day t_{day}	h/a	2543
annual hours of occupancy at night t_{night}	h/a	207
daily operation period of aircon, ventilation, cooling	hours	5am–6pm
annual operation days of each aircon, cooling and heating $d_{op,a}$	d/a	250
daily operation period of heating	hours	5am–6pm
Room conditioning (if provided)		
room set temperature for heating $\vartheta_{i,h,set}$	°C	21
room set temperature for cooling $\vartheta_{i,h,set}$	°C	24
minimum design temperature for heating $\vartheta_{i,h,min}$	°C	20
minimum design temperature for cooling $\vartheta_{i,h,max}$	°C	26
temperature decrease for low occupancy period $\Delta\vartheta_{i,nf}$	K	4
moisture demand	–	with margin
Minimum outside airflow V_o		
person related	m³/(h·person)	60
space related	m³/(h·m²)	6

Mechanical outside air-flow or air exchange	Unit	Quantity
air exchange (general)	h^{-1}	2–3
air exchange (full cooling via supply air)	h^{-1}	4–8
Lighting		
illuminance maintenance value $\bar{E}_m$	lx	500
height of work surface h_{ws}	m	0.8
reduction factor k_A	–	0.93
relative absence C_A	–	0
room index k	–	2.5
reduction factor for building operation period F_t	–	1
Occupant load		
maximum number of occupants	m²/person	low: 12 medium: 10 high: 8
Internal heat sources		
	full use hours h/d	max. specific power output W/m²
persons (70 W/person)	6	low: 6 medium: 7 high: 9
office equipment [1]	6	low: 4 medium: 10 high: 19
heat input per day $q_{i,p} + q_{i,fac}$	Wh/(m²·d)	low: 60 medium: 102 high: 168

[1] low/medium/high correspond to 50/100/150 W per person for office equipment

1.27

Space and volume definitions and their applicability

Term	Symbol	Application	Definition
conditioned volume	$[V_e]$	EnEV	Volume (exterior dimensions) enclosing heated or cooled space. The exact boundary between conditioned and unconditioned space (system boundary) is determined in DIN V 18599.
envelope surface area	[A]	EnEV	The envelope surface area describes the area enclosing the conditioned volume.
conditioned net floor area	$[A_{NFA}]$	EnEV	The conditioned net floor area is calculated according to DIN 277 and then divided into so called zones. The zones, which are each defined by characteristic conditions in accordance with their use, are determined in DIN V 18599-10.
conditioned usable floor area	$[A_{UFA}]$	lease, sale	In comparison to the conditioned net floor area, this assessment only considers the actually usable floor space as a reference value. Circulation zones and staircases are not taken into account, even if they are heated or cooled. In this case, the compactness and functional quality of the floor plan are incorporated in the assessment.
conditioned living area	$[LA_{cond}]$	passive house standard	Area within a dwelling unit that is conditioned. Balconies are not considered in this case. This value is meaningful for reference purposes since it describes the compactness of residential buildings which can then be assessed in comparison to other properties.
living area	[LA]	lease, sale	According to the German Residential Space Ordinance (Wohnflächenverordnung), the living area also includes areas that are not conditioned (balconies, terraces). In order to assess the compactness, the above-mentioned term "conditioned living area" is more meaningful as a reference value.

1.28

1.27 User profile of open-plan offices to assess the energy efficiency accoridng to DIN V 18599-10, table A.4

1.28 Definition of space and volume terms as a basis for the calculation of building compactness

heated basements and the buildings tend to be rather cubic in shape and therefore fairly compact.
Up until the Energy Saving Ordinance (Energieeinsparverordnung, EnEV) came into force in 2009, the geometric ratio of envelope to volume was a criterion used in determining the energy efficiency objectives of a building. This classification of buildings according to their geometric compactness was, however, abolished with the introduction of the EnEV 2009. Since then the EnEV uses an abstract reference area A_n, which relates to the conditioned volume V_e, for the energy assessment of residential buildings ($A_n = 0.32 \cdot V_e$). The assessment method provides no incentive to design compact buildings and layouts of functions, despite the fact that compactness has a strong impact on the energy demand.
Based on the joint effectiveness of DIN V 18599 and EnEV 2009, a direct reference area is introduced for the energy assessment of non-residential buildings applying the conditioned net floor area A_{NFA} according to DIN 277 as a reference factor. The optimisation of a building's design concerning a small proportion of circulation area or other areas not intended for immediate use is not encouraged in this case either.
It is important to note that, by improving the compactness of a building, energy can be saved in two ways: the reduction of envelope area not only minimises the transmission heat loss during the operation of the building (useful energy), it also leads to energy savings in manufacturing the envelope (production energy).
The relation between compactness and urban density is obvious (see adjacent example). Compact, large buildings, such as multi-storey residential buildings, have only up to one third of the envelope area of single family homes with the same living or useful floor area. From an energy efficiency point of view, the single family home is therefore the most absurd form of living.

Building use

The energy demand is dependent on the use of the building. Whereas residential buildings are fairly easy to describe in terms of energy due to their continuous 24-hour use, this is a lot more difficult in the case of non-residential buildings due to the constantly changing conditions. In order to provide at least a certain degree of comparability between the energy demands, DIN V 18599-10 lists forty different usage types that are described by a selection of different criteria.
Figure 1.27 illustrates, by way of example, the expected assessment values for the use of an open-plan office (office with seven or more workplaces). The user profile is categorised according to occupation periods, room temperature, outside air volume flow, lighting, number of occupants and interior heat sources.
Based on these differentiated values, each non-residential building must be subdivided into various zones in accordance with their usage type so that the use-related energy demand can be determined for each zone individually. The sum of the energy requirements of each of the individual zones gives the building's total energy demand. Thus the energy demand changes as soon as the use of the building changes. The usage of the building is therefore an energy-related design parameter.
However, a change of use does not necessarily require the issue of a new energy certificate. This is only the case if structural alterations are made.

Orientation

The orientation of rooms or groups of rooms is extremely important with regard not only to the energy required for heating, but also that needed for cooling. Some measures that have a positive effect in winter are frequently contraproductive in summer. Due to the high internal heat loads, non-residential buildings, in particular, require a differentiated approach regarding the orientation which, among other things, should also consider aspects such as urban design, exposure to noise and the penetration of daylight. To avoid excessive heat gains, the building must be conceived so that external heat loads (solar irradiation) do not occur at the same time as the internal occupancy-related heat loads. The classrooms in a school, for example, which are predominantly used in the morning, should not face east since this is also where solar heat gains occur most during the early hours of the day.
Since non-residential buildings, in contrast to residential buildings, have a large number of internal heat sources, additional heat gains from the outside must be avoided. The objection is frequently raised that the improvements made to summer heat protection might also eliminate the desired heat gains in winter. But in fact these heat gains are so slight that

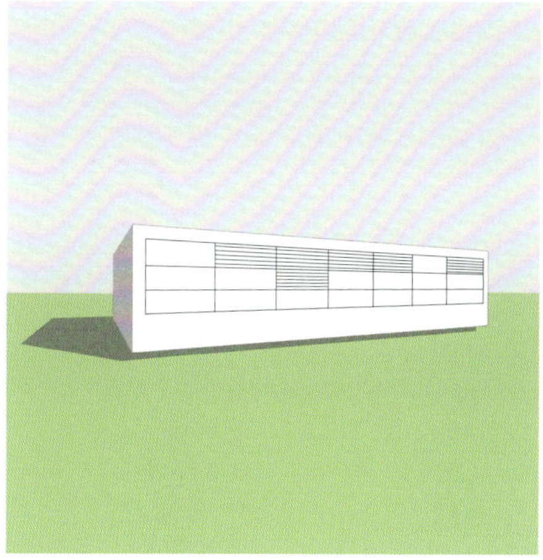

Example: building shape
The office pavilion's conditioned usable floor area of 900 m² (6 cubes of each 150 m²) is arranged in different ways: single storey with a linear arrangement of six separate cubes, single storey with a linear arrangement without gaps, and as a two-storey building. The heated usable floor area in the vertically arranged design is reduced by 10 m² per cube to allow for vertical stair access.

Design 1: linear arrangement with 6 single cubes

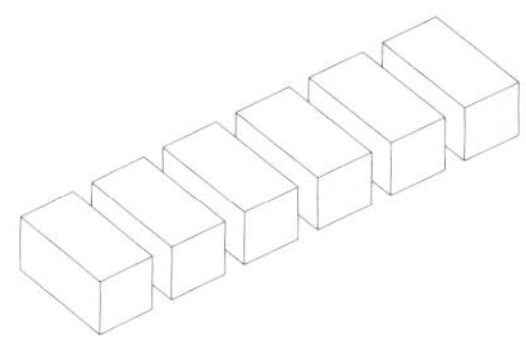

• usable floor area	900 m²
• envelope surface area	4090 m²
• relation usable floor area to envelope surface area	4.5

Design 2: linear arrangement

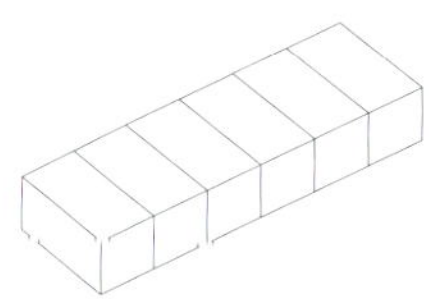

• usable floor area	900 m²
• envelope surface area	3603 m²
• relation usable floor area to envelope surface area	4.0

Design 3: two-storey arrangement

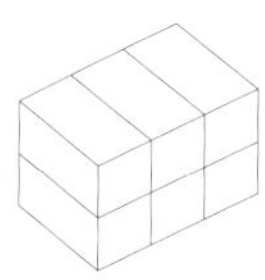

• usable floor area	840 m²
• envelope surface area	2349 m²
• relation usable floor area to envelope surface area	2.8

Comment
Despite the same use, approximately 40 % less envelope surface is required for the most compact arrangement. This design reduces transmission heat loss, the energy required for the construction of the building, and the building costs.

they can be safely neglected in the interest of reducing heat gains in summer. Regarding external heat gains and the orientation of the building, the proportion of window surface area is of crucial importance in improving the energy efficiency since, in well insulated buildings, solar heat gains occur only through transparent surfaces.
However, the window surface area together with the orientation of openings does not only influence the heat gains, it also has an effect on the penetration of daylight and heat loss in winter. From an energy efficiency point of view, the perfect size and orientation of transparent surfaces must therefore be determined individually for each property. Architectural aspects, such as connections to outdoor space or the handling of daylight, must also be included in the considerations.

Energy efficiency refurbishments

An examination whether and to what extent the energy demand can be reduced by conceptual measures should be made at a very early design stage, even in the case of energy efficiency upgrades.

Compactness

It is often believed that compactness has no important role to play in energy efficiency refurbishments since the volume of the building, the envelope and the heated useful floor area have already been determined by the original building. In fact, however, almost every property due for refurbishment offers potential for improving the density.
In terms of conditioned volume, existing properties can be enlarged by extensions and conversions. Conversions can increase the useful floor area within the existing volume. In these cases, the compactness of the property is improved, which in turn has a positive effect on the overall energy efficiency of the building (fig. 1.29).
In some circumstances, it might make sense to reduce the volume of the building by removing certain elements. This can be the case with existing large-scale structures, such as the large-panel system buildings in former East Germany, which can no longer be rented out due a lack of acceptance. These buildings can only meet today's standards by removing some of the volume (fig. 1.30). In this case, the compactness is decreased for more important reasons. The disimprovement this entails in terms of energy efficiency is accepted in the interest of ensuring a sustainable long-term use of the property.

Utilisation

The conditioned floor area can be increased during the course of a refurbishment by improving the layouts. The benefit of such measures must be analysed by studying the space and design concepts of the property.
In the case of extensive refurbishment work, there is often an opportunity to totally rearrange the layout by taking into account all energy efficiency aspects, by pooling similar room functions and therefore improving the building from an energy point of view.

Orientation

It goes without saying that the basic orientation of the building cannot be altered when performing refurbishment measures. However, it is possible to consider whether the floor plans can be revised to, for example, place rooms with large internal heat loads, such as conference rooms, into building zones exposed to little or no solar radiation. The rearrangement of uses can also be performed to achieve a time delay between the exposure to external and internal heat loads. A further measure which can be performed to correct defects concerning the orientation, is to marginally reduce the proportion of window surface area in fully-glazed buildings.

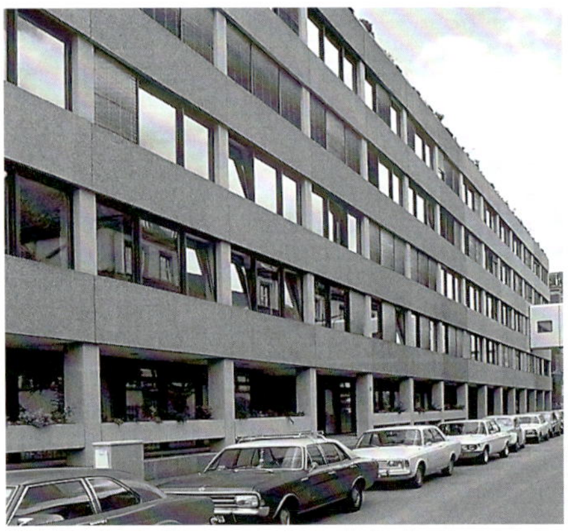
a

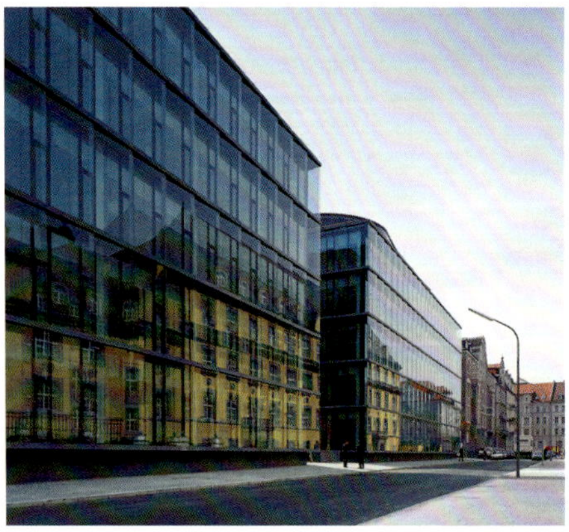
b 1.29

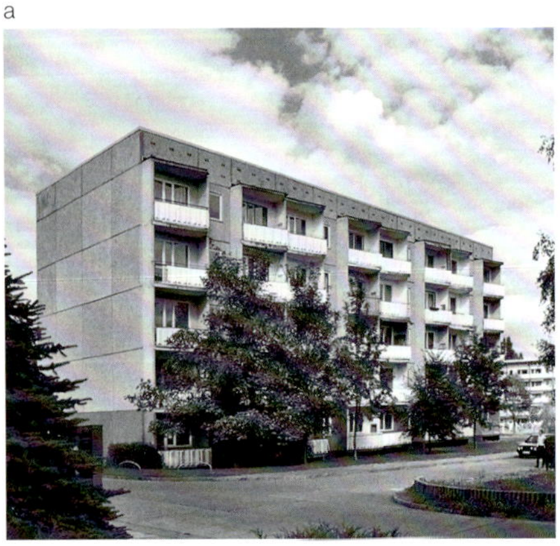
a

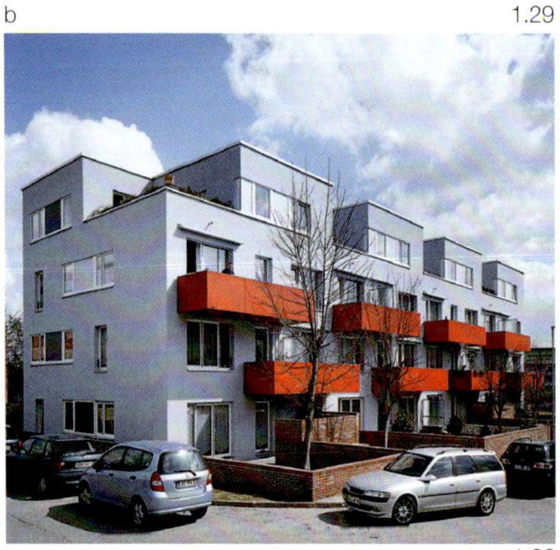
b 1.30

1.29 Extension to Münchener Rückversicherung
Location and completion: Munich, 2001
Architect: Baumschlager Eberle
The existing office building of the Münchener Rückversicherung was refurbished and integrated into the new grounds created during the course of extension work. The selected design concept contributes towards improving the energy demand in two ways: firstly, by reusing the existing building stock, the energy demand to carry out the construction work was reduced, secondly, by increasing the density, the compactness on the site was improved contributing towards further reductions of the energy demand to produce and condition the building.
a former building
b refurbished building

1.30 Refurbishment of large-panel system building "Haus 4 – Goethestraße 25–31"
Location and completion: Leinefelde, 2003
Architect: Stefan Forster Architekten
Increasing the density is not always beneficial. Due to a dwindling demand and higher standards of living, the deconstruction of once compact buildings can provide a basis for ensuring continued use of the remaining substance.
By removing elements, changing floor plans and making small additions, a no longer lettable large-panel system building has been improved to ensure that a considerable proportion of the former building mass is now being used on a long-term basis.
a former building
b refurbished building

Appropriateness

The appropriateness of a building's concept is expressed by the time, effort and expense invested in the selection and configuration of materials to meet the specific user requirements. The choice and quantities of materials is influenced by the compactness of the volume, the efficiency of the structure that transfers the vertical and horizontal loads, and the details used to complete the interior and exterior enclosing surfaces. The type and scope of mechanical and electrical installations are also of considerable importance in this context.

Energy balance

The issues concerning the amount of energy required for the production of building materials, the so-called ecological rucksack, was first mentioned to describe and assess the environmental quality of products in the DGNB Certificate, which was introduced by the German Sustainable Building Council in 2009.
Based on a building materials database (Ökobau.dat in www.nachhaltigesbauen.de) listing the appropriate key figures, it is now possible to assess the total environmental load associated with the manufacturing process of a building product, the finished component and finally the completed building and to define the load as a measurable and therefore comparable factor as early as during the design stage of a building.
The following parameters are considered and can be included in the assessment:

- Global Warming Potential (GWP)
  The global warming potential is referred to as the carbon dioxide equivalent (CO_2 eq) (fig. 1.31, p. 28). The CO_2 equivalent considers all pollutants that contribute towards climate change.
- Ozone Depletion Potential (ODP)
  The emissions induced by humans contribute towards the degradation of the ozone layer in the stratosphere. The ozone layer regulates the quantity of UV radiation reaching the Earth's surface. The ozone depletion potential is referred to as R11 equivalent (R11 eq).
- Acidification Potential (AP)
  The sulphur dioxide and nitrogen dioxide in the air reduce the ph level in rain water. The so-called acid rain harms the ecosystem and, among other things, leads to forest dieback.
  The acidification potential is expressed in terms of its sulphur dioxide equivalent (SO_2 eq).
- Eutophication potential (EP)
  Due to pollutants (fertilisers, air contamination), the food chain is becoming enriched with nutrients that cause harm to the ecosystem and humans. The eutrophication potential is considered in terms of its phosphate equivalent (PO_4 eq).
- Photochemical Ozone Creation Potential (POCP)
  The ozone concentration in lower levels – summer smog in the troposphere – is believed to harm vegetation and, in the case of larger quantities, also humans. In life cycle assessments, the photochemical ozone creation potential is expressed as ethane equivalent (C_2H_4 eq).

Generally discussions on the appropriateness of construction work use the wrong focus, as frequently the energy balance of a component is only considered and assessed from a quantitative perspective. The quality-related issue concerning the quantities of materials required to construct the building is not considered.
Environmentally friendly construction work begins at the design phase when, for example, creating a building with an intelligent layout and a small envelope surface area in proportion to the usable floor area.
The first step in assessing the appropriateness of the design at this early stage is therefore to establish the quotient of surface area to heated usable floor area. The smaller this quotient, the greater the compactness and consequently the better the design from the viewpoint of material consumption. Thus the arrangement of a specific usable floor area on two levels requires less envelope surface area and less material than the arrangement of the same space on a single level.
The use or reuse of existing building stock is also a means of saving energy that would otherwise be required to construct a new build (fig. 1.33, p. 28).
If buildings are, from a structural and installation point of view, designed to allow for changes in use over time, a reduction of the production energy is achieved due to the increase in the building life span.

Environmental assessment

The environmental effort invested in the production of building materials that are eventually intended to save energy is expressed by the two following terms:

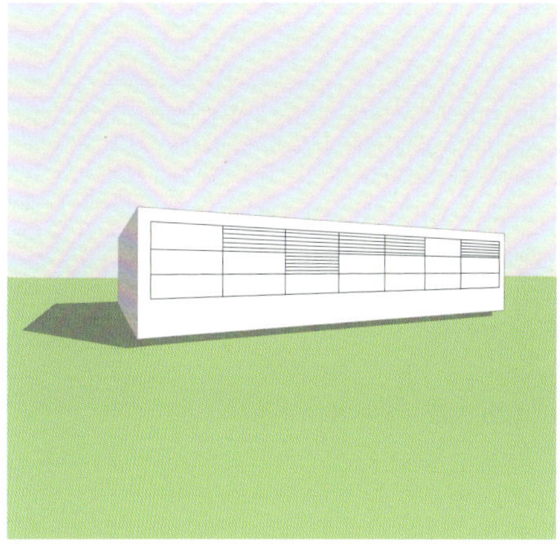

Example: appropriateness
In order to assess the appropriateness of structural decisions, the pavilion is used here to examine whether fitting triple glazing in comparison to double glazing is beneficial from a wholly environmental perspective. This example is representative for many other decisions.

Double glazing
4-16-4 with argon-filled cavity
- g-value 62 %
- U-value 1.1 W/m²K
- τ-value 80 %

Triple glazing
4-16-4-16-4 with argon-filled cavity
- g-value 61 %
- U-value 0.7 W/m²K
- τ-value 73 %

Heating: gas condensing boiler system
observation period: six-month heating period
location: Munich
glass surface area: 26 m²
Sun protection on the south-facing facade is dealt with by fitting exterior-mounted blinds.

Production energy
(Ökobau.dat at www.nachhaltigesbauen.de):
- double glazing 584 kg CO_2
- triple glazing 876 kg CO_2

North-facing glazing
Double glazing
- energy loss -1888 kWh
- energy gain +914 kWh
- energy balance -974 kWh

Triple glazing
- energy loss -1201 kWh
- energy gain +899 kWh
- energy balance -302 kWh

Total balance (30 years)
- CO_2 saving (heating) 4898 kg
- CO_2 excess (production) 292 kg

South-facing glazing
Double glazing
- energy loss -1888 kWh
- energy gain +2468 kWh
- energy balance +580 kWh

Triple glazing
- energy loss -1201 kWh
- energy gain +2428 kWh
- energy balance +1227 kWh

Total balance (30 years)
- CO_2 saving (heating) 4710 kg
- CO_2 excess (production) 292 kg

Comment
Independent of orientation, the fitting of triple glazing always represents an ecologically sensible measure, which can be easily carried out during the course of refurbishment work.

Energy payback time

The energy payback or amortisation time is the period required by a building material or product to harvest the same amount of energy as was used for its production (fig. 1.32).
Photovoltaic panels, for example, have an energy payback time of two years, insulation materials, in contrast, have a payback time of only two to six months.

Energy yield factor

The energy yield factor expresses how much more energy a product saves or yields during its lifetime compared to the amount of energy required to produce it. For a photovoltaic module measuring one square meter, the factor is approximately 8; in the case of insulation material, the factor is 15 to 30. The production of, for example, one square meter of insulation material, a 10-cm-thick mineral wool panel, requires one litre of oil. This same insulation material reduces the amount of energy required to heat the building by one litre of oil per year during the insulation material's lifetime (30 years). The energy payback time is, in this case, one year, the energy yield factor is 30.
In a second step, the envelope surface area must or can be improved in terms of its structural configuration. The aim is to achieve a solution using materials appropriate to the task, i.e. materials with the least environmental impact.

Sustainable building

The DGNB assessment scheme uses a holistic approach to determine and evaluate the quality of buildings. The quality in terms of energy is considered in the subitem "Life cycle assessment". Aspects considered include the environmental load created by producing the selected building materials and the input of resources to heat, cool and ventilate the building, which is rated according to the consumption of primary energy.
However, some aspects that are associated with the assessment system, such as the weighting of individual criteria or the lack of parameters to evaluate the compactness and appropriateness of buildings, require further review.
In particular, the relatively little significance paid to the subitem "Ecology" within the overall assessment is questionable. Because it is fairly easy to accumulate the necessary points for a green label in other sections, these important measures are simply not carried out. The method used to assess a property is presented in figure 1.34.

Global warming potential for the insulation of a brick wall using a variety of products

Insulation material	WLG	λ value [W/mK]	density [kg/m³]	thickness [cm]	CO_2 eq. [kg CO_2/m²]
mineral					
rockwool	035	0.035	180	9	18.9
glass wool	032	0.032	100	8	5.0
mineral foam	045	0.045	115	12	
polyurethane					
PUR- hardfoam	024	0.024	30	6	9.7
polystyrene					
expanded	032	0.032	15	8	3.7
extruded	035	0.035	40	9	8.4
foamglass					
	038	0.038	100	10	10.9
replenishable raw materials					
wood fibre	051	0.051	280	13	-23.7
cellulose	040	0.040	60	11	3.26

1.31

Environmental assessment of exterior wall insulation with different thicknesses

Material thickness	U value [W/m²K]	CO_2 eq. production [1] [kg CO_2/m²]	CO_2 eq. heating [kg CO_2/m²a]		environmental assessment	
			actual value	saving	energy yield	amortisation [a]
existing building	1.35		27.5			
+ 6 cm	0.33	4.13	6.7	20.8	150	0.2
+ 12 cm	0.21	4.13	3.3	3.4	25	1.2
+ 18 cm	0.15	4.13	2.4	0.9	25	4.7
+ 24 cm [2]	0.12	4.13	1.9	0.5	3.7	8.1
+ 30 cm	0.10	4.13	1.6	0.3	2.4	12.5

[1] for each increase of 6 cm
[2] the increase of insulation from 18 to 24 cm (additional emissions due to production: 4.13 kg CO_2/m²) reduces the annual CO_2 emissions by 0.5 (from 2.4 to 1.9 kg CO_2/m²)

1.32

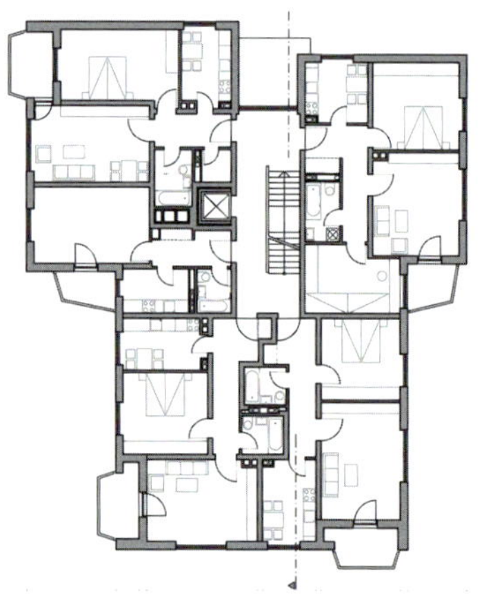
a

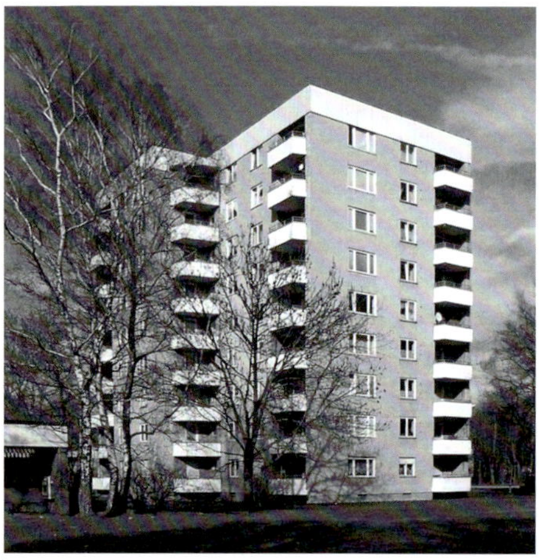
b

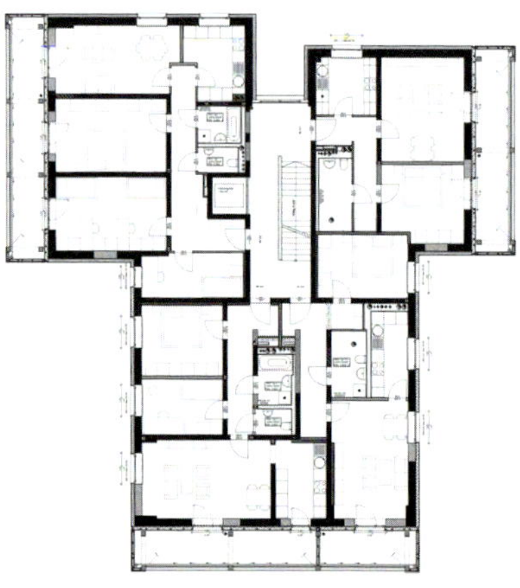
c

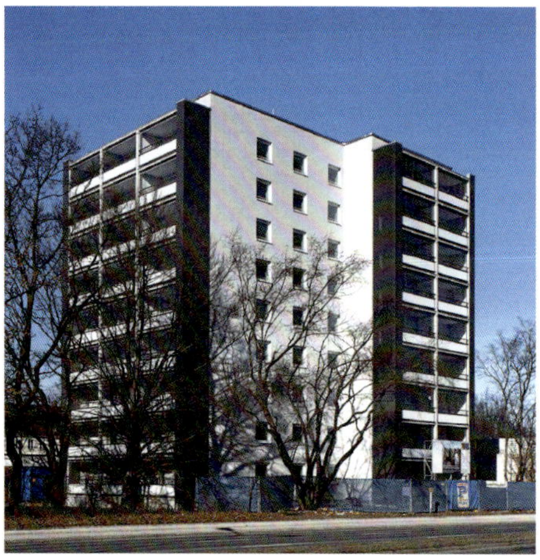
d

1.33

A gold label is awarded for an overall fulfilment of 80 % and at least 65 % in each one of the five criteria groups, a silver label in the case of 65 % and 50 % in each one of the groups, and a bronze label for an overall fulfilment of 50 % and at least 35 % in each one of the criteria groups. The green building certification system has so far only been introduced for new builds. Since 2011, it has also been possible to certify existing and refurbished office buildings. Further occupancy profiles are in preparation.

Energy efficiency refurbishments

In the case of refurbishments, the environmental potential lies in retaining already existing structural components. Every element that is removed, every impetuous and ill-considered decision in this respect, involves the use of material and machinery for its deconstruction, disposal and rebuilding. The appropriateness of refurbishments – and this also applies to energy efficiency refurbishments – is partly determined by the extent to which large parts of existing building stock can be integrated into the newly developed building concept.

1.31 Comparison of the global warming potential involved in the production of insulation materials for the insulation of a masonry wall. Initial situation: wall thickness 36.5 cm plus interior and exterior render of each 2 cm, λ value is 0.7 W/mK for both bricks and render. The aim of the insulation work is to achieve a heat transfer coefficient of 0.3 W/m²K.

1.32 Environmental assessment of insulation thicknesses (mineral woll, WLG 035) increasing in steps of 6 cm. The initial situation is a brick wall (36.5 cm thick plus interior and exterior render of each 2 cm, λ value 0.7 W/mK for both bricks and render). The observation period is 30 years. The building is heated with a gas condensing boiler system.

1.33 Refurbishment of a residential building in Munich
Construction year: 1962
Completion of refurbishment: 2011
Architects: Richarz und Strunz
The aim was to radically replan the apartment layouts (size, barrier-free design), refurbish the building envelope and upgrade the energy efficiency of the building. The nine-storey building was therefore stripped down to the carcass (load-bearing walls, basement, concrete floor slabs); the fit-out was started afresh. Due to the fact that the carcass was almost fully retained, the environmental load was reduced by 2500 t CO_2 compared to a new build.
Due to the annual CO_2 emissions for heating and hot water production of approximately 50 t, the building could, in comparison to a new build, be heated for 50 years carbon free.
a floor plan before the refurbishment
b view of the building before the refurbishment
c floor plan after the refurbishment
d view of the building after the refurbishment

1.34 Criteria and weightings of the DGNB Certificate for the occupancy profile "New office and administration buildings" (2009, extract). The awarded certificate, gold, silver or bronze, depends on the number of points achieved.

Main criteria group	Criteria group	No.[1]	Criterion	max. possible criterion points	significance factor [2]	adjustment factor [3]	max. possible weighted points	max. possible group points	group weighting
Ecological quality	Life cycle assessment	1	Global warming potential (GWP)	10	3	1	30	210	22.5 %
		2	Ozone depletion potential (ODP)	10	1	1	10		
		3	Photochemical ozone creation potential (POCP)	10	1	1	10		
		4	Acidification potential (AP)	10	1	1	10		
		5	Eutrophication potential (EP)	10	1	1	10		
	Impact on local and global environment	6	Risks to the local environment	10	3	1	30		
		8	Sustainable use of resources/ timber	10	1	1	10		
		9	Microclimate	10	1	0	10		
	Utilisation of resources and arising waste	10	Non-renewable primary energy demand	10	3	1	30		
		11	Total primary energy demand and proportion of renewable primary energy	10	2	1	20		
		14	Domestic water consumption and volume of waste water	10	2	1	20		
		15	Area demand	10	2	1	20		
Economic quality	Life cycle costs	16	Building-related life cycle costs	10	3	1	30	50	22.5 %
	Value stability	17	Suitability for third-party use	10	2	1	20		
Sociocultural and functional quality	Health, comfort and user satisfaction	18	Thermal comfort in winter	10	2	1	20	280	22.5 %
		19	Thermal comfort in summer	10	3	1	30		
		20	Indoor hygiene	10	3	1	30		
		21	Acoustic comfort	10	1	1	10		
		22	Visual comfort	10	3	1	30		
		23	Occupants' extent of control	10	2	1	20		
		24	Building-related outdoor quality	10	1	1	10		
		25	Safety and risk prevention	10	1	1	10		
	Functionality	26	Accessibility for disabled persons	10	2	1	20		
		27	Efficient use of space	10	1	1	10		
		28	Suitability for conversions	10	2	1	20		
		29	Accessibility	10	2	1	20		
		30	Convenience for cyclists	10	1	1	10		
	Quality of the design	31	Staging of an architectural competition to ensure creative and urban quality	10	3	1	30		
		32	Artwork	10	1	1	10		
Technical quality	Quality of the technical configuration	33	Fire protection	10	2	1	20	100	22.5 %
		34	Sound insulation	10	2	1	20		
		35	Quality of building envelope's thermal and moisture insulation	10	2	1	20		
		40	Structure's suitability for upkeep and repair	10	2	1	20		
		42	Suitability for deconstruction, recycling and reuse	10	2	1	20		
Process quality	Quality of planning	43	Quality of project preparation	10	3	1	30	230	10.0 %
		44	Comprehensive planning	10	3	1	30		
		45	Optimisation and completeness in the planning approach	10	3	1	30		
		46	Evidence of sustainability aspects in the tendering process	10	2	1	20		
		47	Provision of conditions for perfect use and operation	10	2	1	20		
		48	Construction site/construction phase	10	2	1	20		
		49	Quality of executing firms, pre-qualification	10	2	1	20		
	Quality of construction	50	Quality assurance of workmanship	10	3	1	30		
		51	Systematic commissioning	10	3	1	30		

[1] The numbers missing are categories still in progress. [2] uniform for all occupancy profiles
[3] user-specific, is determined separately for all occupancy profiles

1.34

Economy

Investments made in energy conservation should be balanced by the cost savings they achieve. The repayment (amortisation) must be completed during the life expectancy period of the relevant energy-saving component (fig. 1.39).
In the case of rented property, the person making the investment is not the same as the person benefiting from the savings. It is for this reason that investments are frequently not undertaken.
There are two solutions to this issue: either the refurbishment costs, which are justified from an energy-saving perspective and lead to an improvement of the building stock, are apportioned among the rent payments or an agreement is made with the tenant beforehand guaranteeing that the refurbishment will not have an effect on the all-inclusive rent, i.e. the rent including service charges. As compensation for the lower heating costs, the basic rent excluding service charges increases, thus giving the owner incentive to make the investment.

Determination of investment costs

The savings potential of improvement measures is often underestimated due to incorrect economic assessments, often with the result that potentially beneficial investments are not made.
Cost breakdowns performed according to DIN 276 (fig. 1.42, p. 32) are a good and comprehensive tool for determining the total costs of a construction project. Frequently, the costs of necessary routine maintenance work, such as the replacement of old windows or heating systems, are included in the cost estimations, increasing the amount to be amortised through energy savings. This is incorrect and the effect is to imply that the energy-efficiency measures are uneconomic.
If a differentiated approach to assessing investment costs is taken, the maintenance and repair costs must be determined separately from the costs for the energy efficiency improvements. In the case of the refurbishment of a rendered facade, this would mean that the renewal of the render is classed as "maintenance", whereas the installation of insulation should be included under "energy efficiency improvement". Only the costs for the actual improvement, from a tax point of view these are regarded as construction costs, must then be amortised through energy savings. The maintenance costs are excluded and are also treated differently from a tax point of view.
The investment costs, which are subject to interest, must be repaid over the assumed lifetime of the improvement measure. Interest and capital are paid back in instalments in the form of a fixed total annual amount (annuity). Due to the regular repayments, the interest component decreases from one year to the next, whereas the capital component increases (fig. 1.37). This must be taken into account if it is intended to claim tax relief on interest on the outstanding debt. In 2010, the average interest rate for mortgage loans, depending on the repayment term, ranged between 3.0 and 4.0%. The government encourages construction work that leads to reductions in energy consumption by offering low-interest mortgages.
In early design phases, building costs are often determined by extrapolating from the costs of an already completed property. The Construction Costs Information Centre of the German Chamber of Architects (BKI) has evaluated and published data on numerous completed buildings specifically for this purpose. However, before these reference values can be

Regional BKI factors for building costs

Germany	regional factors 2011
Bad Tölz area	1.21
Berlin	0.939
Frankfurt a. M.	1.047
Cologne	1.076
Leipzig	0.889
Munich	1.187
Zwickau area	0.877
Europe	
Germany	1
Italy	0.738
Norway	1.287
Switzerland	1.2
Czech Republic	0.524

1.35

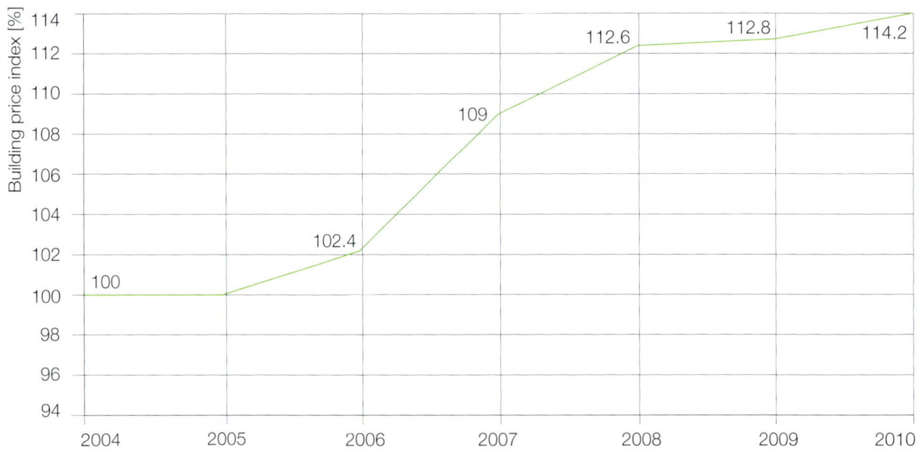

1.36

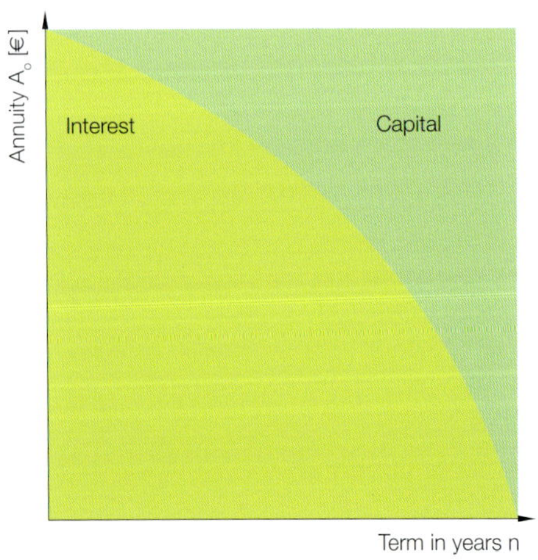

1.37

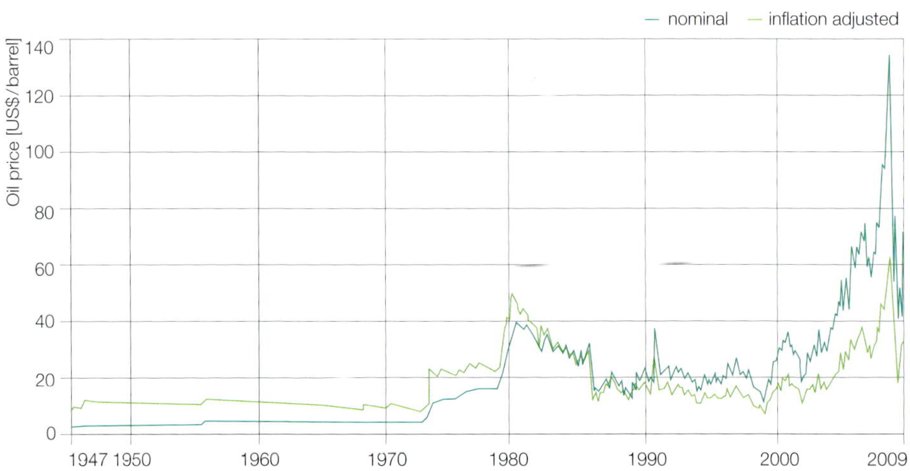

1.38

adopted, it is necessary to determine the year of construction and location of the comparative property and make adjustments for the date and location of the new project by applying the appropriate adjustment factors (fig. 1.35 and 1.36).
If, for example, the refurbishment of a house in Leipzig from the Gründerzeit era (1871 - 1873) in 2004 had cost €1,000,000, a comparable project in Munich in 2011 would have cost €1,522,000. The original total of €1,000,000 has to be adjusted to the different location by applying the quotient of 1.187 and 0.889, and then multiplying it by the factor 1.14 to cater for price changes.
Having estimated and determined the investment sum, the annuity for the investment is calculated according to the following equation:

$$A_o = \frac{K_o \cdot q^n \cdot (q - 1)}{q^n - 1}$$

K_o = amount to be invested
n = term (life expectancy)
q = interest rate (5 % $\rightarrow q = 1.05$)
A_o = annual instalment

Savings
Initially, the savings are determined in relation to the energy demand of the building before the refurbishment. It is very important in this context to make sure that the energy demand before and after the refurbishment refers to the same size of property. Possible extensions of the useful floor area made during the construction work, which inevitably lead to an increase in energy consumption, should be excluded from the comparative analysis.
The actual savings achieved by the reduced energy consumption during the first year after completion of the work are then adjusted according to assumed increases in energy prices over the same term as the annuity. The sum of the adjusted figures gives the total saving over the expected lifetime of the improvement measure. The savings in energy costs therefore rise on an annual basis in line with the expected price development. This is accounted for in the calculations by assuming an average price increase for energy.
Raw materials are subject to considerable price fluctuations. It is therefore necessary, at times of low energy prices, to expect a slightly higher average price increase (approx. 8 %) whereas, at times of high energy prices, a slightly lower price increase (approx. 6 %) may be assumed (fig. 1.38).
The savings achieved through reduced energy consumption can be calculated for an assumed energy price increase and an assumed life expectancy of the improvement measure by using the following formula:

$$K_n = R \, \frac{r^n - 1}{r - 1}$$

K_n = capital saved
R = savings during 1st year
r = price increase of annual savings
n = duration of savings in years

The economic assessment always refers to the year of the investment. Price inflation of everyday goods is not considered. This would only be necessary if provision for the investment costs associated with future replacement measures were to be included from the start in the cost-efficiency analysis.

Repayment
If the sum of the savings with added interest is higher than the costs of the investment calculated on an annuity basis, the investment is worthwhile.
When considering the period for assessing whether an investment is worthwhile, the statistical life expectancy of the investment plays an important role. Technical installations, such as heating systems or solar plants, must amortise over a period of 15 years since it is then that the installations must be replaced due to wear and tear, failure or the introduction of technically more advanced systems (fig. 1.39). The life expectancy of structural measures is almost twice as long. Thus it is appropriate to assume an payback period of around 30 years. The payback period is calculated according to the following formula:

$$TA = \frac{\ln\left(1 - \frac{K_o \, (q - r)}{R}\right)}{\ln\,(r / q)}$$

TA = payback period
q = interest factor for the investment (e.g. $q = 1.05$)
r = price increase of annual savings (e.g. $r = 1.08$)
R = savings during 1st year
K_o = amount to be invested

The invested capital must be restored during the statistically determined life expec-

Life expectancy of selected building components

Component	average life expectancy [a]
Load-bearing structure	
concrete	120
masonry (with plaster)	120
timber (cladded)	
• soft wood	70
• hard wood	100
External wall finish	
render	40
paint	
• emulsion paint	20
• mineral paint	15
composite thermal insulation system	30
rear-ventilated, insulated facade	
• wood sheathing	30
• cement board siding	55
• aluminium panelling	60
• natural stone facing	80
• sheet steel covering	45
• glass	50
Windows	
wooden windows (softwood)	40
wooden windows (hardwood)	50
PVC windows	50
aluminium windows	50
inculation glass	25
sun shading device	
• textile	15
• aluminium	25
Pitched roof (including substructure)	
roof tiles	50
concrete roofing tiles	50
titanium zinc roof sheet	25
copper roof sheet	50
slate	70
flat roof (with insulation):	
without roof covering	20
with roof covering (gravel, planted)	30
Building services	
burner / blower	12
boiler	20
heat pump	12
thermal solar plant	20
heat distribution	40
heat output	
• heaters (radiators, heating convector)	25
• panel heating	25
earth-to-air heat exchanger	60
photovoltaics	20
air-conditioning and ventilation plant	15
air ducts	35
refrigeration plant	15
cable	25

1.39

1.35 Regional factors according to BKI in 2011: Germany (selection) Europe (selection)
1.36 Development of building costs for residential buildings between 2004 and 2010
1.37 Annuity: the capital component increases in accordance with the term, whereas the interest component decreases.
1.38 Nominal and inflation-adjusted development of crude oil price between 1947 and 2009 in US dollars per barrel (1 barrel = 158.98 litres). The diagram clearly shows that there is no significant price increase in comparison to the living costs. Price increases for primary energy resources, which will be much higher than average price increases, are expected in the near future with the rising demand and the more difficult pumping conditions.
1.39 Life expectancy of building components, which influence the power economy of a building

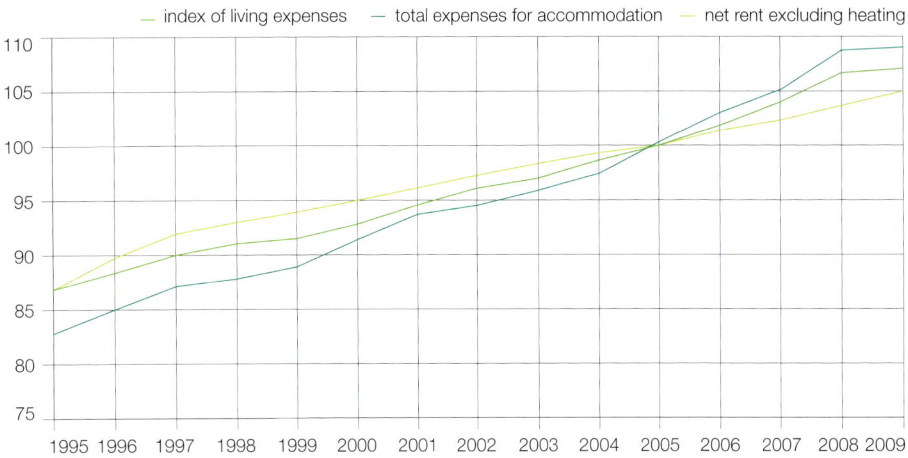

1.40

Distribution of costs according to DIN 18960 (extract)

No.	Cost category
100	**Capital costs**
110	Loan capital
120	Equity capital
130	Depreciation
190	Capital costs, other items
200	**Property management costs**
210	Labour costs
220	Non-labour costs
230	Outside services
290	Property management costs, other items
300	**Operating costs**
310	Provision of public utilities
320	Waste disposal
330	Cleaning and care of buildings
340	Cleaning and care of outdoor facilities
350	Operation, inspection and maintenance
360	Monitoring and security services
370	Charges and fees
390	Operating costs, other items
400	**Repair costs**
410	Structural repair work
420	Repair of building services
430	Repair of outdoor facilities
440	Repair of fit-out
490	Repair costs, other items

1.41

Distribution of costs according to DIN 276 (extract)

No.	Cost category
100	**Site**
110	Site value
120	Incidental costs
130	Site clearance
200	**Preparation and development**
210	**Site preparation**
220	Public development
230	Private development
240	Compensation charges
250	Transitional measures
300	**Building construction**
310	Construction pit
320	Foundation
330	Exterior walls
340	Interior walls
350	Floor slabs
360	Roofs
370	Structural fit-out
390	Other construction-related work
400	**Technical installations**
410	Waste water, water, gas
420	Heating systems
430	Ventilation systems
440	Power installations
450	Communication technologies
460	Transport equipment
470	User-related systems
480	Building automation
490	Other measures related to technical installations
500	**Exterior facilities**
600	**Fit-out and art**
700	**Incidental building costs**
710	Client responsibilities
720	Preliminary project work
730	Architect and engineer services
740	Surveys and consultancy
750	Artistic services
760	Finance costs

1.42

1.40 Development of living costs
1.41 Distribution of operating costs according to DIN 18960: 2008-02, User costs of buildings
1.42 Distribution of building costs according to DIN 276-1: 2006-11, Building costs

tancy of the improvement measure so that the installation or component can be replaced at the end of its life cycle. Thus the payback period should never be longer than the life cycle of the improvement measure being financed. The life expectancy is therefore an important aspect when performing comparative investment analyses. The shorter the payback period of an improvement measure in relation to its life cycle, the more efficient the measure from an economic point of view.

User costs of buildings

In the same way that DIN 276 describes the total costs that accrue when developing a property, DIN 18960 "User costs of buildings" defines the costs which accumulate during the life cycle of a building (fig. 1.41 and 1.42). The systematic recording of all building maintenance costs is a pre-requisite for assessment of the economic efficiency of a property. The income from rents or leases is for this purpose compared with the building maintenance costs.

Cost categories

The user costs of buildings incorporate all the expenses incurred by the property after completion. In the case of residential buildings, a proportion of these costs can be divided among the tenants. In Germany, it is the ordinance on the distribution of running costs (Betriebskostenverordnung, BetrKV) that determines which and how these are divided. Similar schemes for distributing operating costs also apply to commercial properties. In the case of rented properties, all costs that cannot be distributed among tenants may be treated as business related expenses for tax purposes.

The following cost categories are included:

- Capital costs
  The capital costs are in fact the interest charges for the capital, which is required to finance the investment. These costs are fully tax deductible.
- Depreciation
  Depending on the use of the building, a property depreciates over a certain period of time. Examples of typical depreciation periods are 15 years for shopping centres and 40 years for office buildings. Depreciation amounts are fully tax deductible.
- Administration costs
  All expenditure for the administration of properties are treated as expenses and are fully tax deductible.
- Tax
  In most cases, buildings are subject to

property tax. According to the German ordinance on the distribution of running costs (BetrKV), this tax can be divided among the tenants. It does not therefore affect the owner of the property.

- Operating costs
  All operating costs, e.g. costs for cleaning the building, water supply and sewerage, heating and cooling, electricity, service personnel, maintenance and inspection of technical installations and the tending of access ways and planted areas, may be divided among the tenants.
- Building maintenance costs
  These cover all costs for maintenance and repair of the property. They cannot be divided among the tenants, but are fully tax deductible.

Operating costs
The categories and level of operating costs can be explained using the example of a multi-family dwelling, completed in 1980, with a lift. The monthly costs are listed in euro per square metre of rented space for the year 2009. The first figures are mean values for Germany determined by the German Tenants' Association. The figures in brackets refer to Munich and comply with the rental index of 2009.

Property tax	€0.19 (0.24)
Water	€0.39 (0.33)
Heating	€0.90 (0.90)
Hot water	€0.28 (0.28)
Waste disposal	€0.19 (0.26)
General electricity	€0.05 (0.06)
Insurances	€0.13 (0.13)
Aerial/cable	€0.11 (0.11)
Cleaning of building	€0.14 (0.17)
Lift	€0.11 (0.13)
Caretaker	€0.19 (0.34)
Garden upkeep	€0.09 (0.06)
Others	€0.05 (0.05)
Total	€2.82 (3.06)

The impact on operating costs achieved through energy efficiency refurbishments is frequently overestimated.
If, for example, the energy demand for heating and hot water were reduced by 50 % due to the improvement measures of an energy efficiency refurbishment, the operating costs would decrease from €2.82 to €2.23, a reduction of only 21 %. It should be noted that operating costs rise much more rapidly than rental costs. The result is that in regions with low rent levels the operating costs are regarded as a so-called second rent due to the high levels reached (fig. 1.40).

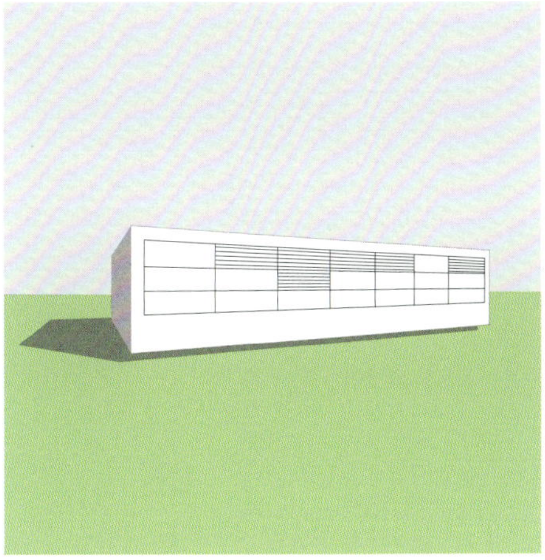

Example: ecology and economy
The presented pavilion is to be refurbished in accordance with the statutory requirements to meet the EnEV standard + 40 %. Different scenarios will be used to study the effect optimising the heating system has on the economic and ecological efficiency of the building.

Pavilion data

• opaque envelope	625 m^2
• transparent envelope	53 m^2
• heated area	196 m^2

The comparative analysis is based on the following figures:

• interest return	4 %
• price increase of energy	6 %
• gas costs	0.70 €/m^3
• electricity costs (grid)	0.22 €/kWh
• electricity costs (heat pump)	0.15 €/kWh

The calculations were made according to DIN V 18599.

Scenario 1: EnEV + 40 % (initial situation)

Parameters

• $\bar{U}_{opaque}$	0.41 W/m^2K
• $\bar{U}_{transparent}$	1.38 W/m^2K
• heating	gas condensing boiler

Energy balance

• useful energy	31,505 kWh
• final energy	39,928 kWh
• primary energy	40,793 kWh
• CO_2 emissions	9742 kg

Scenario 2: window replacement
Improvement measure
• installation of passive-house windows, U_w = 0.8 W/m^2K

Parameters

• $\bar{U}_{opaque}$	0.41 W/m^2K
• $\bar{U}_{transparent}$	0.8 W/m^2K
• heating	gas condensing boiler

Energy balance

• useful energy	29,477 kWh
• final energy	35,667 kWh
• primary energy	36,431 kWh
• CO_2 emission	8667 kg

Economy

• life expectancy of the measure	30 years
• costs	€4857
• savings	€2913
• payback period	upgrade is not amortised

Comment
If one considers the additional costs to manufacture the additional pane of glass at a magnitude of 629 kg CO_2, it is clear that the measure is worthwhile from an ecological point of view, not however from an economic viewpoint (see example of Appropriateness on page 27).

Scenario 3: thermal insulation
Improvement measure
• thermal insulation is increased from 6 to 20 cm

Parameters

• $\bar{U}_{opaque}$	0.2 W/m^2K
• $\bar{U}_{transparent}$	1.38 W/m^2K
• heating	gas condensing boiler

Energy balance

• useful energy	20,970 kWh
• final energy	27,166 kWh (gas)
• primary energy	27,918 kWh
• CO_2 emissions	6628 kg

Economy

• life expectancy of the measure	30 years
• costs (annuity)	€26,023
• savings	€68,462
• payback period	15 years

Comment
The production of the insulation is the cause for 6020 tonnes of CO_2 emissions. Due to the annual savings of 3114 tonnes achieved by the improvement, the ecological payback period of the measure is two years. From an economic point of view, this is also a very efficient measure.

Scenario 4: heat pump
Improvement measure
• installation of a heat pump

Parameters

• $\bar{U}_{opaque}$	0.41 W/m^2K
• $\bar{U}_{transparent}$	1.38 W/m^2K
• heating	water-to-water heat pump

Energy balance

• useful energy	31,505 kWh
• final energy	14,622 kWh (electricity)
• primary energy	38,017 kWh
• CO_2 emissions	9255 kg

Economy

• life expectancy of the measure	15 years
• additional costs (annuity)	€20,236
• savings (including price increases)	€13,989
• payback period	21 years

Comment
The heat pump as a replacement for insulation is not worthwhile, neither from an ecological nor from an economic point of view. It only makes sense to install a heat pump after improving the structural envelope and the heat distribution system (forward flow temperature, panel heating).

Scenario 5: ventilation with HR
Improvement measure
• installation of a ventilation unit with heat recovery

Parameters

• $\bar{U}_{opaque}$	0.41 W/m^2K
• $\bar{U}_{transparent}$	1.38 W/m^2K
• heating	gas condensing boiler

Energy balance

• useful energy	19,283 kWh
• final energy gas	25,201 kWh
• final energy electricity (ventilation)	521 kWh
• primary energy	27,288 kWh
• CO_2 emissions	6478 kg

Economy

• life expectancy of the measure	15 years
• additional costs (annuity)	€20,236
• savings (including price increases)	€21,554
• payback period	14.6 years

Comment
It generally makes sense, both from an economic and an ecological point of view, to install a controlled ventilation system with heat recovery. It is not only beneficial from an energy aspect, but also in terms of hygiene and acoustics.

Energy accounting

In order to calculate the energy demand of buildings in a comparable way, it is absolutely essential that the rules and regulations for performing the calculations are coherent and comprehensive. The consistency of the arithmetic techniques should enable, on the one hand, a clear definition of target values and, on the other hand, a verification of the results.
The basic guidelines concerning the type and scope of energy accounting have been established Europe-wide by the EU in the form of regulations, which are obligatory for all member states.
Based on these framework conditions, each member state is permitted to issue further national regulations. In Germany these rules and regulations are manifested in the Energieeinspargesetz (EnEG) and the complementary regulation for its implementation the Energieeinsparverordnung (EnEV).
The EnEV does not actually define rules for computing, instead it coordinates already available DIN standards and determines calculation methods based on these. The most significant standards for residential buildings are DIN 4108-6 and DIN 4701-10. DIN V 18599 has been generated as a rule in Germany for non-residential buildings to better implement the conditions for energy accounting laid down by the EU. The EnEV defines target values, which must be met by providing the requested calculations. The objectives for implementing renewable energy sources to heat buildings are stated in the Renewable Energies Heat Act (Erneuerbare Energien Wärmegesetz, EEWärmeG). This regulation is intended for new builds, but is also relevant for refurbishments if new useful floor space is created during the course of the improvement measures.

Regulations for saving energy

If the regulations for saving energy are assessed according to measures suitable for existing buildings, there are some first detailed stipulations in the German Thermal Insulation Regulation (Wärmeschutzverordnung, WSVO) of 1984. §8 determines minimum requirements for the thermal protection of certain components, which must be adhered to in the case of renewing the facade, the roof or replacing the windows. The WSVO 1995 adopts the measures mentioned in WSVO 1984, but increases the minimum requirements quite significantly. Since the introduction of the Energy Saving Ordinance (EnEV) in 2002 and its revision in 2004, the technical installations of a building, including the generation of the required useful heat, have become an integral part of every energy assessment. The directive no longer only assesses the heat demand of the building, but also the primary energy consumption. It therefore includes all sources of energy necessary to generate heat in the building as well as all the energy to source and produce the applied energy medium (oil, gas, electricity, wood and other biomass). Hence, the primary energy demand is also a parameter for the CO_2 emissions, which are caused by heating a building.
With this new method, which is applicable to all buildings, it has become possible to evaluate and assess improvement measures to existing buildings. It is for example possible to compare the energy efficiency achieved through improvements to the thermal protection of exterior walls with the replacement of an outdated boiler plant. A new build can now be compared with an existing building, and, for the first time, soundly based decisions can be made on whether to refurbish or demolish the existing building concerned.
Alongside the described assessment of the total energy balance, the EnEV also still enables the application of methods established in 1984 to assess individual components according to their relevant U-values. The energy assessment according to EnEV 2007 has become more detailed and therefore also more complex with the introduction of the European directive "Energy performance of buildings" (2002/91/EG of 16 December 2002). Whereas the procedure for residential buildings has more or less remained the same, for non-residential buildings it is now necessary to assess and evaluate the total primary energy demand for the conditioning of the building in accordance with set target values. In addition to heating energy, now the energy used for the lighting, ventilation and cooling of the building is also taken into account and summed up to form a primary energy parameter. The rules applicable for the assessment are described in the standard DIN V 18599, which have been an integral part of the EnEV since 2007.
The target values have been tightened further in the newly revised EnEV 2009. For the first time, it is now also possible to include electricity generated with photovoltaics into the assessment of the primary energy demand.
The overview on pages 36 and 37 (fig. 1.43) offers an insight into the relevant regulations and highlights the most important minimum values.

Renewable Energies Heat Act (Erneuerbare-Energien-Wärmegesetz , EEWärmeG)

The Renewable Energies Heat Act of 7 August 2008 stipulates that owners of new buildings with a useful floor space greater than 50 m² must cover part of their heating demand with renewable energies. This applies to buildings for which a building application was submitted after 1 January 2009.
The extent of renewable energies is determined as a fixed proportion of the property's total thermal energy demand. In the case of both residential and non-residential buildings, the thermal energy demand includes all sources required for heating, cooling and the provision of hot water.

Use of renewable energies
According to EEWärmeG, the following renewable energies may be used:

- Thermal solar energy
  The use of thermal solar energy is permitted only for residential buildings. The collectors must cover at least 15 % of the thermal energy demand. This requirement is fulfilled if a collector area (only the glass area of the collector) of 0.04 m² is provided for each square metre of useful floor space in residential buildings with up to two living units and a corresponding area of 0.03 m² for each square metre of useful floor space in residential buildings with more than two living units. The useful floor space includes the heated or cooled space as stipulated by the EnEV. A residential building with a floor area of 800 m² therefore should have a collector surface area of 24 m².
- Solid and liquid biomass
  The use of wood or liquid biomass must cover the thermal energy demand by at least 50 %. The boiler efficiency must be at least 86 % (88 % for small plants under 50 KW)
- Gaseous biomass
  In the context of a combined heat and power generation, the use of gaseous biomass must cover at least 30 % of the thermal energy demand.
- Geothermal and environmental heat
  The use of ground heat, groundwater or air must cover at least 50 % of the thermal energy demand. The heat pumps required to make use of this heat, must have an annual performance factor of at least 3.5.

Permitted compensatory measures
The EEWärmeG also allows for the aforementioned renewable energy sources to be replaced by other measures, so long

as these cover at least 50% of the thermal energy demand.
The following additional methods are permitted:
- use of exhaust heat
  The heat pumps applied for this purpose, must have an annual performance factor of at least 3.5.
- block-type thermal power station
- district heat, so long as it is generated by renewable energy sources or from the exhaust heat recovered from power generation plants

An improvement of the construction and technical conditions of the building may act as a substitute for the use of renewable energies. In the case of residential buildings, the EnEV requirements for transmission heat loss of the building envelope (H'_t, Heat sinks on page 43) must be undercut by 15%. In the case of non-residential buildings, the requirements refer to the average U-values ($\bar{U}$) for the transparent and opaque surfaces of the building envelope as well as to the primary energy demand (Q_p).

Energy efficiency refurbishments
Despite the fact that these regulations only refer to new builds, there are in fact aspects of relevance to the improvement measures of refurbishments. When, for example, dealing with the development of new space (e.g. an attic conversion) the requirements of the EEWärmeG must be observed.
The difference between a refurbishment and a new build is not always clear in this context. According to § 16 of the EnEV, an energy certificate must be drawn up if the heated or cooled useful floor space is extended by more than 50% or if several of the improvement measures according to Part 3 of the EnEV are carried out and the technical installations are replaced. In this case the requirements of the EEWärmeG must also be complied to.
Fact is, however, that there is still room for interpretation when it comes to the question of when a refurbishment is a new build and which boundary values apply to the energy matters. There are plans to integrate the requirements of the EEWärmeG into the EnEV and also incorporate refurbishment measures into the revised edition of the EnEV in 2012.

Future requirements for buildings
The targets to reduce the energy demand of buildings have been determined by the European Union for all member states.
The EU directive 2010/31/EU of 19 May 2010 dealing with the energy performance of buildings has set out the framework conditions, which must in future be observed by the member states in their national regulations.
Furthermore, the national rules and regulations must be adjusted at least every five years to incorporate state of the art technology.
A so-called ultra-low energy standard, which still requires further detailing, but which will roughly compare with passive house standards, will apply to all private new builds at the latest as of 1 January 2021 and to all public new builds at the latest as of 1 January 2019. The remaining primary energy demand for the conditioning of these new builds will then have to be covered by using renewable energies.

Future requirements for refurbishments
The requirements concerning existing buildings cannot be defined as precisely as those for new builds:
In the case of major refurbishment work, the national regulations will in future include minimum requirements which take the price-earning ratio into account. A refurbishment is considered as a major refurbishment when the total costs amount to more than 25% of the building's value (excluding the price for the land), or when at least 25% of the building envelope's surface is renewed.
The minimum requirements can either be determined in total or according to the individual building components. This procedure has been implemented in Germany since the introduction of the EnEV 2002.
According to the EU directive 2010/31/EU, more stringent requirements also apply when replacing or retrofitting individual components to the envelope if these have a considerable impact on the building's energy demand.
The same applies to technical installations, which are modernised, replaced or retrofitted.
The intention is to introduce special financial subsidies as an incentive to perform refurbishments according to the ultra-low energy standards applicable to new builds.
Energy certificates will presumably become more significant in future because they will have to be presented automatically, and not just on request, when selling, renting or leasing a property. The legal significance of the certificates will increase due to the fact that the "indicator" stated in the certificate will have to be included in the offers for sale or rent.
Furthermore, the revised EU directive requests that random checks be made of the certificates by authorised experts from regulatory bodies. These measures will underline the seriousness of the energy situation and enhance the approach towards environmental issues in the real estate business. Revised versions of the energy saving directive will be responsible for implementing the above-mentioned aspects in Germany.
Accounting performed according to the standard DIN V 18599, which has been providing the corresponding parameters since 2007, will be the only accounting system in future. The current possibility to carry out the accounting of residential buildings according to the standard DIN 4108-6 in combination with DIN 4701-10 will be withdrawn in the short to medium term.

A broader understanding of sustainability
A look into the future and the topical analysis of relevant research projects clearly presents that the planned revision of the EnEV for 2012 will not yet complete the energy efficiency assessments for construction work. It could, for example, be fairly simple to determine the resource requirements which are necessary for the production, maintenance and disposal of a building material if the corresponding standardised and therefore generally applicable values were available. If all products were recorded in tender preparation software with the data concerning the production energy, it would be possible to determine the consumption of resources for a building component and eventually the total building and hence optimise the outcome at the same time as preparing the tender documents.
Furthermore, the aim of all energy related considerations should not only be to take account of the measurable quantity of resources required for the production and conditioning of buildings, but to understand these topics as partial aspects of a greater perception of sustainability. This is the only way of avoiding an isolated approach to energy-related questions and, for example, to ensure that the thickness of thermal insulation is not the only feature regarded as relevant for assessing a building's quality in terms of energy.

References

[1] Jacobson, Mark Z.; Delucchi, Mark A.: Evaluating the Feasibility of a Large-Scale Wind, Water, and Sun Energy Infrastructure. Stanford University, University of California at Davis, USA 2009
[2] GEMIS = Globales Emissionsmodell integrierter Systeme; www.gemis.de
[3] Heyer, Ernst: Witterung und Klima – Eine Einführung in die Meteorologie und Klimatologie. Wiesbaden 1988

Date	Name	Basis	Specifications for structural measures	Specifications for technical installations
July 1952 Aug. 1969 Oct. 1974 Aug. 1981 July 2003	Minimum thermal protection (Mindestwärmeschutz)	DIN 4108: 1952–07 TGL 35424 (former GDR) today: DIN 4108-2: 2003–07	The aim is to avoid damage caused by moisture through the application of structural heat protection measures and to protect the building from overheating in summer. According to the currently valid regulation, the following U-values may not be exceeded: exterior wall 0.73 W/m^2K roof to outdoors 0.73 W/m^2K floor over exterior space 0.51 W/m^2K floor over ground 0.93 W/m^2K floor over unheated space 0.90 W/m^2K	no specifications
1 Nov. 1977	1st Thermal Insulation Regulation (Wärmeschutzverordnung)	EnEG 22 July 1976	Winter thermal protection: • $k_{m,max}$ dependent on SA/V ratio (surface area/heated volume, e.g. SA/V = 0.8 → $k_{m,max}$ = 0.85 W/m^2K) • $k_{m,ew+win}$ < 1.85 W/m^2K Refurbishments: • no specifications Summer thermal protection: • minimum thermal protection according to DIN 4108-T2 (10.74)	no specifications
1 Jan. 1984	2nd Thermal Insulation Regulation (Wärmeschutzverordnung)	EnEG	Winter thermal protection: • $k_{m,max}$ dependent on SA/V ratio (e.g. SA/V = 0.8 → $k_{m,max}$ = 0.66 W/m^2K) • $k_{m,ew+win}$ < 1.2 W/m^2K • joint permeability coefficient at windows limited to 1.0 m^3/mh ($daPa^{2/3}$) Refurbishments (recently introduced): • building component-related minimum requirements (wall 0.6 W/m^2K; floor over exterior space 0.45 W/m^2K; basement ceiling 0.7 W/m^2K; window: double/insulating glazing) Summer thermal protection: • minimum thermal protection according to DIN 4108-2 (8.81)	no specifications
1 Jan. 1995	3rd Thermal Insulation Regulation (Wärmeschutzverordnung)	EnEG	Winter thermal protection: • heat demand ($Q_t + Q_v - Q_s - Q_i$) dependent on SA/V ratio Refurbishments: • building component-related minimum requirements (wall 0.40 W/m^2K; floor over exterior space 0.30 W/m^2K; basement ceiling 0.5 W/m^2K; window 1.8 W/m^2K) Summer thermal protection: • minimum thermal protection according to DIN 4108-2 (8.81) with differentiated additions	no specifications
1 Feb. 2002	Energy Saving Ordinance (Energieeinsparverordnung, EnEV) and its revision in 2004	EnEG (changed on 10 Nov. 2001) DIN standards	Winter thermal protection: • the specific transmission heat loss H'_t, which is related to the heat transferring surface of the building envelope, is limited in accordance with the SA/V ratio and the window area. Refurbishments: • building component-related minimum requirements (wall 0.35 W/m^2K; pitched roof to outdoors 0.30 W/m^2K, flat roof 0.25 W/m^2K; basement ceiling 0.4 W/m^2K /0.5 W/m^2K new layer; window 1.7 W/m^2K). Total accounting can be performed in the same way as for a new builds; permissible values H'_t and Q_p' (m^2 related) and Q_p'' (m^3 related) may be exceeded by 40%. Summer thermal protection: • minimum thermal protection according to DIN 4108: 2001-03; since the revised version of 2004 according to DIN 4108-2: 2003-3 • determination of the solar insolation in relation to the net floor area of a room according to the values $s_{present}$ (present solar input value) and s_{max} (maximum solar input value) including a balancing of the two factors. The solar insolation is in this case determined according to the window area, the glass quality and the sun protection device and related to a target value, which is made up of different values derived from the building's conditions. Evidence must only be provided if the window area is greater than 30%.	Total energy balance (recently introduced): • primary energy demand Q_p (heating and provision of hot water) including the plant technology is limited in relation to the SA/V ratio • different requirements for the total accounting of residential and non-residential buildings (Q_p)

Date	Name	Basis	Specifications for structural measures	Specifications for technical installations
2007	Energy Saving Ordinance (Energieeinspar-verordnung, EnEV)	EnEG (changed on 8 July 2005) DIN standards (new: DIN V 18599)	Winter thermal protection: • residential buildings: calculations according to DIN 4106-6 and DIN 4701-10 The specific transmission heat loss H'_t, which is related to the heat transmitting surface of the building envelope, is limited in accordance with the SA/V ratio. • non-residential buildings: calculations according to DIN V 18599 The specific transmission heat loss H'_t, which is related to the heat transmitting surface of the building envelope, is limited in accordance with the SA/V ratio. Refurbishments: • residential and non-residential buildings: building component-related minimum requirements (wall 0.35 W/m²K; pitched roof to outdoors 0.30 W/m²K; flat roofs 0.25 W/m²K; floors and walls adjoining unheated space or ground 0.4 W/m²K with insulation on the cold side, 0.5 W/m²K for insulation on the warm side; window 1.7 W/m²K). Total accounting can be performed in the same was as for a new build. Permissible values H'_t and Q_p may not exceed 40%. • Summer thermal protection: see EnEV 2002. In the case of non-residential buildings, proof must be provided for each individual zone. The regulation that evidence must only be provided if the window area is greater than 30% is no longer applicable. Evidence must be provided for all SA/V ratios.	Total energy balance: • residential buildings: The primary energy demand Q_p for heating and the provision of hot water is limited in relation to the SA/V ratio. • non-residential buildings: The primary energy demand Q_p for heating, the provision of hot water, cooling, lighting, ventilation and humidification is limited to that of a reference building.
2009	Energy Saving Ordinance (Energieeinspar-verordnung, EnEV)	EnEG (changed on 28 March 2009)	Winter thermal protection: • residential buildings: calculations either according to DIN 4106-6 and DIN 4701-10 or according to DIN 18599. The specific transmission heat loss H'_t, which is related to the heat transmitting surface of the building envelope, is limited in accordance with the shape and the size of the building (SFH 0.4 W/m²K; MFH detached 0.50 W/m²K; semi-detached houses 0.45 W/m²K; other residential buildings 0.65 W/m²K) • non-residential buildings: calculations according to DIN V 18599. The average U-value is limited according to the following aspects: average value of opaque surfaces ≤ 0.35 W/m²K; average value of transparent surfaces ≤ 1.9 W/m²K Refurbishments: • residential and non-residential buildings: building component-related minimum requirements (wall 0.24 W/m²K, pitched roof to outdoors 0.24 W/m²K, flat roofs 0.20 W/m²K; floors and walls adjoining unheated space or ground 0.3 W/m²K with insulation on the cold side, 0.5 W/m²K for insulation on the warm side, window 1.3 W/m²K). Total accounting can be performed in the same was as for a new build. Permissible values H'_t and Q_p may not exceed 40%. Summer thermal protection: • see EnEV 2007	Total energy balance: • all buildings: Electricity generated by renewable energy sources is taken into consideration for the first time and deducted from the primary energy demand in the assessment. The power must be generated in the immediate vicinity of the building and used exclusively by the building. • residential buildings: The primary energy demand Q_p for heating and the provision of hot water is limited to that of a reference building. • non-residential buildings: The primary energy demand Q_p for heating, the provision of hot water, cooling, lighting, ventilation and humidification is limited to that of a reference building.

1.43 Development of German regulations to limit energy consumption in buildings with average inside temperatures.

- Heat sinks
- Heat sources
- Ventilation
- Daylight

Heat sinks

The chapter "Construction-related measures" addresses specific approaches to improving the energy balance of existing buildings by making changes to the construction. It is subdivided according to the accounting system of the Energy Saving Ordinance (Energieeinsparverordnung, EnEV) and the standard DIN V 18599 into the topics "Heat sinks" (heat loss), "Heat sources" (heat gain) as well as "Ventilation" and "Daylight".
The technical installations required to operate a building are dealt with in the chapter "Building services measures".

Minimum thermal protection according to DIN 4108

Winter thermal protection according to DIN 4108-2: 2003-07 is intended to ensure a temperature of the surfaces enclosing the room that is sufficient to avoid damage caused by moisture. According to the standard, surface temperatures of over 12.6 °C are regarded as sufficient because, at a normal room usage with a temperature of 20 °C and a relative air humidity of 50 %, the air humidity close to the cooler wall surfaces reaches a maximum value of 80 %. If however the surface temperature of the walls falls below 12.6 °C for a longer period of time, dew condensation and mould growth are likely to occur on the affected building components.
For planar, room enclosing surfaces, it is therefore necessary to observe the values for minimum thermal protection included in table 3 of DIN 4108-2 (fig. 2.1). If these principal conditions (room temperature, air humidity) are observed, there is no danger of mould growth and no further evidence is required.
These minimum thermal protection measures according to DIN 4108-2 must be observed by every room enclosing surface. It is therefore especially important to check building components which lack homogeneity caused by changes in material or by point-supported fixing elements. The specific issues concerning these point-supported or linear disturbed components are dealt with in the section "Thermal bridges" (pp. 40ff.).

Additional requirements

In addition to the minimum values for the thermal resistance of individual room-enclosing surfaces, DIN 4108-2 also includes a variety of supplementary aspects for thermal protection in winter. These are explained in the following sections:

Transparent components

According to the minimum thermal protection standards, all windows must have at least double or insulating glazing. The frame must have a heat transfer coefficient U_{frame} (formerly known as k-value) that complies with frame material group 2.1 (DIN V 4108-4:2002-02). The use of the rating "frame material group" is no longer applicable, but is found still in a lot of older documents, which are frequently referred to when performing a survey of a building prior to commencing refurbishment work.
In figure 2.2, the former values are matched with the relevant U_{frame} values for windows today in accordance with the valid standard. These requirements for minimum thermal protection according to DIN 4108-2 have been replaced by some of the regulations contained in the Energy Saving Ordinance (EnEV), such as the replacement of windows as a sole improvement measure. If however, in the case of major refurbishment work, the observance of the EnEV values is proven in an overall energy balance ($H'_{T,}$ Q_P), it is also permissible to install a frame with inferior thermal characteristics. It is in this case that the above mentioned minimum requirements according to DIN 4108-2 apply.

Roller shutter boxes

There are three possible ways of including roller shutter boxes into the energy balance of a building:

- The integrated shutter boxes are treated as a panel component with their appropriate U-value (fig. 2.4 b, p. 40).
- In the case of integrated roller shutter boxes, the shutter boxes can also be ignored and included in the calculations as part of the wall. The result is that the shutter box acts as a thermal bridge and must be considered when accounting for thermal bridges. In this case, the wall surface extends down to the top of the window frame (fig. 2.4 a, p. 40).
- The same applies in the case of exterior-mounted roller shutters. However, in this solution, the window area reaches up to the top of the shutter box (fig. 2.4 c, p. 40).

The inspection lid on the inside of the integrated roller shutter box must have a thermal resistance of $R \geq 0.55$ m²K/W (see p. 60).

Wall cavities

In the case of rear-ventilated wall constructions, DIN EN ISO 6946 must be referred to in order to calculate the insulation effect of non-ventilated, slightly ventilated and highly ventilated cavities. Table 2 in the standard lists the thermal resistance values which apply to the various depths of the non-ventilated cavities (fig. 2.3).
The ventilation openings of a non-ventilated cavity may be no greater than 500 mm² per linear metre in the case of facades and no greater than 500 mm² per square metre of roof area in the case of

roofs. When dealing with slightly ventilated cavities, the facade and roof ventilation openings must be between 500 mm² and 1500 mm². The standard ascribes these cavities a thermal resistance R which is half that of non-ventilated cavities. If the ventilation openings exceed the sizes applicable for non-ventilated and slightly ventilated cavities, the cavity is regarded as highly ventilated and therefore without any insulating effect, as is the case for all rear ventilated facade constructions which comply with DIN standards. However, it is then permitted to apply the more favourable R_{si}-value (0.13 m² K/W) for the outer thermal resistance factor R_{se} when calculating this building component.

Protruding slabs
DIN 4108–2 does not permit the construction of protruding slabs, which are either not insulated or not thermally separated, such as balcony slabs, roof parapets, fascias or columns, if these extend from the heated space into the exterior.

Basement walls
The exterior insulation of sealed basement walls, so called perimeter insulation, is frequently subject to the penetration of moisture. Nevertheless, the calculations can include the insulation so long as the boards are fitted without joints and are made of extruded polystyrene (XPS) in accordance with DIN 18164-1.

Inverted roofs
The calculations for the thermal resistance R of flat roofs consider only the layers beneath the waterproofing. The inverted roof is an exception – in accordance to DIN 18164-1, the insulation consisting of XPS boards is placed on top of the waterproofing and covered with gravel, for instance. The surface of the insulation boards must allow for sufficient drainage, however, the boards may become wet for short periods of time.
It is for this reason that an increment is added to the U-value of the total roof construction, which depends on the thickness of the insulation layers positioned above and below the waterproofing (fig. 2.6, p. 41).
In the case of light-weight inverted roofs with a total area-specific mass of less than 250 kg/m², the thermal resistance value R of the layers below the waterproofing must add up to at least 0.15 m² K/W. This value is achieved by incorporating a 1 cm-thick panel of insulation with a λ-value of 0.04 W/mK.

Minimum values for the thermal resistance of building components according to DIN 4108-2:2003-07, table 3

<table>
<tr><th colspan="2">Building component</th><th>Thermal resistance R [m²K/W]</th></tr>
<tr><td colspan="2">Exterior walls; walls in habitable rooms adjoining attic space, passageways, open corridors, garages, ground</td><td>1.2</td></tr>
<tr><td colspan="2">Walls between spaces with different users; party walls</td><td>0.07</td></tr>
<tr><td rowspan="2">Stair enclosure walls</td><td>to staircases with much lower indoor temperatures (e.g. indirectly heated staircases); room temperature $\theta_i \leq 10\,°C$, but at least without the danger of frost</td><td>0.25</td></tr>
<tr><td>to staircases with indoor temperatures $\theta_i \geq 10\,°C$ (e.g. administrative buildings, offices, school buildings, hotels, restaurants and residential buildings)</td><td>0.07</td></tr>
<tr><td rowspan="2">Party floors, floors between two offices with different users; floors beneath rooms that are between insulated roof pitches or jamb walls in the case of habitable attic space</td><td>general</td><td>0.35</td></tr>
<tr><td>in centrally heated office buildings</td><td>0.17</td></tr>
<tr><td rowspan="2">Lower finish of basementless habitable space</td><td>immediately above the ground not exceeding a room depth of 5 m</td><td rowspan="4">0.90</td></tr>
<tr><td>above a non-ventilated cavity adjoining the ground</td></tr>
<tr><td colspan="2">Floors beneath non-habitable attic space; floors beneath crawl space or even lower areas; floors beneath ventilated rooms between roof pitches and jamb walls in the case of habitable attic space, insulated roof pitches</td></tr>
<tr><td colspan="2">Basement ceilings; ceilings adjoining enclosed, unheated hallways or similar</td></tr>
<tr><td rowspan="2">Ceilings (also roofs) which enclose a habitable space adjoining the exterior</td><td>below, adjoining garages (also heated ones), passageways (also those that can be closed off) and ventilated crawl space</td><td>1.75</td></tr>
<tr><td>above, e.g. roofs according to DIN 18530, roofs and floors beneath terraces; inverted roofs
In the case of inverted roofs, ΔU must be calculated by using the heat transfer coefficient U according to DIN EN ISO 6946 and the correction values from DIN 4108, table 4 (fig. 2.6, p. 41)</td><td>1.2</td></tr>
</table>

2.1

Thermal protection standards of window frames according to the old and new regulations

Frame material group DIN 4108-4:1985-12		Heat transfer coefficient k DIN 4108-4:1985-12	Heat transfer coefficient U_f DIN EN ISO 10077-1:2006-12
1	wood, PVC, wood-aluminium	$k \leq 2.0$ W/m²K	$U_f \leq 2.2$ W/m²K
2.1	insulated metal or concrete frames with corresponding test certificates	$2.0 < k \leq 2.8$ W/m²K	$2.2 < U_f \leq 3.0$ W/m²K
2.2	see 2.1	$2.8 < k \leq 3.5$ W/m²K	$3.0 < U_f \leq 3.7$ W/m²K
2.3	see 2.1	$3.5 < k \leq 4.5$ W/m²K	$3.7 < U_f \leq 4.7$ W/m²K
3	non insulated frame made of steel, aluminium or concrete	4.5 W/m²K < k	4.7 W/m²K < U_f

2.2

R-values of non-ventilated cavities [W/m²K]

Cavity depth [mm]	Direction of heat flow: up	horizontal	down
0	0.00	0.00	0.00
5	0.11	0.11	0.11
7	0.13	0.13	0.13
10	0.15	0.15	0.15
15	0.16	0.17	0.17
25	0.16	0.18	0.19
50	0.16	0.18	0.21
100	0.16	0.18	0.22
300	0.16	0.18	0.23

2.3

2.1 Minimum values for the heat transfer coefficients of room-enclosing building components according to DIN 4108-2:2003-07, table 3. If these values are observed, no further evidence is required and there is no danger of mould growth.
2.2 Thermal protection standards for window frames according to the old and new regulations
2.3 Thermal resistance of non-ventilated cavities according to DIN EN ISO 6946 (intermediate values can be interpolated)

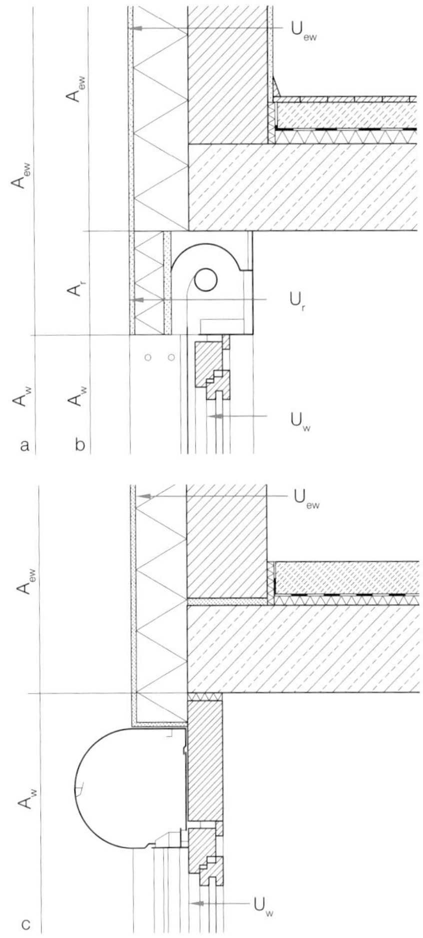

A_w window area
A_r roller shutter area
A_{ew} exterior wall area
U_w heat transfer coefficient window
U_r heat transfer coefficient roller shutter
U_{ew} heat transfer coefficient exterior wall

2.4

2.4 Determination of U-value for roller shutter box according to DIN 4108-2:2003-07
a definition of areas when including the integrated roller shutter box into the exterior wall area
b definition of areas for integrated roller shutter box with own area and own U-value
c definition of areas when including the exterior-mounted roller shutter box into the window area

2.5 Interior thermal resistance (R_{si}) and exterior thermal resistance (R_{se}) for the calculation of the surface temperature (moisture protection) and heat loss (thermal protection) in the case of thermal bridges

2.6 Increments for the heat transfer coefficient ΔU in the case of inverted roofs according to DIN 4108-2:2003-07, table 4

2.7 Residential block in Ingolstadt
Architects (refurbishment): Adam Architekten
Year of completion: 1968/2003
The nine-storey residential building was refurbished whilst inhabited. A 12-cm-thick composite thermal insulation system and new windows reduce the heat loss of the building envelope. The balcony slabs, which are not thermally separated, have been changed into thermal buffer zones by mounting a simple sliding glass element to the exterior of the building. The old windows could be retained in these zones. This allowed the danger of mould growth at the protruding balcony slabs to be reduced significantly.
a building before the refurbishment
b building after the refurbishment
c vertical section, scale 1:20

Airtightness
The airtightness of buildings is not only regulated in the EnEV, but also in the guidelines for minimum thermal protection. It can therefore be determined either according to DIN EN 12114 or DIN EN 13829. The joint permeability coefficient for component junctions may, if the specified pressure conditions apply, not exceed the value of 0.1 m^3/mh. The values in DIN 18055 apply for windows. External doors may not exceed a value of 2.0 m^3/mh.

Energy efficiency refurbishments
In the planning of a new building, the demands of minimum heat protection no longer represent a major problem in terms of design or construction, as the requirements of the Energy Saving Ordinance (EnEv) are in any case more stringent. However, the specification that the minimum heat protection must be provided for every part of the building causes difficulties for point-fixed components, such as projecting roofs and balconies, if their mounting brackets are not thermally separated.
One of the major issues in the case of refurbishments is to ensure that "panel components" meet the minimum heat protection standards. If the assessments reveal that the minimum heat protection standards are not met, the planner is obliged to consult the client and offer professional advice on the measures required to make the necessary improvements. The planner must explain that the rules applied at the time of the building's development have become more stringent and complex. Furthermore, the changing lifestyles and comfort needs of the occupants are also often a cause for structural damage in buildings which were originally built according to standards and codes, such as damage caused by moisture. This especially applies when old windows are replaced without insulating the exterior walls and these walls then no longer comply with today's minimum heat protection standards. Due to the greater airtightness of the new windows, the relative humidity tends to rise in the rooms. The risk of mould growth increases on the still uninsulated walls, which are, according to today's standards, too cold.
Architects are therefore well advised to check whether and to what extent minimum heat protection is ensured for the building concerned prior to commencing the planning for the refurbishment work, possibly in cooperation with a building physicist.

Thermal bridges
The heat transmission through a component can increase in certain places due to geometric or structural irregularities. Depending on the type of disturbance, a difference is made between geometric and structural thermal bridges. Geometric thermal bridges occur at all building corners and in all places where there is a greater heat transmission due to a discrepancy between the heat absorbing and heat dissipating surfaces, such as the outside corners of a building, window reveals or protruding floor slabs.
Structural thermal bridges are a result of changes in material, for example a concrete pillar in a brick wall, the tie-in areas of floor slabs and interior walls, window joints, the brackets for protruding roofs or balcony constructions and even the fastening devices for insulation boards.
Thermal bridges can affect a single point (one-dimensional), a linear area (two-dimensional) or a spatial configuration (three-dimensional).
Thermal bridges lead to a greater heat loss in specific areas and, as a result, a reduction in the temperature of the room-enclosing surfaces in the areas concerned. The first effect described leads to a greater energy demand of the building, the second to the risk of condensation

Thermal resistance in m^2K/W

	Calculation of the surface temperature (moisture protection) DIN 4108-2		**Calculation of the heat loss (thermal protection) DIN EN ISO 6946**	
	R_{si}	R_{se}	R_{si}	R_{se}
Opaque exterior components				
facade (horizontal heat flow) [1]	0.25	0.04	0.13	0.04/0.13 [2]
roof (vertical heat flow)	0.25	0.04	0.10	0.04/0.10 [2]
floor (vertical heat flow)	0.25	0.04	0.17	0.04
Transparent exterior components	0.13	0.04	–	–

[1] these values also apply to facades with inclinations of up to ± 30° in relation to the vertical
[2] this value also applies to rear ventilated facades and roofs

2.5

and mould growth in the area of the thermal bridge. It is therefore always necessary to consider these two separate aspects when thermal bridges are concerned:

- calculation of the surface temperature in the area of the thermal bridge
- calculation of the thermal bridging heat loss by applying the ψ-value in W/mK

Surface temperature
The surface temperature of a building component determines whether the formation of mould is encouraged. According to DIN 4108-2:2003-7, for a room-enclosing surface temperature θ_{si} of 12.6 °C, a room air temperature θ_i of 20 °C, a relative air humidity of 50 % and an outside temperature θ_e of -5 °C, the following temperature factor (f_{Rsi}) is determined as a boundary value and should not be undercut:

$$f_{Rsi} = \frac{(\theta_{si} - \theta_e)}{(\theta_i - \theta_e)} \geq 0.7$$

If the temperature factor is equal or greater than 0.7, mould growth is not to be expected so long as the mentioned "normal" conditions prevail. The relative humidity in the vicinity of the wall surface, however, is close to 80 % , which is a boundary value as regards encouraging mould growth (fig. 1.10, p. 14).
In this case it is necessary to understand that a different thermal resistance must be applied for the internal surface (R_{si}) and the external surface (R_{se}) to calculate the surface temperatures and the heat loss depending on whether moisture protection or heat protection is the main issue to be considered (fig. 2.5).
So long as the boundary conditions for minimum heat protection according to DIN 4108-2:2003-07, table 3 are observed, the protection against moisture is given for a described situation. This is the actual intention of the minimum heat protection standard.

Increment factors for inverted roofs

Proportion of thermal resistance on the inside of the waterproofing compared to the total thermal resistance [%]	Increment factor U [W/m² K]
below 10	0.05
from 10 to 50	0.03
above 50	0

2.6

Thermal bridging heat loss
DIN 4108-2 does not only consider the issue of condensation through thermal bridges, but also the effect of transmission heat loss. In order to perform an energy balance according to EnEV, the thermal bridging heat loss can either be included by using default values or by performing individual calculations.

Default values for thermal bridges
According to DIN 4108-6:2000-11, the thermal bridging heat loss can be included in the energy balance by using one of the following default values:

- without providing further evidence by simply using the thermal bridging factor 0.1 W/m²K for each component of the building envelope or
- by observing the values listed in DIN 4108, supplement 2 for thermal bridges and applying a default factor 0.05 W/m²K

Providing the evidence according to supplement 2 is generally fairly straightforward. The fact that all corners and areas where interior walls and floor slabs are tied into the exterior walls are ignored simplifies matters considerably. However, this rule only applies to floor slabs if a continuous board of insulation with a thermal resistance of R = 2.5 m²K/W is placed to the face of the floor slab at the wall support. It is then only necessary to prove that the construction or the attributed values are comparable with those described in supplement 2 (fig. 2.8, p. 42).
If the construction is not identical to the one in supplement 2, individual calculations must be performed to prove that the values of supplement 2 can also be achieved with the alternative detail solution. These numerical methods are explained in the following sections.
If the thermal conductivity of the insulation material used deviates only slightly from the requirements in supplement 2, which always assumes a value of 0.04 W/mK for insulation material, simple comparative calculations can provide the necessary proof.
For example, the construction described in supplement 2 achieves a thermal resistance of 2.5 m²K/W with 10-cm-thick insulation. This value can also be met with 8-cm-thick insulation when using an insulation material with a thermal conductivity of 0.32 W/mK.

Individual calculations
To perform individual calculations of the thermal bridging factor according to

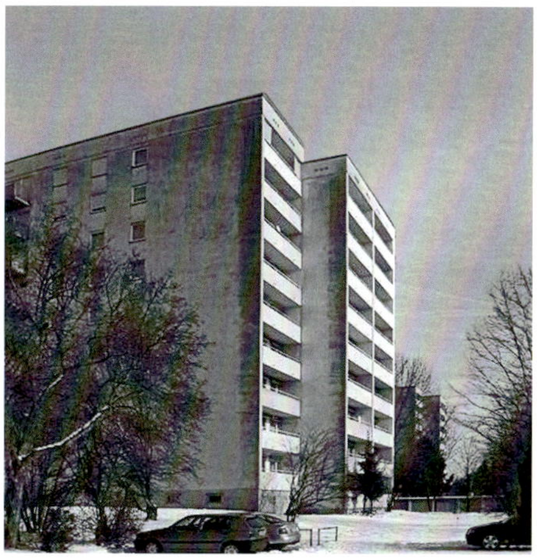
a

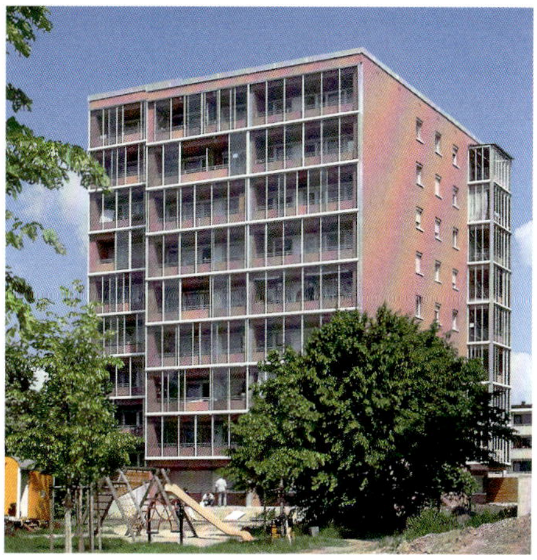
b

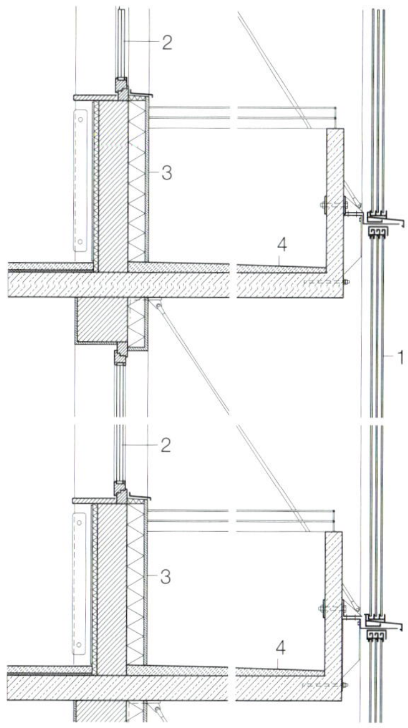

c

1 sliding element made of 12 mm single pane security glass
2 wooden frame window (existing)
3 composite thermal insulation system
decorative silicone resin render, 3 mm grain
5–10 mm mineral base coat
100 mm insulation
150 mm masonry wall (existing)
30 mm coated insulation boards (existing)
4 40 mm bonded concrete screed, sloped (existing)

2.7

DIN EN ISO 10211-1 and -2, the ψ-value (linear thermal bridging heat loss coefficient) of the thermal bridge concerned must be determined by performing a thermal bridge simulation. The ψ-value is then multiplied by the length of the thermal bridge to determine the heat loss in W/K for that one specific thermal bridge. According to DIN 4108-6:2004-06, the following thermal bridges need to be taken into consideration:

- building edges
- reveals to all windows and doors
- areas where walls and floor slabs tie into the exterior walls
- floor slab supports
- thermally separated balcony slabs

In contrast to the calculations performed according to DIN 4108, supplement 2, with the individual verification method not only selected cases but all linear thermal bridges must be recorded and calculated. Point thermal bridges are not incorporated in the energy balance of thermal bridges. The sum of all calculated thermal bridges gives the building's total thermal bridging heat loss. This sum divided by the building envelope surface area produces the individual thermal bridging factor for the building in W/m²K.

In the case of accurately planned constructions, this calculated factor is frequently lower than the default value – the fairly elaborate numerical method is therefore justified.
The surface temperature of a building component and the heat loss at the inspected part of the building are determined by using complex computer programmes.

Point thermal bridges
The devices for fixing composite thermal insulation systems, for the substructure of rear-ventilated facades or other point-fixed elements are taken into account by using default values in accordance with DIN EN ISO 6946:1996, enclosure D. The approvals of the specific facade systems provide the relevant data regarding the type and number of anchor pins. This extra value is arrived at by multiplying the number of pins per square metre by the thermal conductivity of the anchor pin. The value is only considered if the result is greater than 0.04 W/m²K, which would, for instance, be the case if there were more than ten fixings per square metre using stainless steel anchor pins with plastic heads (0.004 W/K). The extra value can be ignored for composite thermal insulation systems, if plastic dowels with insulated caps are used. Special attention must be paid to the thermal bridges that result from mounting the aluminium substructure for rear-ventilated facades. These details must be planned and executed with utmost care to avoid producing, depending on the type of fixing device used, extra values of around 0.05 W/m²K. These extras must then be added to the U-value calculated for the wall construction.

Thermal bridges in passive houses
The thermal bridging analysis used in passive houses, in which the energy standards voluntarily far exceed the values required by DIN or EnEV, i.e. without the pressure of standardisation, limit the heat loss of each individual thermal bridge to a maximum of 0.01 W/mK.
This leads, for example, to the use of 5 cm of insulation over the window frames at the joint with the reveals rather than the 3 cm required by DIN 4108, supplement 2.

Energy efficiency refurbishments
Compared to today's construction

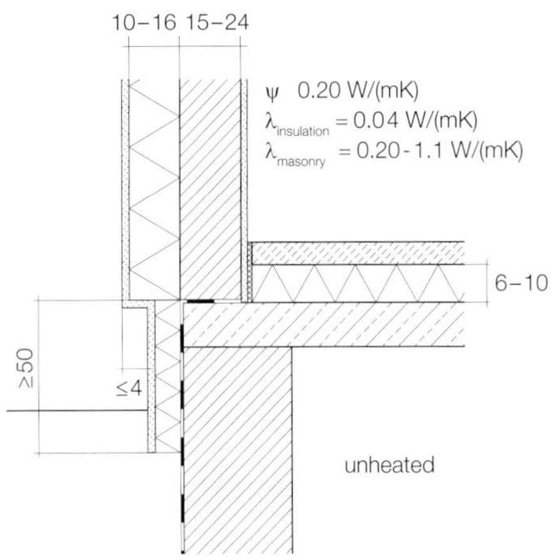

a

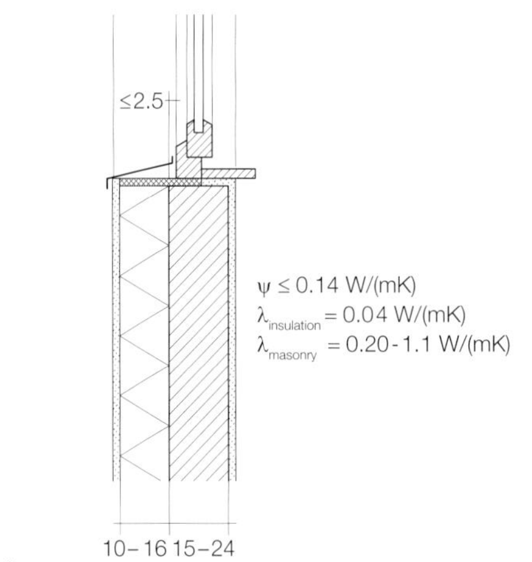

b

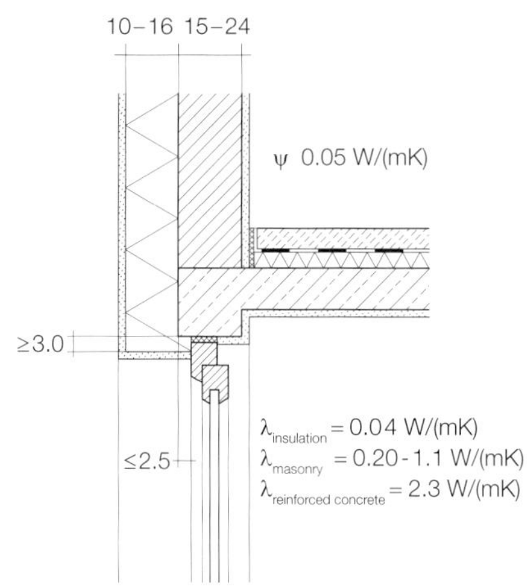

c

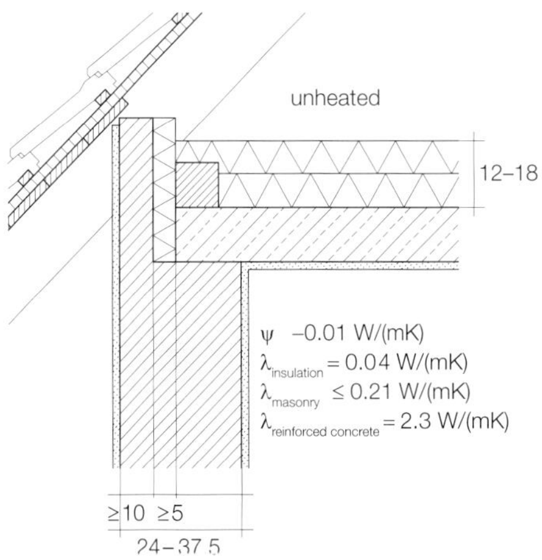

d

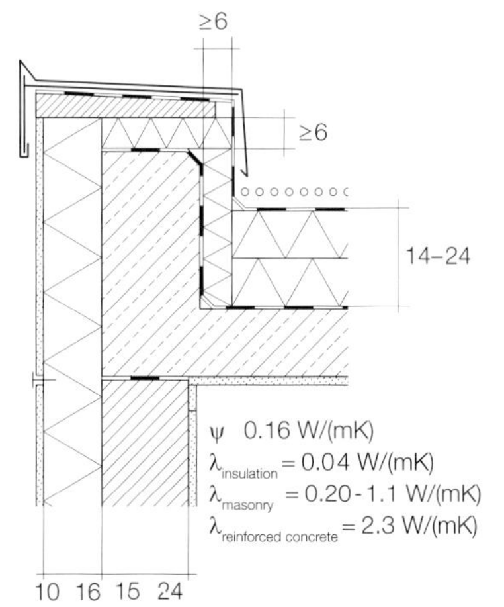

e

2.8

2.8 Constructions without thermal bridges according to DIN 4108, supplement 2:2006-03 (selection, all dimensions in cm)
a plinth, unheated basement
b bottom window detail
c top window detail
d highest floor slab, unheated attic space
e parapet, flat roof

standards, there are numerous thermal bridges in existing buildings. Reflecting the time at which they were built, many buildings do not even satisfy the necessary thermal resistance of $R \geq 1.2$ m²K/W for minimum thermal protection in their homogenous, planar components. The R-value is then even lower at thermal bridges, such as areas where interior walls or floor slabs are tied into exterior walls or balcony slabs protrude.
Further weak points include the detailing at the base of the building, around windows and the junctions at the roof. In the case of individual calculations, the thermal bridging factor often adds up to around 0.2 W/m²K in old buildings. For a building with a heated volume of approximately 1000 m³ and an envelope area of 700 m², this amounts to a heat loss of approximately 11,000 kWh per year, which is almost 20 % of the building's total heat loss. For construction-related reasons, it is not always possible to eliminate all of the thermal bridges in existing buildings. In such cases it is particularly important to at least reduce the thermal bridging heat loss by making some improvements. The variations for optimising construction details should therefore always be accompanied by corresponding simulation calculations.
The primary aim when remedying thermal bridges is to provide evidence for the elimination of mould growth; the second aim is to reduce heat loss. In some circumstances, the heat loss of individual thermal bridges is actually accepted, so long as numerical proof has been provided that guarantees the elimination of mould growth. This is especially justified if the financial expense involved in removing the thermal bridge bears no relation to the energy savings achieved, such as would be the case if protruding balcony slabs were to be fully insulated. Mould growth can already be eliminated by insulating the exterior wall (see Protruding balcony slab, pp. 66ff.).

Energy balance according to DIN V 18599
Based on the new standard DIN V 18599: 2007-02, the Energy Saving Ordinance (EnEV) now requires that heat flow is considered not only for the heating period, but for the whole year. This has the effect that the energy demand to cool the building is included in the total energy balance in the same way as the energy demand to heat the building. The abstract numerical method for the accounting procedure is based on the terms "heat sinks" (Q_{sink}) for heat loss and "heat sources" (Q_{source}) for heat gain. Both of these values are recorded on a monthly basis and added together (fig. 2.9, p. 44).
Typical heat sinks include the normal transmission heat loss through the building envelope, the ventilation heat loss at low outside temperatures and the heat loss that occurs from building use through production processes or the integrated plant technology, e.g. cold water pipes. Typical heat sources, on the other hand, include solar radiation, ventilation at high outside temperatures as well as heat gains caused by the use of the building through residents, machines, lighting or the installed plant technology, such as the distribution systems for heating and hot water. Heat sources tend to be uneven, i.e. they occur irregularly, which means that on some occasions, especially in the months of spring and autumn, there are often more heat gains than are actually required to cover the heat sinks. These circumstances are considered in the energy balance by calculating the monthly utilisation factor of the heat gains (η). If the total of heat sinks is greater than the total of heat sources, the building must be heated. This means that the determination of the heating period is one of the first steps when performing the balancing of heat loss and gain. The heating period can be calculated in five steps and is then incorporated in the determination of the heat demand in a sixth step. The example on pages 45 and 46 describes the practical implementation of this method.

Step 1: Determination of heat sinks
The heat sinks (Q_{sink}) are assessed separately according to transmission heat loss and ventilation heat loss.

Transmission heat loss
The heat loss through the building envelope enclosing and separating the heated volume from the exterior is expressed in the heat transfer coefficient, the U-value, in W/m²K.
The U-value is calculated by assigning each individual layer of the exterior wall an appropriate thermal conductivity λ. The layer's thermal resistance is determined by dividing the layer thickness s by its thermal conductivity λ. The sum of all layer-related thermal resistance values gives the global thermal resistance R of a building component. The total of R and the inner and outer thermal resistance values R_{si} and R_{se}, which express the heat transfer caused by the airflow at the building component, lead to the thermal resistance R_t of the component. The

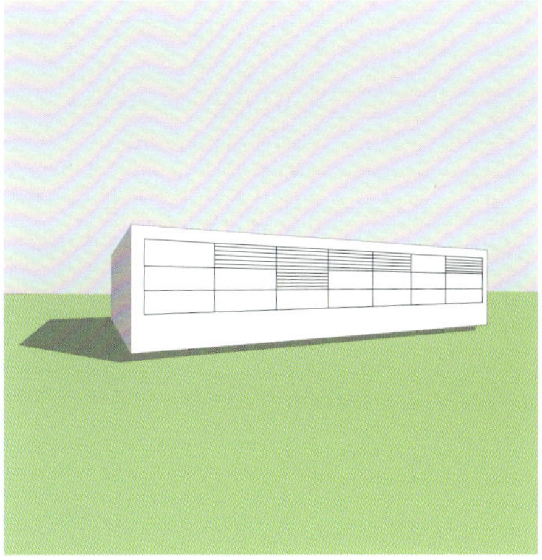

Example: thermal bridges
The pavilion (here with a residential use) presents the impact the calculation method for thermal bridges has on the specific transmission heat loss H'_T and the heat demand.

Condition	Surface area A_i [m²]	U-value U_i [W/m²K]
total surface area A	680.6	
• transparent surface	42.6	1.4
• floor area [1]	225	0.3
• roof area	225	0.3
• wall area	188	0.3
[1] temperature correction factor F_x		0.5

Calculation of the specific transmission heat loss according to DIN 4108-6, table D 1:
$H_t = \sum (F_{xi} \cdot U_i \cdot A_i) + \Delta U_{tb} \cdot A$

Calculation of the specific transmission heat loss in relation to the heat transferring envelope area H'_t according to EnEV, enclosure 1:
$H'_t = H_t \div A$

According to EnEV, enclosure 1, the maximum value H'_t for single-family detached homes with $A_u \leq 350$ m² is **0.4 W/m²K.**

Default thermal bridging factor
The calculation of H'_t is performed by using the fixed thermal bridging factor ΔU_{tb} of 0.1 W/m²K.
It follows that **H'_t = 0.42 W/m² K.**

Building without thermal bridges according to DIN 4108, supplement 2
If the details are constructed according to supplement 2 or evidence is provided that the ψ-values assigned to these constructions are observed, the thermal bridging factor ΔU_{tb} can be reduced to 0.05 W/m²K.
It follows that **H'_t = 0.37 W/m² K.** This means that the maximum value prescribed in enclosure 1, table 2 of the EnEV (0.4 W/m²K) is observed.

Calculated thermal bridging factor
All thermal bridges must be calculated by means of simulation. By improving specific details (e.g. using 5 cm rather than 3 cm of insulation over the fixed window frame at the reveals), the calculated thermal bridging heat loss is reduced as follows:

Thermal bridge	length l [m]	ψ-value (calculated) [W/mK]	Energy loss $l \cdot \psi$ [W/K]
plinth detail	61	-0.04	-2.44
outside corners	15.2	-0.14	-2.128
roof edge	61	0.16	9.76
window surround			
• top	16.8	0.06	1.008
• bottom	16.8	0.01	0.168
• side	4.5	0.02	0.09
Total			6.458

Calculated thermal bridging factor: 0.01 W/m² K
It follows that **H'_t = 0.33 W/m² K**.
The maximum EnEV value is undercut by 18 %.

reciprocal of the thermal resistance gives the heat transfer coefficient U in W/m²K. The U-value defines the amount of heat lost between the inside and outside per square metre of surface area at a temperature difference of one Kelvin.
The smaller the U-value, the lower the heat loss. The total specific heat loss of the building envelope (H_t) in W/K is determined by multiplying the individual U-values by the area of the specific components and finally adding these values together (fig. 2.10). The specific heat loss is characteristic of the building envelope's energy efficiency.

Ventilation heat loss
Ventilation heat loss occurs when a space is aired and the warm inside air is exchanged for cold outside air. The calculation of ventilation heat loss is dependent on the air volume that must be provided to a certain space due to its use. In the case of indoor space, the fresh air demand per person is 30 m³/h. The required fresh air volume is therefore a variable factor, which depends on the number of occupants using the room (see Ventilation, p. 76).

It is for this reason that a mean hourly air exchange rate n is used to calculate the energy balance. The value expresses how often the volume of a room or space must be exchanged to provide a sufficient supply of fresh air for a particular use and the relevant number of occupants. An air exchange rate of 1/h means that the total air volume of a room is exchanged once per hour.
The calculations for the energy balance are always performed using an average air exchange rate, in some cases this could of course be higher or lower. In order to obtain the monthly balance, in accordance with DIN V 18599, the average ventilation heat loss H_v is at first determined for the occupancy and non-occupancy periods on a daily basis and then added up to a monthly total. The monthly ventilation heat loss can then be calculated according to the air exchange rates (n) of the appropriate use (zones) (see example on p. 45):

H_v [Wh] =
V [m³] · n [1/h] · 0.34 Wh/m³K · Δθ [K] · t [h]

Here V is the net room volume, 0.34 Wh/m³K is the thermal capacity of the air and Δθ is the average temperature difference during the time period t concerned.

Step 2: Determination of heat sources
The numerical assessment of heat input through solar radiation and the use of the building is explained in the example opposite.

Step 3: Use of heat sources
The aim of an energy upgrade is to use the heat gains to cover the heat demand. However, the numerical assessment of this process is difficult since the heat sources and heat sinks vary in their timing and level depending on the weather and the intensity of their use. The impact of these fluctuations also depends on the storage capacity of the building and the proportion of transparent surfaces within the building envelope. The numerical approach of DIN V 18599, which has been adopted from DIN 4108-6, is aimed at defining a percentage to describe the utilisation of the balanced heat sources.
The energy balance for the months between the heating period and the non-heating period is extremely imprecise, which is why a detailed calculation is necessary to determine the heat sources' degree of utilisation (η) for these months:

$$\eta = \frac{1 - \gamma^a}{1 - \gamma^{a+1}} \quad \text{for } \gamma \neq 1$$

The utilisation of the heat sources is particularly dependent on the ratio of heat gains to heat losses. The ratio (γ) is calculated as follows:

$$\gamma = \frac{Q_{source}}{Q_{sink}}$$

The value a as a numerical parameter takes account of the building's storage capacity. It is a function of the building's specific heat loss H (W/K). In the case of a monthly balance, the parameter is calculated as follows:

$$a = 1 + \frac{C_{effect}}{H \cdot 16\,h}$$

According to DIN EN ISO 13786:1999-12, the following values can be applied for the effective storage capacity C_{effect} of a building:
- for buildings without solid components (lightweight building) 50 Wh/(m²K)

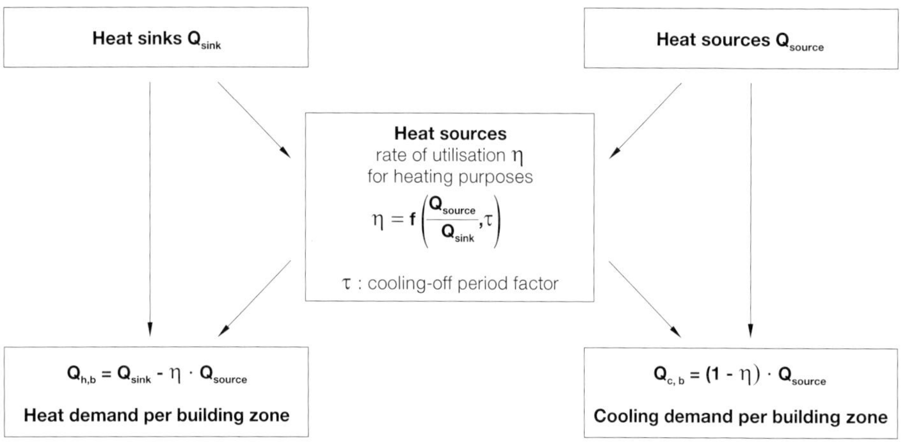

2.9

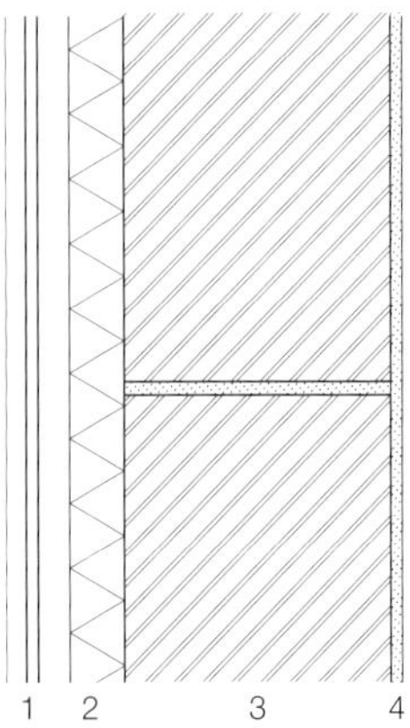

Calculation of U-value for building component

Layer	Thickness s [m]	Thermal conductivity l [W/mK]	s/l [m²K/W]
1 cladding, cavity	–	–	–
2 insulation	0.16	0.4	4.00
3 sandlime brick (1600 kg/m³)	0.24	0.79	0.30
4 plaster	0.02	0.35	0.06
Total 1 (thermal resistance R)			*4.36*
+ R_{si} *(thermal resistance inside)*			0.13
+ R_{se} *(thermal resistance outside)*			0.13
Total 2 (thermal resistance R_T)			*4.62*
Reciprocal provides the heat transfer coefficient U [W/m²K] U = 1/R_t			0.22

2.10

2.9 Calculation of a building zone's heating and cooling demand according to DIN V 18599-2:2007-02

2.10 Example for the calculation of a component's U-value (cavity wall)

- for buildings with solid ceilings and floors (moderate building)
  90 Wh/(m²K)
- for buildings exclusively made of solid components (heavy building)
  130 Wh/(m²K)

Step 4: Determination of accountable heat sources

The usable heat gains, which can be included into the balance, are calculated by multiplying the monthly heat gains Q_{source} determined in Step 2 with the monthly degree of utilisation factor η determined in Step 3:

$$\eta \cdot Q_{source}$$

Step 5: Assessment of monthly heating days

Especially in spring and autumn not every day is a heating day. The calculation of the monthly heating period is therefore, in a similar way to the utilisation of heat sources, important to be able to make a differentiated assessment.

The monthly heating period is determined by comparing the heating system's mean rate of utilisation ($\beta_{h,\,use}$) with a boundary value $\beta_{h,\,boundary} = 0.05$. If the heating system's mean rate of utilisation is less than the boundary value, then the month considered is not a full heating month and the monthly heating days must be assessed.

$$\beta_{h,\,use} = \frac{Q_{h,\,b,\,use}}{Q_{h,\,max,\,res} \cdot 24\ h}$$

The following aspects need to be considered when calculating $\beta_{h,use}$:

- The monthly heat demand $Q_{h,b,use}$ is determined using the usual process. The transmission heat loss, ventilation heat loss, solar heat gain and heat input from the building's utilisation are put in relation to the monthly climate data (solar radiation, outside temperature).
- When calculating the maximum heat demand $Q_{h,max,res}$ according to DIN V 18599-2, enclosure B2, only the transmission and ventilation heat losses are taken into account. All months are based on an outside temperature of -12 °C and an inside temperature of +20 °C. The monthly values only differ slightly due to the different number of days in each month.

In order to calculate the number of heating days per month, the number of days in that month in question are multiplied by the quotient $\beta_{use} \div \beta_{boundary}$.

Example: heating period

The assessment of the heating period is described using the following example. The period when no heating is required must be examined to determine whether the building requires cooling for the provision of comfortable room conditions and a maximum interior temperature of 26 °C. This is the case when the available heat sources exceed the available heat sinks.

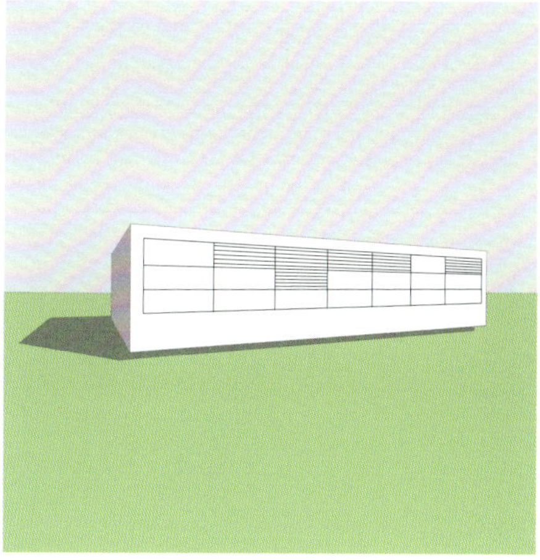

Step 1: Determination of heat sinks

1.1. *Transmission heat sinks*

Boundary conditions:		
	H'_t	0.3 W/m²K
	total building envelope	681 m²
	inside temperature	21 °C
	average monthly outside temperature	DIN 18599-10, table 7 (reference climate: Germany)

Month	days [d]	· hours [h]	· ΔT [K]	H'_t [W/m²K]	· A [m²]	= transmission heat sinks Q_t [kWh]
March	31	24	16.9	0.3	681	2568.79
April	30	24	11.5	0.3	681	1691.60
May	31	24	8.1	0.3	681	1231.19

1.2 *Ventilation heat sinks*

Boundary conditions:		
	accountable area	196 m²
	air volume V	680 m³
	air exchange rate n	0.6/h

Month	days [d]	· hours [h]	· ΔT [K]	· c_{air} [Wh/m³K]	· V [m³]	· n [1/h]	= ventilation heat sinks Q_v [kWh]
March	31	24	16.9	0.34	680	0.6	1744.21
April	30	24	11.5	0.34	680	0.6	1148.60
May	31	24	8.1	0.34	680	0.6	835.98

1.3 *Heat sinks Q_{sink} ($Q_t + Q_v$)*

month	[kWh]
March	4313.00
April	2840.21
May	2067.18

Step 2: Determination of heat sources

2.1 *Solar radiation through transparent surfaces*

Boundary conditions:		
	average monthly radiation intensity according to DIN 18599-10, table 7	
	g-value of the glass	0.7
	reduction due to shading	0.9
	· reduction due to pollution	0.9
	· reduction due to frame	0.7
	= total reduction	0.567

South-facing window

Month	area [m²]	· radiation [W/m²]	· days [d]	· hours [h]	· reduction (see above)	· g-value	= heat sources from transparent surfaces Q_s [kWh]
March	37.8	80	31	24	0.567	0.7	892.97
April	37.8	137	30	24	0.567	0.7	1479.88
May	37.8	119	31	24	0.567	0.7	1328.29

2.2 *Heat input due to utilisation*

Boundary conditions: interior heat sources according to DIN 18599-10, table A2 no. 2

Month	source [Wh/day]	· office area [m²]	· days [d]	· use [h/d]	= heat sources from use $Q_{i,source}$ [kWh]
March	72	120	31	6	1607.04
April	72	120	30	6	1555.20
May	72	120	31	6	1607.04

2.3. *Total of heat sources ($Q_{source} + Q_{i,source}$)*

March	2500.01
April	3035.08
May	2935.33

Step 3: Use of heat sources

3.1 *Calculation of the heat gain/loss conditions γ*

Month	heat sources [kWh]	÷ heat sinks [kWh]	= γ (sources ÷ sinks)
March	2500.01	4313.00	0.58
April	3035.08	2840.21	1.07
May	2935.33	2067.18	1.42

Example (continued from page 45)

3.2 Heat loss H

a) transmission heat loss H_t

H'_t [W/m²K]	· envelope surface area A [m²]	= transmission heat loss H_t [W/K]
0.3	681	204.3

b) ventilation heat loss H_v

air volume V [m³]	· air exchange rate n [1/h]	· c_{air} [Wh/m³K]	= ventilation heat loss H_v [W/K]
680	0.6	0.34	138.72

c) heat loss H (H_t + H_v) [W/K]
343.02

3.3 Calculation of the degree of utilisation η

Month	γ	C_{effect}	H	a	η
March	0.58	90	343.02	4.2	0.96
April	1.07	90	343.02	4.2	0.78
May	1.42	90	343.02	4.2	0.65

Step 4: Determination of accountable heat sources

Month	heat sources (see 2.3)	· degree of utilisation η	= accountable heat sources [kWh]
March	2500.01	0.96	2400.00
April	3035.08	0.78	2367.36
May	2935.33	0.65	1907.96

Step 5: Assessment of monthly heating days

5.1 Calculation of the monthly heat demand ($Q_{h,b,use}$)

Month	sinks (see 1.3)	– sources (see Step 4)	= heat demand [kWh]
March	4313.00	2400.00	1913.00
April	2840.21	2367.36	472.85
May	2067.18	1907.96	159.22

5.2 Maximum monthly heat demand ($Q_{h,max,res}$)

Boundary conditions: outside/inside temperature - 12 °C/20 °C

Month (see 3.2)	specific heat loss H [W/K]	· days [d]	· hours [h]	· Δ θ [K]	= $Q_{h,max,res}$ [kWh]
March	343.02	31	24	32	8166.62
April	343.02	30	24	32	7903.18
May	343.02	31	24	32	8166.62

5.3 Determination of $β_{use}$ and comparison with $β_{max}$

Month	$β_{use}$	$β_{max}$	$β_{use} ÷ β_{max}$ ($β_{use} < β_{max}$)	heating days
March	0.223	0.05	–	31
April	0.060	0.05	–	30
May	0.019	0.05	0.38	11.78

Step 6: Calculation of the heat sinks and the monthly heat demand

6.1 Calculation of the heat sinks with the determined heating period

a) transmission

Month	days [d]	· hours [h]	· Δ θ [K]	· H'_t [Wm²K]	· envelope surface area [m²]	= transmission heat sinks [kWh]
March	31	24	16.9	0.3	681	2568.79
April	30	24	11.5	0.3	681	1691.60
May	11.78	24	8.1	0.3	681	467.85

b) ventilation

Month	days [d]	· hours [h]	· Δ θ [K]	· c_{air} [Wh/m³K]	· V [m³]	· n [1/h]	= ventilation heat sinks [kWh]
March	31	24	16.9	0.34	680	0.6	1744.21
April	30	24	11.5	0.34	680	0.6	1148.60
May	11.78	24	8.1	0.34	680	0.6	317.67

6.2 Total heat sinks

Month	transmission (see 6.1 a) [kWh]	+ ventilation (see 6.1 b) [kWh]	= total heat sinks [kWh]
March	2568.79	1744.21	4313.00
April	1691.60	1148.60	2840.20
May	467.85	317.67	785.52

6.3 Determination of heat demand

Month	sinks (see 6.2)	– sources (see Step 4)	= heat demand [kWh]
March	4313.00	2400.00	1913.00
April	2840.20	2367.36	472.84
May	785.52	1907.96	–1122.44

Comment:
According to the presented data concerning use, climate and building, May is the month in which the heat gain compensates for the heat loss. The non-heating period, or the period when it might be necessary to cool the building, starts in this month.

Step 6: Calculating the heat demand
The last step in the calculation determines the monthly heat sinks by multiplying the number of heating days by the average temperature difference between the inside and outside during that specific month (number of degree days).
The heat sinks caused by transmission and ventilation are then multiplied by the number of degree days. The result defines the heat sinks for the in-between season months (either the last or the first heating month of the year). The heat sources determined in Step 3 are deducted from this amount. The remaining heat demand required for this month can then be incorporated into the overall calculation.
These individual steps are presented in numerical format in the example opposite.

Energy efficiency refurbishments
The differentiated approach to calculating the heating period is a significant improvement on the heating period method practised in Germany up until 2002, which irrespective of the building's structural quality simply assumed a fixed heating period of 185 days.
Due to the differentiated calculations, the reduction of the heating demand achieved through construction-related measures, for example increasing the heat sources by installing more suitable windows or reducing the heat sinks by improving the insulation, can now be presented in a much more comprehensible way. Furthermore, user-related heat sources, such as heat generating production processes, can be included into the calculations so that, in the case of substantial heat gains, a reduction of the heat losses does not necessarily have to be pursued.
Whereas old unrefurbished buildings in Germany (Würzburg is the place of reference) can have up to 300 heating days, well-insulated passive houses only have approximately 120 heating days due to their very low transmission heat loss.
By dividing the year up into heating days and non-heating days, it is furthermore possible to use the total of non-heating days to establish the number of cooling days. Thus a differentiated assessment method has been developed, which allows for a structural assessment of buildings with a variety of uses featuring different climate conditions throughout an entire year.

Structural survey

A structural survey to define the initial situation is one of the greatest challenges facing the planner, since it is the survey which is used as a basis for assessing the energy-saving potential achieved by the various insulation and refurbishment measures. In many cases, it is impossible to establish the exact nature of components, such as walls or intermediate floors without using destructive inspection techniques. Avoiding destruction requires experience, intuition and a good understanding of construction methods.

The following aspects provide invaluable support in such circumstances:

- original drawings from when the building was completed
- building authority records
- invoices from contractors and other tradespeople
- the DIN standards valid at the time of completion provide information about typical building materials and constructions

The German Energy Agency, dena, has compiled a table with heat transfer coefficients each typical for a certain period of time (fig. 2.11). Since 30 July 2009, when the Federal Ministry for Transport, Building and Urban Development (BMVBS) issued "rules to adopt and use data for residential building stock", it has been permitted to use the listed data when performing a verification procedure according to EnEV. However, the fixed rates listed in the rules may only be used if there is no precise information due to a lack of planning documents. BMBVS has also compiled corresponding data for non-residential building stock.

Typical U-values [W/m²K] for components in existing buildings

	Exterior wall			Topmost floor/ flat roof			Pitched roof			Floor over basement/ ground floor		
pre-1918	Brick or rubble masonry	S M	2.2[1]	Timber joist floor with straw loam infill	S M	1.0	Without insulation, plaster on reed mats or lathing	S M	2.6[1]	Timber joist floor with straw loam infill	S M	1.0
	Timber frame with loam infill	S M	2.0[1]				Straw loam infill between rafters, plaster to underside	S M	1.3[1]	Stone floor on soil or vaulted basement	S M	2.9[1]
1919–1948	Brick wall 25–38 cm	S M L	1.7[1]	Timber joist floor with false floor and loam infill	S M	0.8	Without insulation, plaster on reed mats or lathing	S M	2.6[1]	Timber joist floor with false floor and loam infill	S M	0.8
	Single-leaf masonry 38–51 cm or two-leaf wall	S M	1.4[1]				Straw loam infill between rafters, plaster to underside	S M	1.3[1]	Solid brick arch floor	S M L	1.2
1949–1968	Lightweight masonry with hollow blocks, honeycomb bricks, aerated concrete	S M	1.4[1]	Concrete slab, ribbed slab, hollow block floor	S L	2.1[1]	3.5 cm woodwool panel, plastered	S M	1.4[1]	Concrete slab, ribbed slab, hollow block floor with minimum impact sound insulation	S L	1.5[1]
	Masonry with pumice concrete blocks	S M	0.9	Timber joist floor with false floor	S M	0.8	Pumice concrete blocks between rafters	S M	1.4[1]	Timber joist floor with false floor	S M	0.8
							5 cm insulation between rafters	S M L	0.8			
1969–1978	Lightweight masonry with porous brick and normal-weight mortar[2]	S M	1.0	Concrete slab with 5 cm topside insulation	S M L	0.6	3.5 cm woodwool panel, plastered	S M	1.4[1]	Concrete slab with 2 cm impact sound insulation	S M L	1.0
	Precast concrete elements with core insulation or lightweight concrete	M L	1.1	Flat roof: concrete slab with 6 cm topside insulation (cold roof)	S M L	0.5	Pumice concrete blocks between rafters	S M	1.4[1]			
	Timber stud walls with 6 cm insulation	S	0.6	Timber joist floor with 4 cm insulation (timber structure/prefab house)	S	0.8	5 cm insulation between rafters	S M L	0.8			
1979–1983	Masonry with lightweight or perforated brick and lightweight mortar	S M	0.8	Flat roof: concrete slab with 8 cm topside insulation	S M	0.5	8 cm insulation between rafters	S M	0.5	Concrete slab with 4 cm impact sound insulation	S M L	0.8
	Masonry with aerated concrete	S M	0.6	Flat roof: concrete slab with 8 cm topside insulation (warm roof)	M L	0.5						
	Precast concrete elements with core insulation or lightweight concrete	M L	0.9	Timber joist floor with 8 cm insulation (timber structure/prefab house)	S	0.5						
	Timber stud walls with 6 cm insulation	S	0.5									
1984–1994	Masonry with lightweight or perforated brick and lightweight mortar	S M	0.6	Flat roof: concrete slab with 12 cm topside insulation	S M L	0.3	12 cm insulation between rafters	S M L	0.4	Concrete slab with 5 cm impact sound insulation	S M L	0.6
	Masonry with aerated concrete	S	0.5	Timber joist floor with 12 cm insulation (timber structure/prefab house)	S	0.3						

S = single family residence, M = multi-family residence, L = large multi-family residence

[1] a global U-value of 1.0 W/m²K can be used for retrofitted, at least 20-mm-thick insulating boards

[2] this construction is based on the examples presented on the following pages

2.11

Insulation of opaque surfaces

The main task when performing an energy efficiency upgrade to a building is the addition of insulation to the non-transparent surface areas of the facade. This is, with an insulation thickness of up to approximately 16 cm, the most effective construction-related measure to reduce heat loss. If meeting the desired energy levels requires an insulation thickness of over 16 cm, it is worth considering alternative, less expensive measures to achieve the desired saving potential. Possible alternatives include improving the windows by using triple thermal insulation glazing instead of double glazing or installing a solar thermal plant.
In the case of all insulation materials, it is advisable to use materials with a low thermal conductivity. The slightly higher price of the insulation amortises within a short period of time due to the increased savings in energy consumption. The choice of material should also always take ecological factors into consideration.

Pitched roof

The space between the rafters is generally used for insulation purposes (rafter insulation, usually using clamping felt made of mineral wool). Most mineral wool products have a thermal conductivity rating of WLG 032, which corresponds to a thermal conductivity of 0.032 W/mK. If there are no constraints in terms of fire protection, it is also possible to use insulation boards made of regrowable raw materials or blow-in insulation made of cellulose fibres. The latter requires the installation of panelling on both sides of the rafters to form a hollow cavity for the loose material. However, these insulation products all only have a thermal conductivity rating of WLG 040 and therefore have clearly lower insulation properties than mineral wool.
Above-rafter insulation is installed on top of sheathing, which is fitted above the rafters; prefabricated, stable elements made of rigid polyurethane foam (WLG 024) are best suited to this solution. For below-rafter insulation, mineral wool mats are used, which are clamped between battens fixed below the rafters.
When retrofitting insulation in roofs, it is important to bear in mind the correct fitting of the vapour barrier, especially in areas which are not so easily accessible, such as at the eaves. Furthermore, it is necessary to determine precisely the required sd-value of the vapour barrier by performing building physics calculations in order to select the most appropriate product.

Flat roof

The energy efficiency upgrade of sound flat roofs with additional insulation fitted above the roof sealing (inverted roof) is described on page 56. When performing insulation measures to cold roofs, it is important to ensure the necessary air circulation.

Exterior wall insulation

The energy saving potential of solid exterior walls can be improved by using insulation rendering, external thermal insulation composite systems (ETICS), rear-ventilated facades or facing blocks using insulation brickwork. In the case of two-leaf constructions, it is possible to fill the cavity between the two wall layers with insulation. Insulation rendering has the advantage that it is easy to apply to rough wall surfaces. It is also suitable to even out more severe irregularities within the wall surface by adding different thicknesses of the render, possibly in several steps. However, the efficiency of the insulation is moderate due to the low thermal conductivity rating of WLG 060.
Mineral or stone wool products with a thermal conductivity rating of up to WLG 035 are used in external thermal insulation composite systems (ETICS) if fire safety requirements must be complied with. Better insulation values can be achieved by using rigid polyurethane foam boards with WLG 024 (rigid resole foam board: WLG 022). Thermal insulation composite systems with wood fibre boards or insulating boards made of mineral wool are very environmentally friendly. However, these boards also only reach a thermal conductivity rating of WLG 040. Two thick layers of render can be applied to thermal insulation composite systems using mineral wool boards to protect them better from vandalism and woodpeckers.
Because the thickness of insulation in older composite systems is often insufficient, it may be necessary, in the case of an energy efficiency refurbishment, to increase the thickness of the existing insulation. If the insulation is damaged, no longer stable or moist, it is advisable to replace the entire system. For thermal insulation composite systems that are still in good condition, manufacturers now offer systems approved for use by construction authorities.
Special insulating blocks for facing brickwork have only recently been introduced to the market. They are easy to use and have a robust surface that can be rendered with conventional products, which is a clear advantage. However, due to their thermal conductivity rating of WLG 060, a far greater thickness must be used than with comparable thermal insulation composite systems.
In rear ventilated facades, the separation of the building's weather-tight skin from the insulation by means of a cavity is, from a building physics point of view, ideal and is moreover an extremely durable, low-maintenance construction. The insulation boards (rock or glass wool) are usually fixed to the wall using plastic dowels, whereas the wind and dead load of the exterior cladding is diverted into the existing load-bearing exterior wall by using an aluminium substructure.

Interior wall insulation

If it is not possible to insulate the walls from the outside, the insulation must be fitted on the inside. However, in this case, it is very important to carefully analyse the physical properties of the wall in terms of thermal bridges and moisture content. Furthermore, it should be ensured that the wall structure is resistant to water penetration from the outside.
Interior wall insulation fitted as a dry construction requires the addition of a vapour barrier, which complicates the installation of building services, in particular the electrical wiring. By fixing a substructure of battens on top of the vapour barrier, a cavity is created especially for these services. Mineral fibre boards with a thermal conductivity rating of WLG 035 are generally used in this situation.
Systems with insulation boards that also function as a vapour barrier can be fixed directly to the wall and plastered. Rigid foam glass or polyurethane boards are most suitable in this case.
Advances are being made in the area of capillary active insulation systems. These products can quickly absorb large amounts of moisture, for example when the room is in full use, and release the stored moisture back into the room once the intensity of use has subsided. These insulation boards are best used in combination with a controlled room ventilation system.
The product range includes silicate panels, mineral foam panels, loam plaster systems as well as products made of mineral wool with integrated moisture absorbing materials (aerogels). Hollow facing bricks filled with perlite can also be included into this category.

Topmost floor

In a similar way to the floor above the basement, walk-on insulation boards with a finished wearing surface are most suitable for the topmost floor. However, if it is necessary to even out irregularities in the floor surface, which is frequently the case when dealing with timber floors, adjustable constructions with sleepers, infill insulation and boarding are more suitable.

Floor above basement
Even when not required by the building regulations, the floor above the basement should be insulated using insulating materials classified as building material class A (non-flammable). It makes sense to use insulation boards with a finished surface in order to avoid the additional work process of adding plaster or cladding. All insulation manufacturers offer suitable finished systems.

Innovations
In recent years, new building materials with improved or adjustable insulation properties or ones that are easier to use have further broadened the product range.

Vacuum insulation
Vacuum insulation panels, also called vacuum isolation panels (VIP), consist of a special rigid core panel inside a gas-tight envelope from which air has been evacuated. Despite being fairly expensive, these prefabricated insulation elements with thicknesses ranging between 10 and 50 mm and WLG 007 are ideal for situations in which high thermal resistance is required but space is limited, such as on roof terraces and balconies to avoid steps, to insulate thermal bridges or as thin panels in curtain wall facades. Concerning the practical application, the high sensitivity of the vacuum envelope and, particularly in refurbishments, the fact that the elements cannot be cut to size are a clear disadvantage.

Switchable insulation
By supplying a current (5 W/m²) to an approximately 2-cm-thick panel that has been fixed to a solid wall, hydrogen is desorbed/adsorbed into the material which changes the pressure conditions inside the panel and hence the thermal conductivity. The switchable insulation therefore enables the solid wall to store solar insolation in winter. When no solar insolation is available, the thermal conductivity is reduced so that, depending on the wall structure, U-values of approximately 0.2 W/m²K are achieved. During the heating period, the heat gain capability of the panel surface is up to 150 kWh/m².

Nano-cellular foams
Due to the small cellular structure, the thermal conductivity of foamed insulation (e.g. polyurethane) is almost halved. These highly efficient insulation materials are however not yet available on the market.

Composite building materials
The number of composite building materials available on the market is increasing steadily. By combining different material properties, these products are of better quality than homogenous building materials. They include composite elements for the exterior insulation of buildings consisting of vacuum insulation boards enclosed by traditional insulation materials, which function as protection and lathing. The combination of vacuum insulation with 4-cm-thick plywood panels enables the prefabrication of large facade elements. In the case of interior insulation, silicate panels with a high water-absorbing capacity are combined with rigid PUR foam boards. These panels have the advantage of being extremely vapour tight.

Environmental product declarations according to ISO 14095
In view of the great diversity of available insulation materials, the environmental product declarations according to DIN EN ISO 14025 provide transparency to the marketplace. The declaration lists the technical data of the materials in an objective, neutral way including the manufacturing energy and the environmental impact caused by the production process itself (see Appropriateness, pp. 27ff.)
The examples of refurbishments documented on the following pages clearly show that individual measures lead to a reduction of energy transmission through the upgraded building component, but that thermal bridges can frequently only be removed by also considering the adjoining components (e.g. window surrounds). Hence, in the case of refurbishments, it is necessary to take a holistic approach to problematic issues and to develop a differentiated strategy to possible alterations.
In order to establish the energy balance for the presented situations, the U-values from the dena table (fig. 2.11, p. 47) for the building age group 1969–1978 have been applied. The savings in heating energy requirements and heating oil consumption achieved through the upgrade are presented per square metre of useful floor area and year.

Insulation materials for exterior walls (selection)

Insulation material	Thermal conductivity λ [W/mK]	Unit weight [kg/m³]	Supply (thickness) [cm]	Fire protection (DIN 4102 T1) 1998-05	Global warming potential (Ökobau.dat) [kg CO_2/m³]	Required thickness for U-value of 0.24[1] [cm]
Exterior insulation to outside air						
composite systems						
• insulating rendering	0.07	200	up to 12	B1	288	14
• mineral wool	0.032	40	up to 20	A1	68.8	6
• expanded PS (EPS)	0.035	20 (PS 20)	up to 30	B1	59.2	8
• mineral foam	0.045	115	up to 20	A1	127.6	10
• vacuum insulation	0.007–0.008	170–210	1–5	B2	n.s.	2
• wood fibre board	0.045	190–220	16	B2	-1.37	10
• rigid PU foam	0.028	32	20	B2	148	6
multilayered facades / rear-ventilated facades						
• mineral wool	0.032	40	up to 20	A1	68.8	6
• cellulose	0.4	60	any	B2	0.66	8
thermal insulation blocks	0.06	400	12/18	A1	93.1	12
Exterior insulation to ground						
extruded PS (XPS32)	0.035	32	2–16	B1	93.7	8
Interior insulation to outside air						
insulation with vapour barrier						
• mineral wool						
moisture active systems	0.032	40	up to 20	A1	68.8	6
• calcium silicate panels						
• composite of mineral	0.06	200–240	2.5–5	A1	504	12
wool and aerogel	0.019	40	2–4	A1	n.s.	4
• thermal insulation blocks	0.06	400	12 a. 18	A1	n.s.	12
mineral foam	0.045	115	20	A1	128	10
insulation as a vapour barrier						
• foam glass	0.038	100	14	A1	127.6	8
• rigid PU foam	0.28	32	20	B2	148	6

[1] Initial situation is a 36 cm brick wall, 2 cm plaster on both sides, with λ = 0.7 W/mK for plaster and brickwork

2.12

2.12 Insulation materials for exterior walls (selection)
The table includes typical constructions suitable to achieve an improvement in energy efficiency.

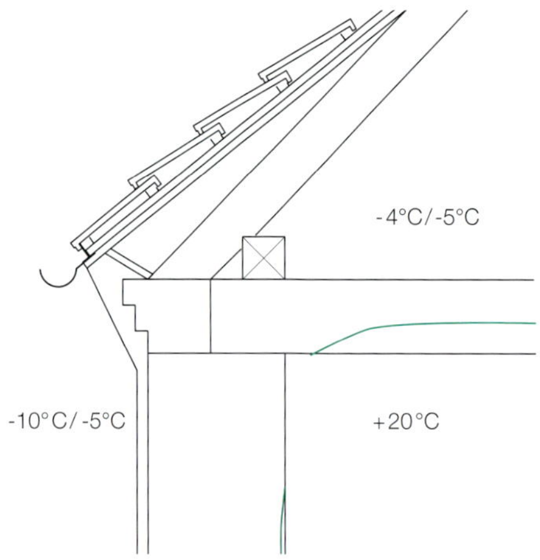

a

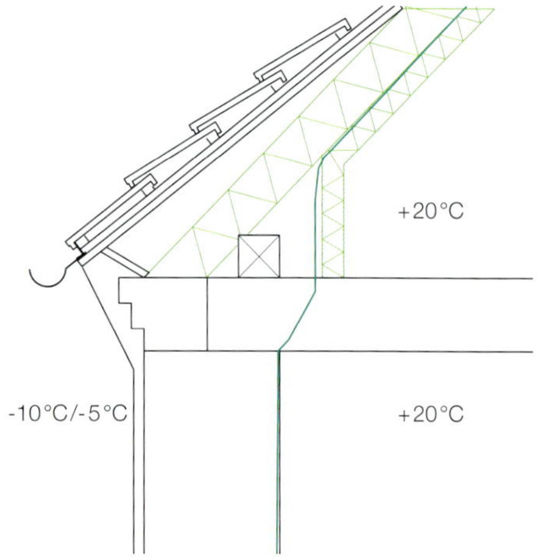

b

2.13

Requirements acc. to EnEV (enclosure 3, table 1):

Pitched roof after refurbishment:

max. U-value: 0.24 W/m²K

The requirements of the EnEV are regarded as satisfied if, in the case of already habitable attic space, the maximum possible thickness of insulation has been fitted between the rafters. If the roof storey has not yet been converted, the depth of the rafters must be increased on the underside so that sufficient insulation can be installed to meet the requirements.

U-value calculation:

Rafter [1]:	t [m]	λ [W/mK]	R [m²K/W]
R_{se}			0.100
roof sheathing	0.024	0.130	0.185
rafters	0.120	0.130	0.923
mineral fibre	0.060	0.032	1.875
gypsum board	0.012	0.210	0.060
R_{si}			0.100
Thermal resistance R_T			3.243
U-value of rafter [W/m²K]			**0.31**

Infill panel [2]:	t [m]	λ [W/mK]	R [m²K/W]
R_{se}			0.100
roof sheathing	0.024	0.130	0.185
mineral fibre	0.120	0.032	3.750
mineral fibre	0.060	0.032	1.875
gypsum board	0.012	0.210	0.060
R_{si}			0.100
Thermal resistance R_T			6.070
U-value of infill panel [W/m²K]			**0.16**

Total U-value [3]:	Share of area	U-value	Share of U-value
rafter	0.13	0.31	0.04
infill panel	0.87	0.16	0.14
Total U-value [W/m²K]			**0.18**

[1] width w = 0.10 m
[2] width w = 0.65 m
[3] Simplified calculation compared to DIN EN ISO 6946:2003. The U-value determined corresponds to the maximum value according to DIN and is sufficiently accurate to perform an energy audit.

Pitched roof: insulation is fitted to the underside whilst the existing roof covering is retained

When insulating roofs, it is necessary to decide whether the roof covering can be retained or whether the entire roofs needs replacing. The conditions for retaining the existing roof covering are as follows:

- good condition of the existing roof covering
- the presence of a properly functioning waterproofing system (underroof) which will protect the new insulation reliably against moisture from the outside
- access to the roof structure from the inside (in the ideal case an attic storey that has not yet been converted into habitable space)

Advantages:

- no costs for new roof covering
- no scaffolding needed
- roof stays watertight during the insulation work (independent of weather conditions)
- no change to external appearance

Disadvantages:

- if the rafters are not sufficiently deep for the required thickness of the insulation, the depth of the rafters must be increased by fixing battens underneath them (loss of internal space and height)
- if the existing waterproofing system (underroof) is vapour tight (bituminous felt, PE sheeting, etc.), the compatibility, in terms of building physics, of the new insulation must be checked thoroughly

☐ a

The water vapour diffusion resistance factor of the waterproofing system (underroof) has a decisive influence on the sequence of layers on the inside. It frequently occurs that a very vapour-tight material is used for the underroof, e.g. bituminous felt or PE sheeting. Any moisture that has infiltrated the insulation through these materials evaporates only very slowly to the outside. In such a case some of the possible insulation thickness must either be sacrificed in order to include an additional ventilation cavity below the waterproofing, or a vapour barrier must be added and installed very carefully on the room-side of the insulation so that no moisture from the interior air is able to penetrate the insulation.
A further option is to use a vapour barrier with a variable water vapour diffusion resistance. The barrier effect of these moisture adaptive vapour barriers is dependent on the average ambient moisture level. The diffusion resistance increases in winter so that only very small quantities of water vapour can infiltrate the insulation, whereas in summer the barrier effect decreases, which allows the insulation to dry out by dissipating the moisture to the interior space. In any case, it is essential to assess the physical performance of the intended roof construction.

☐ b

Rigid insulating boards are not suitable for use between rafters. On the one hand, it is very time-consuming to cut the boards to fit accurately between the existing rafters without any gaps, on the other hand, they are unable to compensate for the natural swelling and shrinkage behaviour of the timber roof structure. It is for this reason that clamping felt materials made from natural or mineral fibres should be installed between the rafters. Loose infill insulation, such as cellulose flakes, are a possible alternative. In the case of loose infill insulation, it is necessary to make sure that all cavities are filled to prevent the insulation from sagging.
The roof insulation should provide for a high specific heat storage capacity (or in other words, a high gross density) in

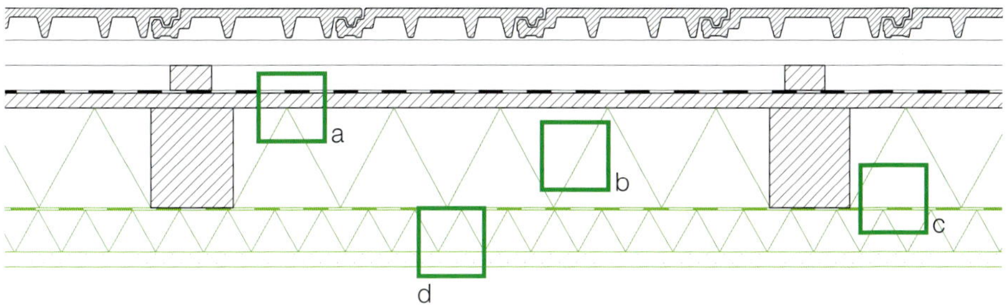

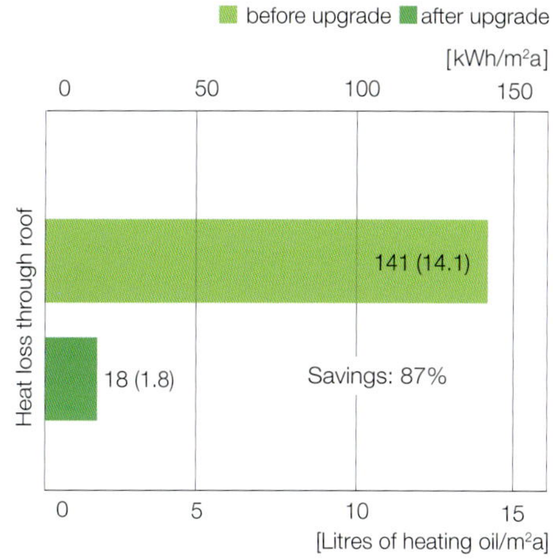

2.14

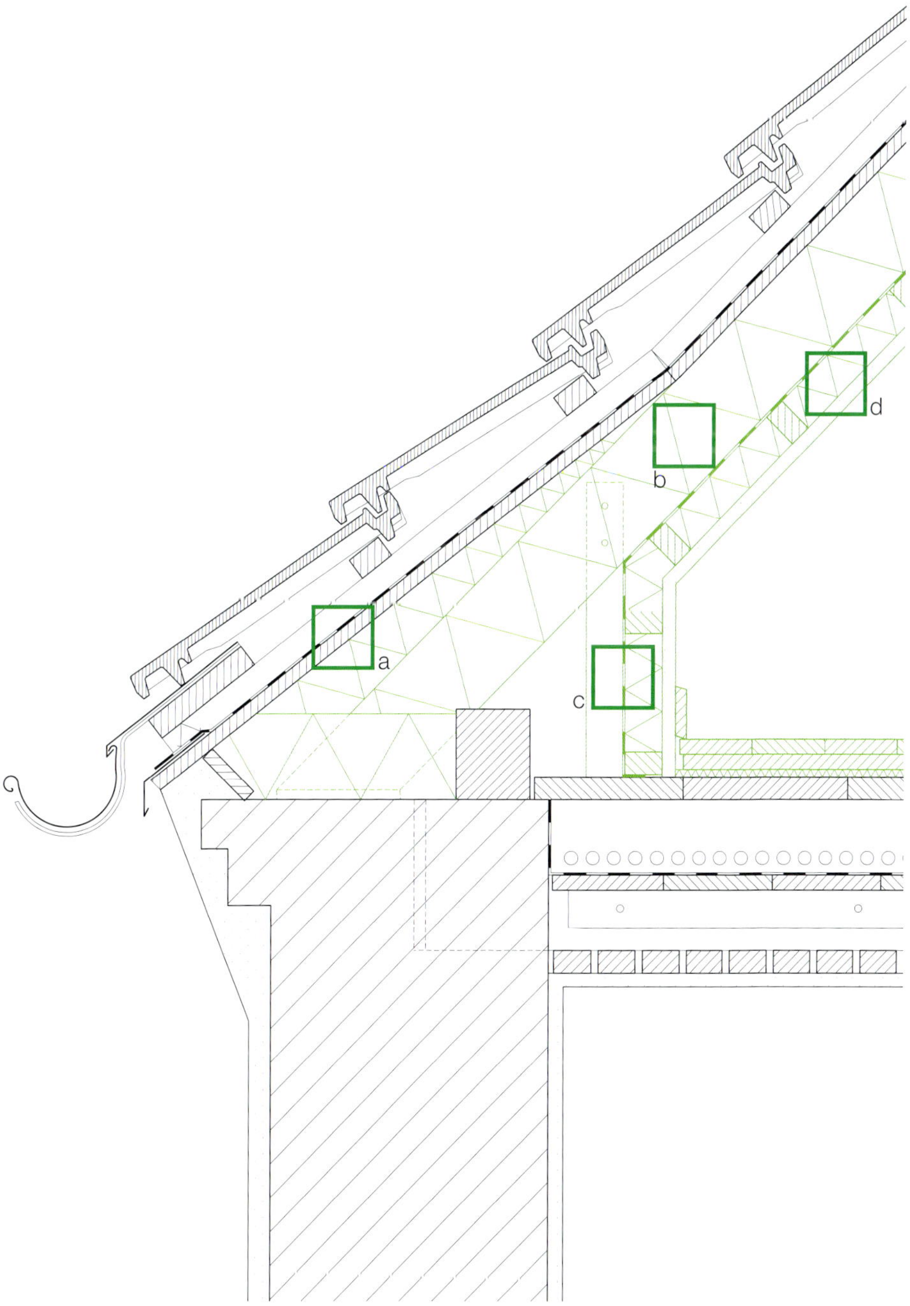

order to ensure a better thermal phase shift and hence prevent the roof space from overheating during the summer months.

☐ c

The vapour barrier also functions as an airtight membrane. Careful laying and bonding at all joints and junctions is vital to prevent the convection of interior air to the outside. The aim is to avoid heat loss and a high level of moisture being transported into the insulation. The vapour barrier should not be interrupted and is therefore best laid below the purlins, collar beams, etc. Thus designs with exposed roof structures are not recommended for this type of energy efficiency upgrade.

☐ d

A second layer of insulation installed between battens mounted at a 90° angle to the rafters increases the insulation thickness and eliminates the thermal bridging effect of the rafters. In addition, this solution provides an installation layer for any electric cables or heating pipes and protects the airtightness achieved by the vapour barrier. On the inside, the layer of insulation can be covered with gypsum plasterboard, gypsum fibre board or timber boarding.

2.13 Isothermal lines before (a) and after (b) the upgrade under standard conditions according to DIN 4108-2 (outside temperature -5 °C, room temperature +20 °C, relative room air humidity 50 %)
— 12.6 °C isotherm (risk of mould growth)

2.14 Annual heat demand in kilowatt hours and in litres of heating oil per square metre of roof area in the existing building and after the upgrade (calculations performed with 84 degree days (DD), U-value of existing structure 1.4 W/m²K, system loss 20 %)

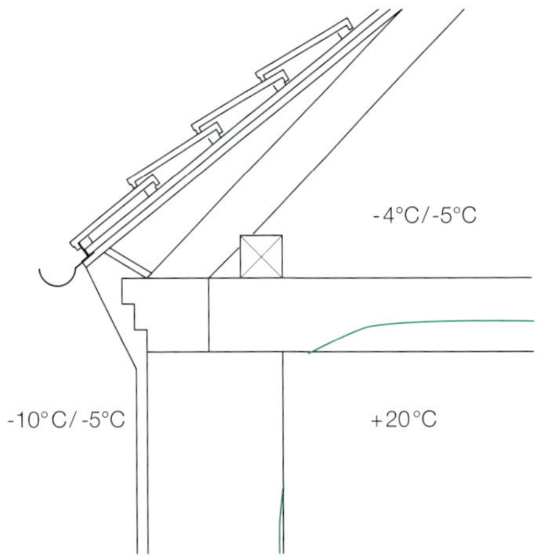

a

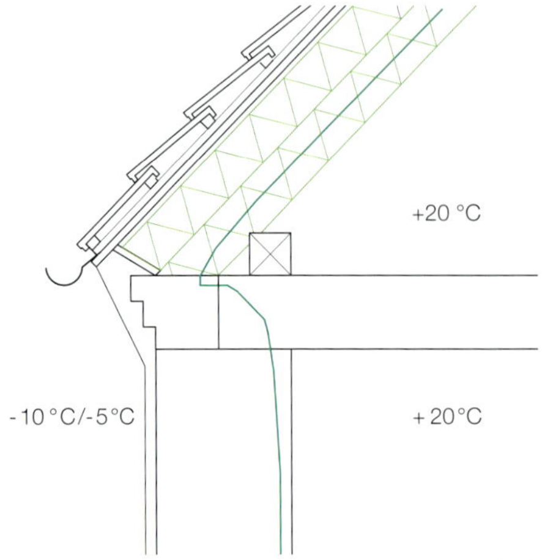

b

2.15

Requirements acc. to EnEV (enclosure 3, table 1):

Pitched roof after refurbishment:

max. U-value: 0.24 W/m²K

The requirements of the EnEV are regarded as satisfied if, in the case of already habitable attic space, the maximum possible thickness of insulation has been fitted between the rafters. There is no obligation to increase the depth of the roof structure above the rafters. However, if the roof space has not yet been converted, the full thickness of insulation required must be installed.

U-value calculation:

Rafter [1]**:**	t [m]	λ [W/mK]	R [m²K/W]
R_{se}			0.100
roof sheathing	0.024	0.130	0.185
mineral fibre	0.120	0.032	3.750
rafter	0.120	0.130	0.923
woodwool panel	0.030	0.065	0.462
lime cement plaster	0.015	1.000	0.017
R_{si}			0.100
Thermal resistance R_T			5.537
U-value of rafter [W/m²K]			**0.18**

Infill panel [2]**:**	t [m]	λ [W/mK]	R [m²K/W]
R_{se}			0.100
roof sheathing	0.024	0.130	0.185
mineral fibre	0.120	0.032	3.750
mineral fibre	0.120	0.032	3.750
woodwool panel	0.030	0.065	0.462
lime cement plaster	0.015	1.000	0.017
R_{si}			0.100
Thermal resistance R_T			8.364
U-value of infill panel [W/m²K]			**0.120**

Total U-value:	Share of area	U-value	Share of U-value
rafter	0.13	0.18	0.02
infill panel	0.87	0.11	0.10
Total U-value [W/m²K]			**0.12**

[1] width w = 0.10 m
[2] width w = 0.65 m

Pitched roof: insulation is fitted from the outside, the existing roof covering is replaced

If the roof covering is in need of replacement or the waterproofing system (underroof) is damaged, it is advisable to carry out the insulating measures from the outside.

Advantages:
- habitable roof space is not affected by the work
- the insulation can be perfectly protected against moisture from the outside

Disadvantages:
- higher costs (insulation, roof covering, flashings)
- scaffolding is required
- if the rafters are not deep enough for the thickness of insulation required, the depth of the rafters must be increased by adding battens to the upper side of the rafters (problems concerning the changed appearance and planning permission)

☐ a

The interior finish of the existing habitable roof space consists of reed matting or woodwool panels, which are plastered on the inside. If the plaster lathing is nailed to the battens, frequently countless numbers of nails penetrate into the space between the rafters. In such a case, it is at first necessary to install rigid insulating boards (e.g. EPS) – at least as thick as the longest nails – between the rafters in order to provide a flat backing for the vapour barrier and one without nails that will prevent it from being damaged by nails.

☐ b

The vapour barrier is installed from the outside. The foil is laid parallel to the roof structure and pressed into the gaps between the rafters so that the vapour barrier stands up at the sides of the timber. The two adjoining sheets are joined at the top of the rafters and bonded in an airtight manner. It is important that the vapour barrier is not stretched taut between the rafters, but rather laid with some slack so that, once the insulation is installed, the vapour barrier presses closely up against the rafters. The aim is to prevent warm, moist interior air from infiltrating into cavities between the vapour barrier and the rafters and then moving into cooler parts of the roof construction. If this is where the temperature drops below the dew point, the moisture condenses and, can, in the worst case, lead to an unacceptable moisture content in the load-bearing timber structure.
The problem can be solved by installing a further layer of insulation on top of the rafters, which ensures that the temperature on the upper side of the rafters is also much higher.

☐ c

It is very important that the material used for the insulation between the rafters is not a rigid insulating material, but rather compressible clamping felt or loose infill insulation. Only these materials can ensure that the vapour barrier is pressed up against the sides of the rafters.

☐ d

A variety of materials can be used for the second layer of insulation above the rafters. In the example illustrated here, horizontal battens have been fitted on top of the rafters and the intermediate gaps have been filled with clamping felt. The insulation is covered with timber sheathing and a roofing membrane that is open to diffusion. If non-compressible insulating materials are used, such as rigid EPS foam boards or wood fibre insulating boards, the battens and sheathing are not

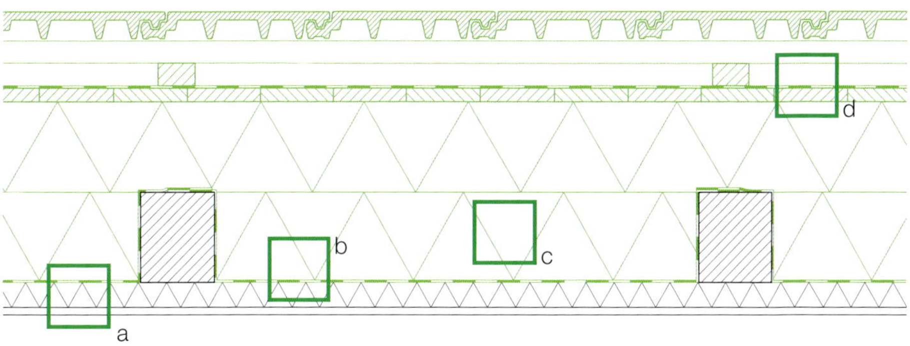

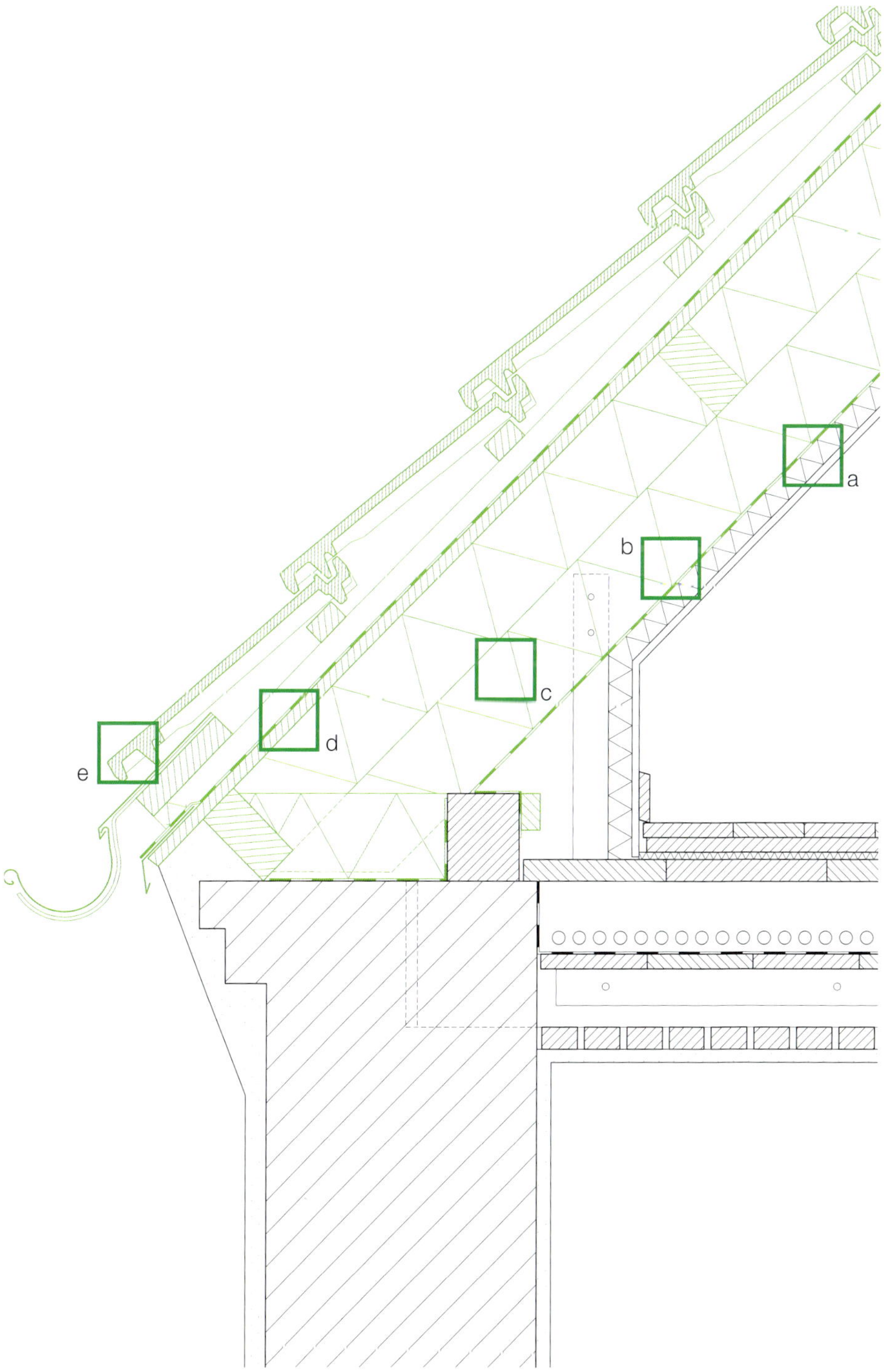

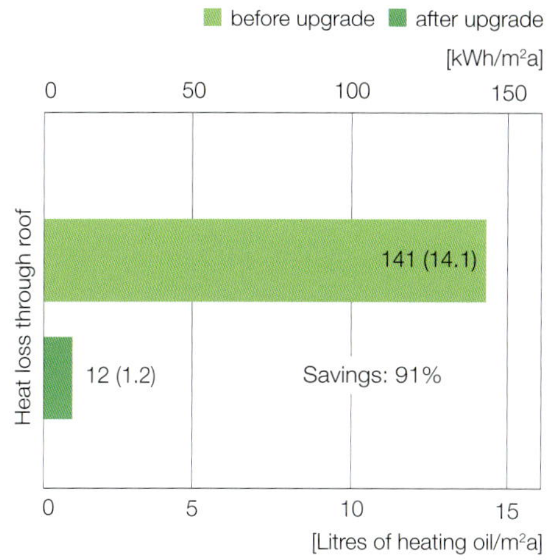

2.16

required. Even the diffusion-open roofing felt can be omitted if specially profiled, bituminous wood fibre boards or rigid foam system boards are used. To be on the safe side, the layering of the roof structure should be checked thoroughly in terms of its physical performance, especially when using rigid, extremely vapour-tight foam boards.

□ e

If, as in this example, the roof structure is raised by adding a second layer of insulation, it is essential to rethink and redesign all the edge and corner details (verge, eaves, etc.). This is especially important in the case of terrace and semi-detached houses, where awkward and unattractive joints can occur in the roof surfaces. It is also important to check whether raising the ridge and eaves levels requires the approval of the local building authority.

2.15 Isothermal lines before (a) and after (b) the upgrade under standard conditions according to DIN 4108-2 (outside temperature -5 °C, room temperature +20 °C, relative room air humidity 50 %)
— 12.6 °C isotherm (risk of mould growth)

2.16 Annual heat demand in kilowatt hours and in litres of heating oil per square metre of roof area in the existing building and after the upgrade (calculations performed with 84 degree days (DD), U-value of existing structure 1.4 W/m²K, system loss 20 %)

Requirements acc. to EnEV (enclosure 3, table 1):

Requirements to be satisfied by intermediate floors adjoining unheated interior space with finishes fitted on the cold side, i.e. the top side of the topmost floor:

max. U-value 0.30 W/m²K

Structure of the floor on the warm side:

max. U-value 0.50 W/m²K

Floors adjoining outside space with insulation on the cold side:

max. U-value 0.24 W/m²K

Thermal resistance of roof space:

When calculating the U-value of the topmost floor, it is necessary to know the thermal resistance R of the unheated roof space. This thermal resistance is specified in DIN EN ISO 6946, table 3, as follows:

Roof description	R [m²K/W]
1. Tiled roof without sheathing or the like	0.060
2. Slab roof or tiled roof with roofing felt, sheathing or the like below the roofing tiles	0.200
3. Same as 2, but with low emissivity on the underside of the roof (e.g. aluminium foil)	0.300
4. Roof with sheathing and roofing felt	0.300

U-value calculation (insulation, reinforced concrete slab):

	t [m]	λ [W/mK]	R [m²K/W]
R_{se}			0.040
roof space of tiled roof			0.060
fibre board	0.022	0.130	0.169
composite PU panel	0.120	0.024	5.000
concrete floor slab	0.160	2.300	0.070
gauged plaster	0.015	0.700	0.021
R_{si}			0.100
Thermal resistance R_T			5.460
U-value [W/m²K]			**0.18**

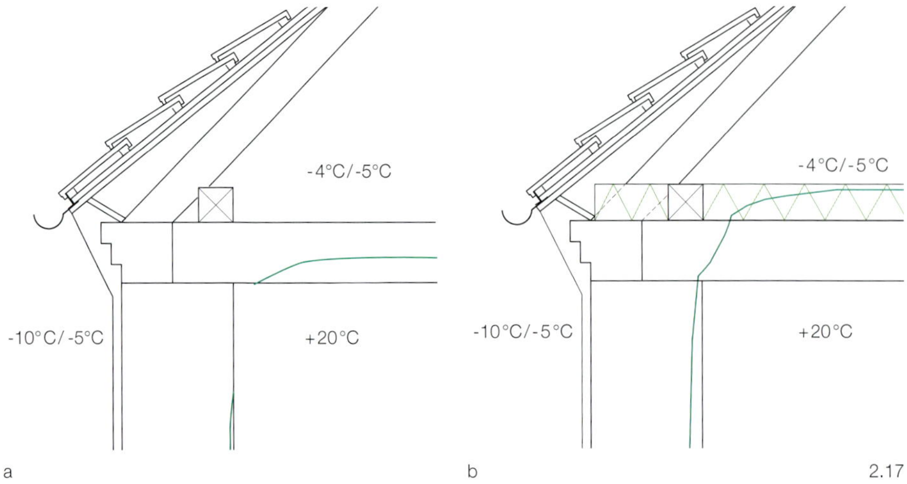

2.17

Pitched roof: insulation of topmost floor

Adding insulation to the topmost floor, between the habitable space and attic space, is one of the few mandatory construction upgrade measures specified in § 10 of the EnEV. A U-value of at least 0.24 W/m²K is specified for the floor construction of roof space that is accessible, but not yet insulated. Residential buildings with no more than two housing units, one of which must be occupied by the owner, are excluded from this upgrading obligation.

This regulation has applied for some time to attic space that is accessible, but not suitable for foot traffic. According to the EnEV, this kind of roof space (accessible, but not suitable for foot traffic) does not qualify for a future conversion into living accommodation owing to its layout. However, it can, nevertheless be used for storage purposes. Such a roof space is normally accessed through a hatch with a folding loft ladder, and not via a permanent stairway.

During the course of the planning work for the floor insulation, special attention must be paid to the access situation of the attic space. The importance here is not so much the insulation value of the door leaf or hatch, but rather its airtightness. Due to the stack effect within the connected interior air volume, large amounts of warm air can escape into the attic space without causing any noticeable draughts, and cold air is continuously drawn into the lower storeys due the untight conditions of older components.

In the following, a distinction is made between adding a new floor construction with clamping felt or loose infill insulation and a floating floor construction supported by rigid insulating material.

□ a

The existing timber floorboards are removed in order to access the existing cavities between the timber joists for insulating purposes. The cavity above the loose fill insulation on the sound boarding is then filled to the top of the timber joists with clamping felt made from mineral or natural fibres or a loose fill material, e.g. cellulose flakes. Either plastic sheeting or building paper with glued joints serves as a vapour barrier and an airtight membrane. Battens are then laid at a 90-degree angle to the joists and the space between the timber is once again filled with insulating material. The floor structure is then covered with simple wood boards or other flooring grade boards.

□ b

The vapour barrier and airtight membrane is fixed tightly to the top of the peripheral masonry wall by using contact pressure strips and pre-compressed sealing tape.

□ c

In the case of a timber joist floor, it is necessary to make sure that outside air does not penetrate the void beneath the sound boarding through gaps at the joist bearing points. This can easily occur if the joints between the masonry and timber joists have not been detailed in an airtight manner. If this is the case, the thermal insulation above the sound boarding is useless in very windy conditions. An additional strip of sheeting must therefore also be added here between the masonry and the eaves purlin. Finally, the top of the masonry wall is covered with loose fill insulation material so that the thermal bridging effect at the top of the wall is reduced and the ends of the timber joists are no longer subject to damage caused by condensation.

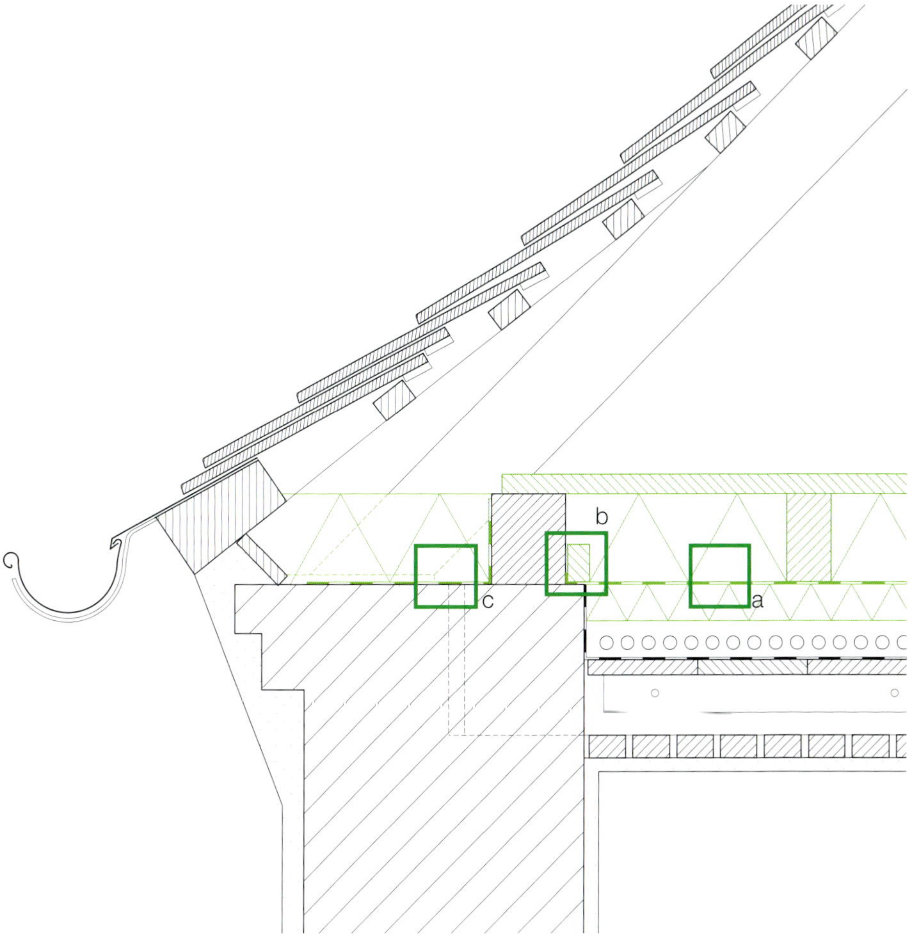

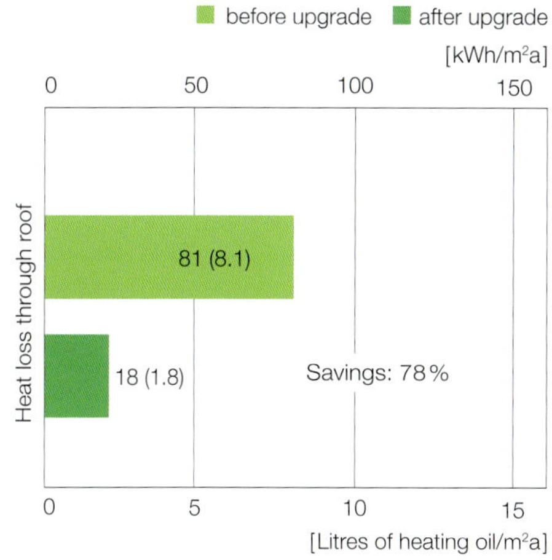

2.18

□ d

If the topmost floor, between the habitable space and attic space, is a reinforced concrete floor slab, a vapour barrier and airtight membrane is not required. If a thick layer of insulation is required, it is advisable to install the insulation boarding in two layers. This has two advantages: firstly, the thinner boards are easier to handle and fit; secondly, the joints between the boards can be staggered to minimise convection through these. Flooring grade boards can simply be laid on top of the rigid insulation with glued joints (floating floor).

□ e

In the case of a concrete floor slab, the top of the masonry wall must also be covered with loose fill insulation material in order to minimise the geometrical thermal bridge along the top edge of the room below.

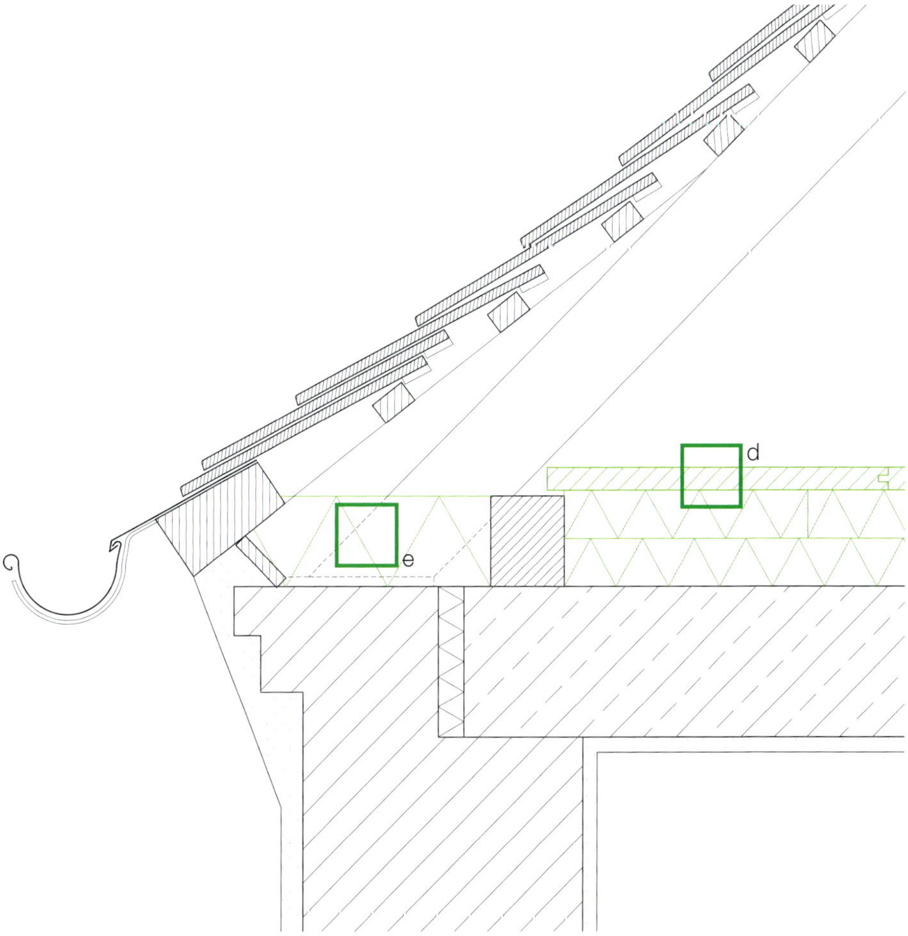

2.17 Isothermal lines before (a) and after (b) the upgrade under standard conditions according to DIN 4108-2 (outside temperature -5 °C, room temperature +20 °C, relative room air humidity 50 %)
— 12.6 °C isotherm (risk of mould growth)

2.18 Annual heat demand in kilowatt hours and in litres of heating oil per square metre of roof area in the existing building and after the upgrade (calculations performed with 84 degree days (DD), U-value of existing structure 0.8 W/m²K, system loss 20 %)

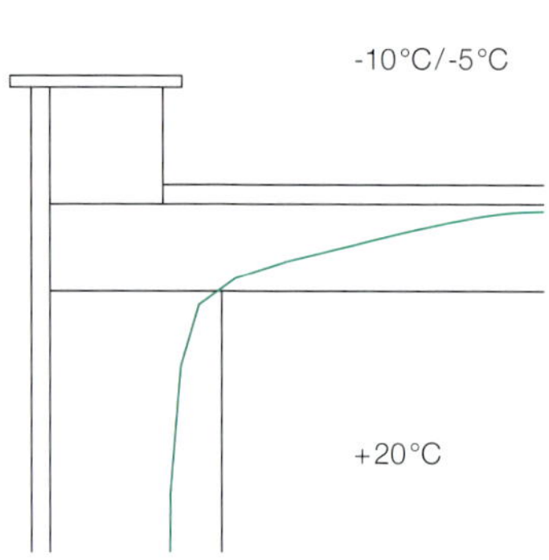

a

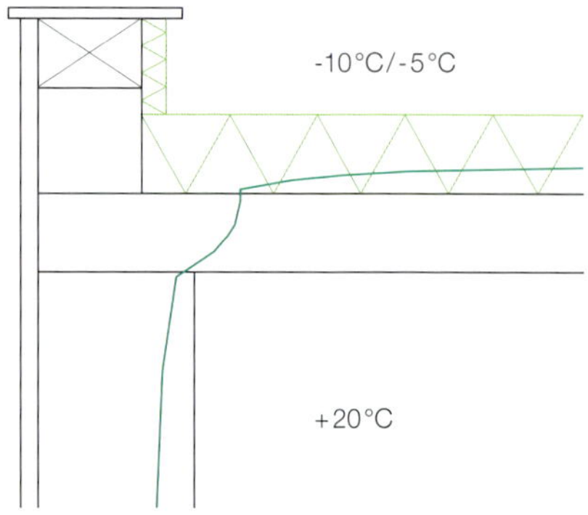

b 2.19

Requirements acc. to EnEV (enclosure 3, table 1):

Flat roof after refurbishment:

max. U-value 0.20 W/m²K

When tapered insulation is used, the U-value must be determined according to DIN EN ISO 6946, appendix C. Minimum thermal protection according to DIN 4108-2 with a U-value of 0.75 W/m²K must also be ensured at the lowest point with the smallest thickness of insulation.

Inverted roof

Correction factors according to DIN 4108-2, see fig. 2.6, p. 41

U-value calculation:

Flat roof:	t [m]	λ [W/mK]	R [m²K/W]
R_{se}			0.040
extruded PS foam	0.120	0.035	3.529
waterproofing	0.008	0.170	0.047
rigid PS foam	0.060	0.030	2.000
vapour barrier	0.005	0.170	0.029
reinforced concrete	0.240	2.100	0.114
R_{si}			0.100
Thermal resistance R_T			5.859
U-value of flat roof [W/m²K]			**0.170**
Correction factor (share$_R$ = 31%) [W/m²K]			0.03
U-value of inverted roof [W/m²K]			**0.200**

Warm roof

U-value calculation:

Flat roof:	t [m]	λ [W/mK]	R [m²K/W]
R_{se}			0.040
waterproofing	0.008	0.170	0.047
rigid PS foam	0.18	0.030	6.000
vapour barrier	0.005	0.170	0.029
reinforced concrete	0.240	2.100	0.114
R_{si}			0.100
Thermal resistance R_T			6.330
U-value of warm roof [W/m²K]			**0.160**

Flat roof

The most stringent requirements of the EnEV apply to the refurbishment of flat roofs, because this is where relatively low-cost insulation measures can be employed to realise a high energy-saving potential. In the case of existing warm roof constructions (non-ventilated roof structure), the first step before commencing the refurbishment is to establish whether the roof structure is in good order and merely the thermal aspects need improving, or whether the entire roof structure, including the waterproofing, must be replaced.
In the case of an existing cold roof construction (ventilated roof structure), the ventilation cavity is usually not high enough to accommodate additional insulating material and still reliably satisfy the ventilation requirements. An analysis of the existing roof structure is therefore required to determine whether the roof is still acceptable from a building physics point of view if the ventilation cavity is completely filled with insulating material. If this is the case, the roof can be upgraded at low cost by blowing cellulose flakes into the cavity. The alternative is that the roof structure including the waterproofing must be replaced completely.

☐ a

If the waterproofing of the warm roof construction can be retained, a closed-pore insulation material can be installed above the existing waterproofing, e.g. XPS, if possible using interlocking insulation boards. The construction of the so-called inverted roof is straightforward from a building physics point of view. The insulation protects the waterproofing against mechanical damage and also reduces the impact of temperature fluctuations. The actual drainage level is now on top of the new insulation, which requires gravel to hold it down and prevent uplift.

☐ b

It is usually necessary to raise the height of the parapet in order to meet the requirements of the flat roof guidelines. The reference level for the parapet height is the top of the gravel; the distance must be at least 15 cm. Problems with the building authorities could arise if the increased height of the parapet has an effect on the distance space.

☐ c

To avoid thermal bridging heat loss, it is essential to insulate both sides of the concrete parapet and, depending on the height, also the top. Insulation on the inside of the parapet is frequently missing in older buildings.

☐ d

It makes sense, in the case of damaged waterproofing, to replace the entire roof system including the old layers of insulation since it is nowadays possible to meet the increased requirements of the EnEV with a minimum thickness of insulation by using innovative products.
It may in some circumstances be possible to retain the old layers of insulation since most of the plastic sheeting available today is to a certain degree open to diffusion, which enables a previously moist roof structure to gradually dry out again once the refurbishment work has been completed. By leaving the existing roof system in place, it is possible to save the costs of removal and disposal. Furthermore, once the existing materials have dried out, they also contribute towards the thermal insulation of the building.

☐ e

Owing to its mass, the addition of rooftop planting to a refurbished flat roof evens out temperature fluctuations and hence improves the summer heat protection of

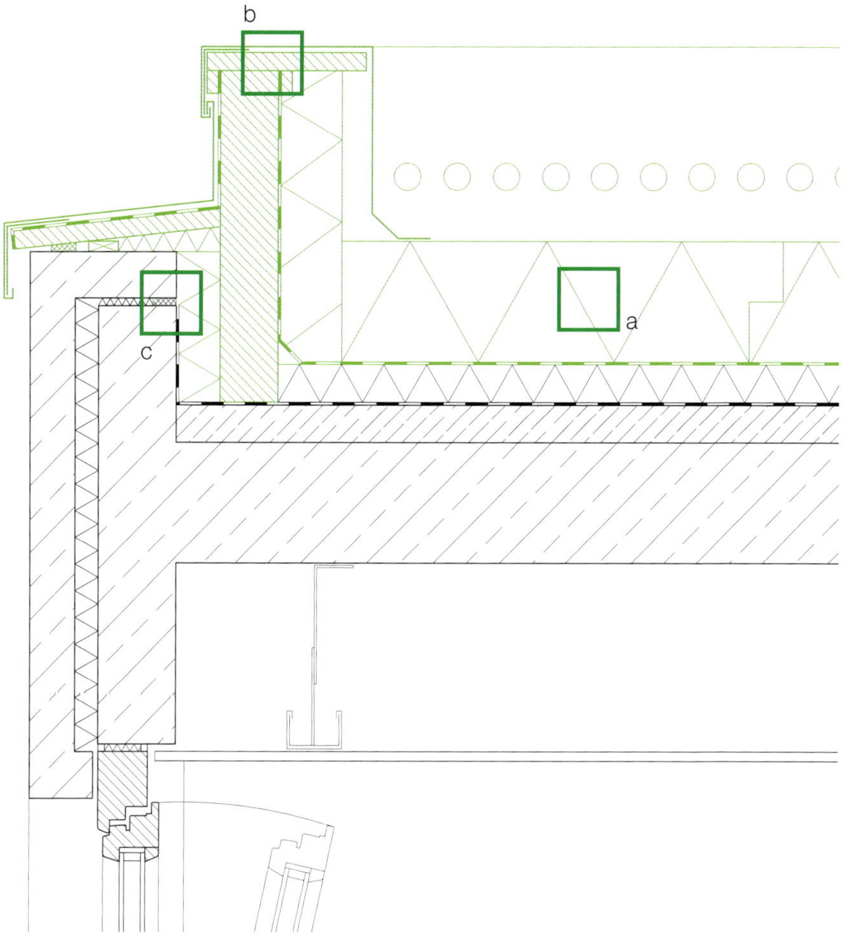

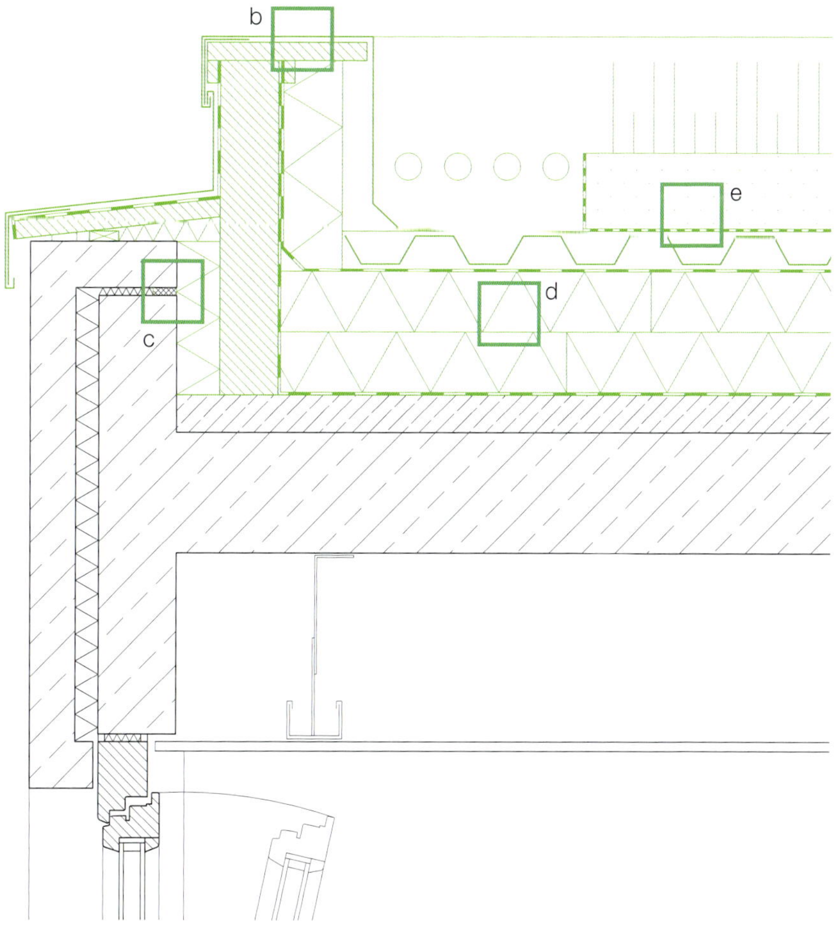

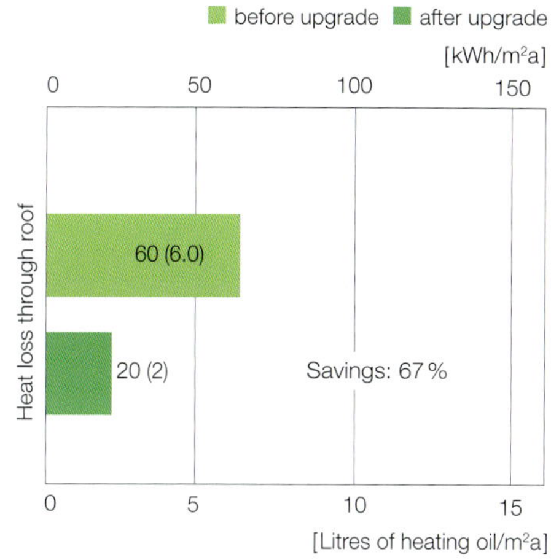

2.20

the building. At the same time, the waterproofing is protected against excessive thermal fluctuations, which can extend its useful life expectancy. It is essential to clarify in advance whether the roof slab is sufficiently dimensioned and able to bear the additional load. In green roofs, there is a major distinction between:

- intensive green roofs
  this is a two-layered system with a drainage, filter and plant-bearing layer
- extensive green roofs
  this is a single-layer system without a drainage layer, but with a water-permeable plant-bearing layer and exclusively low-maintenance sedum varieties

Intensive green roof systems can also be installed on flat roofs without slopes providing that an adequately deep drainage layer is included. This can be a cost advantage for refurbishments since neither sloping screed nor tapered insulation is required.
The weight of the lightest extensive green roof system is less than 100 kg/m² in a water saturated condition and the additional height above the waterproofing is no more than 90 mm. However, owing to the lack of a drainage layer, this simple system requires a slope of 2 %.
Alongside the constructional and thermal advantages, rooftop planting is extremely beneficial in terms of ecological functions, particularly in urban environments.

2.19 Isothermal lines before (a) and after (b) the upgrade under standard conditions according to DIN 4108-2 (outside temperature -5 °C, room temperature +20 °C, relative room air humidity 50 %)
— 12.6 °C isotherm (risk of mould growth)

2.20 Annual heat demand in kilowatt hours and in litres of heating oil per square metre of roof area in the existing building and after the upgrade (calculations performed with 84 degree days (DD), U-value of existing structure 0.6 W/m²K, system loss 20 %)

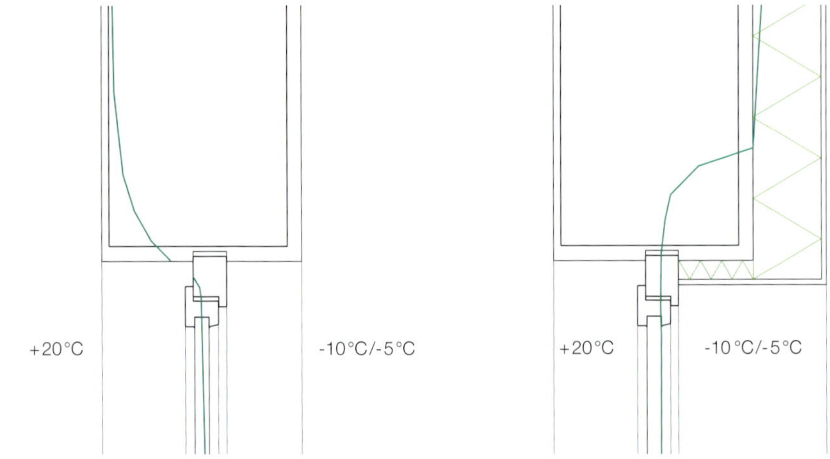

a b 2.21

Requirements according to EnEV (enclosure 3, table 1):

Requirements for exterior walls with cladding or layers of insulation added to the outside:

max. U-value 0.24 W/m²K

If the render is renewed on an existing exterior wall with a U-value greater than 0.9 W/m²K, the owner is obliged to perform insulation measures and meet the U-value of 0.24 W/m²K.

U-value calculation:

Exterior wall:	t [m]	λ [W/mK]	R [m²K/W]
R_{se}			0.040
polymer render	0.005	0.700	0.007
PS boards	0.120	0.032	3.750
lime-cement render	0.025	0.870	0.029
masonry	0.300	0.610	0.492
gauged plaster	0.015	0.700	0.021
R_{si}			0.130
Thermal resistance R_T			4.469
U-value of exterior wall [W/m²K]			**0.22**

2.21 Isothermal lines before (a) and after (b) the upgrade under standard conditions according to DIN 4108-2 (outside temperature -5 °C, room temperature +20 °C, relative room air humidity 50 %)
— 12.6 °C isotherm (risk of mould growth)

2.22 Annual heat demand in kilowatt hours and in litres of heating oil per square metre of envelope area in the existing building and after the upgrade (calculations performed with 84 degree days (DD), U-value of existing structure 1.0 W/m²K, system loss 20 %)

Thermal insulation composite system without replacing windows

Insulating the exterior of an opaque facade by using a thermal insulation composite system (ETICS) is usually one of the most effective and economic insulating measures. On the one hand, the opaque parts of the facade in residential buildings almost always represent the largest share of the building envelope, on the other hand, the U-values of the existing monolithic exterior walls are usually much lower than the multilayered structure of the roof or intermediate floor constructions.

Advantages:
- the solid exterior wall with its heat storage mass remains accessible for the interior space
- the majority of structural thermal bridges can be eliminated; the geometric thermal bridges minimised substantially
- the interior space benefits from the heat loss from, for example, poorly insulated heating pipes incorporated in the external walls

Disadvantages:
- the wall thickness increases (problems related to appearance and planning permission)
- all junctions between the facade and other components must be redesigned and adapted
- the facade becomes more susceptible to mechanical damage and the formation of algae
- replacing the windows at a later date is more difficult

It is for this reason that it makes sense to replace the windows at the same time as adding the thermal insulation composite system. In terms of thermal performance, the simultaneous refurbishment of windows and walls allows for the best results.

The following aspects however may give reason to install new windows in the original position or to even retain the existing windows:
- the windows are in very good condition and can be improved in terms of their thermal properties with some simple measures
- the windows have already been replaced
- the existing roller shutters and roller shutter boxes are to remain in use
- elaborate plastering work around the windows and the replacement of window sills in the interior is to be avoided

☐ a

If the window frames are still in good condition, the windows can be upgraded by exchanging the glass only. In this case, the new window panes must meet a U_g-value of 1.1 W/m²K.

☐ b

The window frame is often too narrow to allow sufficient insulation to be fitted into the reveal and cover the frame. This is especially the case if metal fittings for window shutters are attached to the frames, which are to remain in use. In this situation, it is necessary to remove the render in the reveal and replace it by a very thin layer in order to create sufficient space for at least 2 cm of insulation.

☐ c

The insulating boards and rendering system form a unified whole and must be coordinated as a composite system in correspondence with the accreditation of the system provider. The fire protection requirements have an immediate impact on the choice of insulating material and render. The permissible building material class of a thermal insulation composite system depends on the height h of the topmost floor above the surrounding ground level:

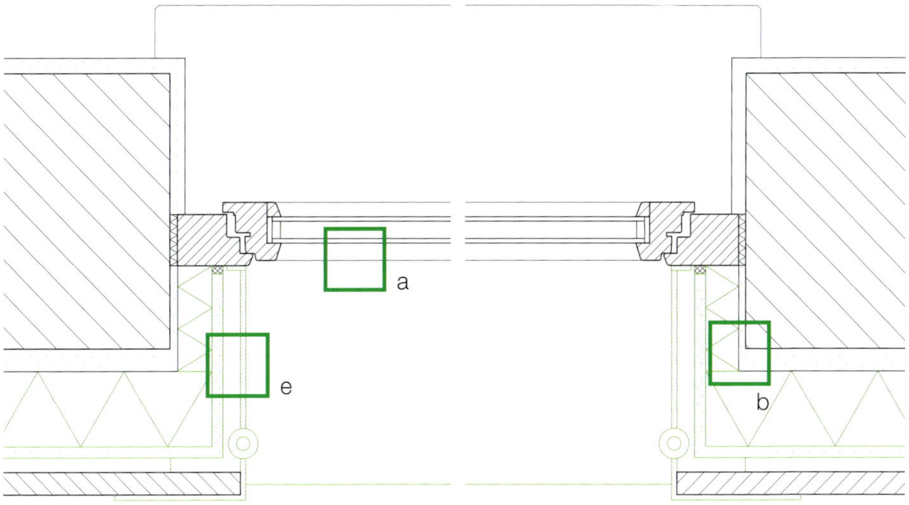

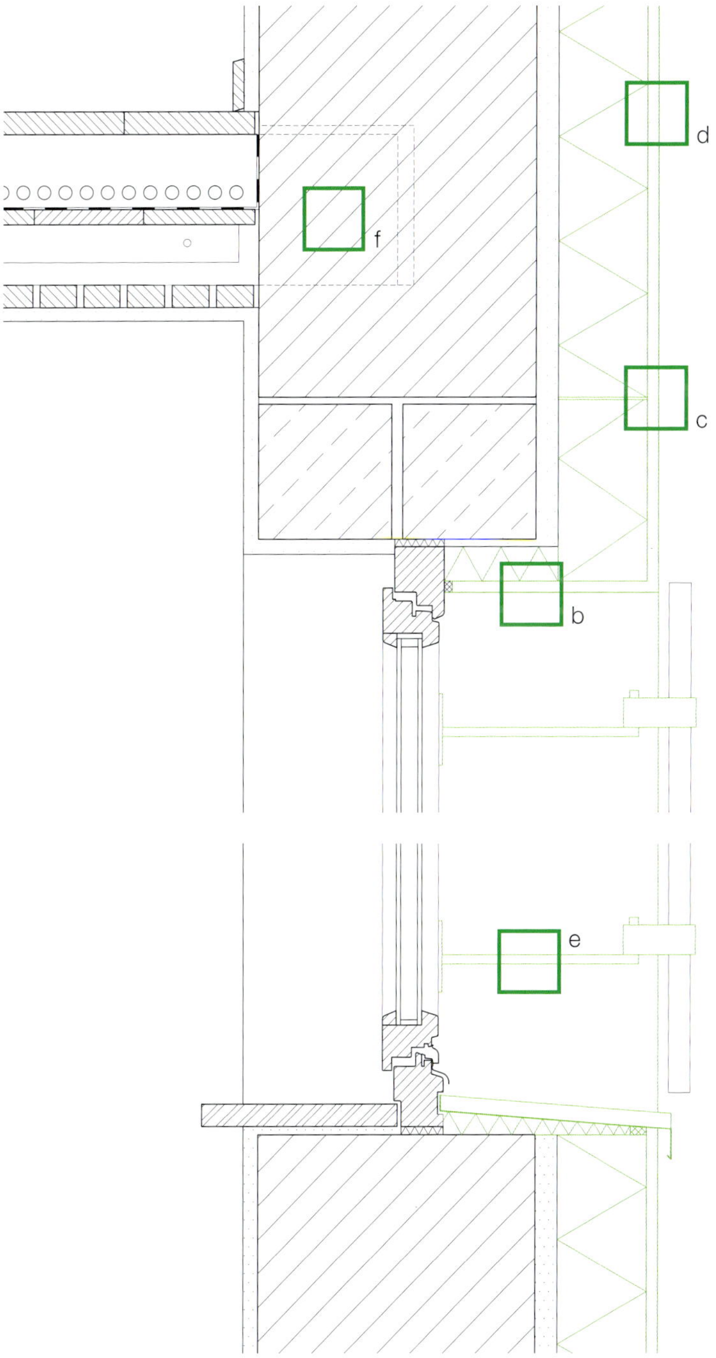

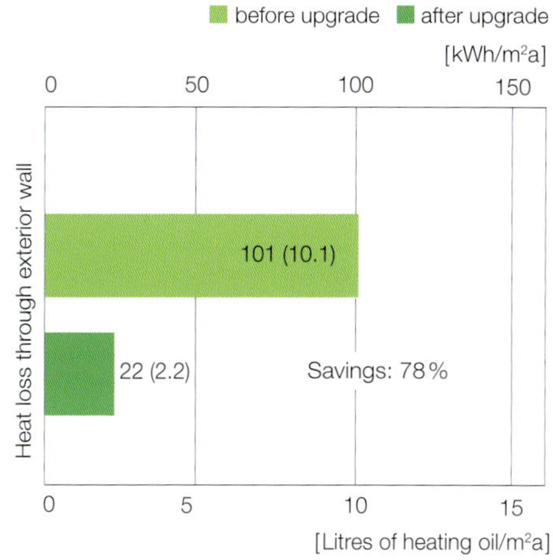

2.22

- at least B2 (flammable) for $h < 7$ m
- at least B1 (not readily flammable) for $h = 7–22$ m
- A (incombustible) for $h > 22$ m

Thermal insulation composite systems with rigid PS or PUR foam boards and an insulation thickness of more than 10 cm represent a special case in terms of fire protection. Strips of A-class insulation must be fitted immediately above the window lintels. These must be at least 20 cm high and extend 30 cm to both sides of the opening. The purpose of these strips is to protect the escape route through the window from melting insulating material in the event of fire. Alternatively, it is possible to install peripheral rings of mineral wool board insulation strips in every second storey.

☐ d

Some systems also use a thick coat render finish. In this case, the lathing and final coat are applied to a 15 to 20-cm-thick undercoat, which must dry out for approximately three weeks before the final coat can be added. Thick coat render systems are less susceptible to acts of vandalism and damage by woodpeckers. Furthermore, damage caused by cracking is comparably low.

☐ e

Existing window shutters will require new fittings and may need to be adapted to fit the slightly smaller clear openings of the new windows.

☐ f

The exterior thermal insulation has a positive effect on the bearing points of timber joist floors. On completion of the insulation measures, the wall temperatures are much higher and the formation of condensation is no longer an issue in these areas.

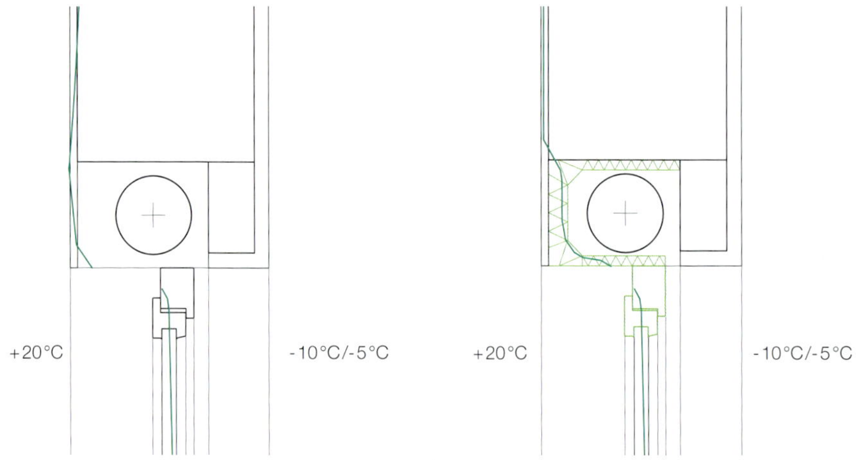

a b 2.23

Requirements acc. to EnEV (enclosure 3, table 1):

Window replacement:

max. U_w-value 1.30 W/m²K

This value must be met if 20 % or more of the window area is replaced in a single building.

Requirements for roller shutter boxes:

The EnEV does not specify any special requirements for roller shutter boxes. However, when calculating the transmission heat loss, the facade areas of the roller shutter boxes must be shown separately and considered with their own U-value if the EnEV's global thermal bridging heat loss coefficient ψ of 0.10 W/m²K is applied.
According to the minimum requirements of DIN 4108-2, the U-value for roller shutter boxes is:

U-value (replacement after 1995)	0.60 W/m²K

If the optimised thermal bridging heat loss coefficient of 0.05 W/m²K is assumed or the thermal bridges are assessed individually, the roller shutter boxes may be added to the surface area of the external wall (see pages 38 and 40). If the intention is to use the optimised thermal bridging heat loss coefficient, the design rules according to supplement 2 of DIN 4108 must be satisfied. This means that approximately 6 cm of thermal insulation with a thermal conductivity rating of WLG 040 must be fitted at all contact points between the roller shutter box and the interior and between the roller shutter box and the floor slab above the roller shutter box.

The U-value calculation of the window is performed according to DIN EN ISO 10077-1.
The example presented here is based on a window measuring 1.20 m by 2.20 m (masonry dimensions) with a frame proportion of 35 % (window area: 2.64 m²). The frame is an upgraded wooden frame with core insulation.

U-value calculation:

Window:	A [m²]	U [W/m²K]	[W/K]
wooden frame (upgraded)	0.79	1.00	0.790
glass	1.85	0.60	1.110
	ψ [W/mK]	length [m]	[W/K]
thermal bridge	0.04	6.00	0.240
total			2.140
U-value [W/m²K] (divided by 2.64 m²)			**0.810**

2.23 Isothermal lines before (a) and after (b) the upgrade under standard conditions according to DIN 4108-2 (outside temperature -5°C, room temperature +20°C, relative room air humidity 50%) —12.6°C isotherm (risk of mould growth)

2.24 Annual heat demand per square metre of window area in the existing building and after the upgrade (calculations performed with 84 degree days (DD), U-value of existing structure 2.7 W/m²K, system loss 20 %)

New windows, upgrade of existing roller shutter boxes

In the majority of cases the windows are replaced at the same time as installing a thermal insulation composite system. This allows the windows to be set in the best position from an energy performance point of view, i.e. flush with the outer face of the original masonry (see pp. 66f.). However, in numerous situations, it is not possible to insulate the building from the outside for reasons concerning the construction and planning permission. In these cases, the only way to improve energy efficiency is to replace the windows.
When new windows are fitted without insulating the exterior wall, the critical surface temperature is always undercut at the window/wall junction. The result is an increased danger of mould growth within these specific areas. The danger can be reduced by installing the windows in an airtight and technically correct manner according to RAL (quality assurance association for windows and doors) and by installing a simple ventilation unit (exhaust air unit) to support the air circulation in the affected rooms. The regular air exchange ensures that the relative humidity always remains below 50 % in winter.

☐ a

When replacing windows, the width of the frame should be chosen in such a way that a thermal insulation composite system, that may under certain circumstances be added at a later date, can be fitted according to the technical requirements at all junctions.
If the windows are replaced and the facade is insulated in a single operation, the position of the windows in their corresponding openings should be considered carefully. From an energy performance and light incidence viewpoint, the window is best mounted on the outer edge of the reveal, flush with the exterior wall surface. Whenever the position of the window in the reveal is changed, the window sill on the inside and outside and possibly also the roller shutter boxes will also need replacing.

☐ b

The installation of the window in the reveal must be performed, according to RAL, in an airtight manner. It is important in this context to ensure that the inner joint is more airtight than the outer one. In detail this means that the reveal must be plastered evenly before installing the window creating an opening which is approximately 2 cm larger than the frame. The space between the window and reveal is then filled with mineral wool and sealed with pre-compressed sealing tape all the way round until airtight. The reveal on the outside should then be rendered. On the inside, the joint can be sealed by using foil strips, which are glued to the reveal. These should then also be covered with a new layer of plaster.
If the windows are to be replaced without damaging the inner reveal, the required airtight seal can also be achieved by using permanent elastic grout or an approved pre-compressed sealing strip. A wood strip at this connection point inside can cover up the sometimes slightly unattractive joint.

☐ c

Depending on the requirements, the choice of glass is made according to following criteria:

- winter heat protection (U-value)
- summer heat protection (g-value)
- daylight transmittance (τ-value)
- noise protection (R_w-value)
- special requirements, e.g. fire protection

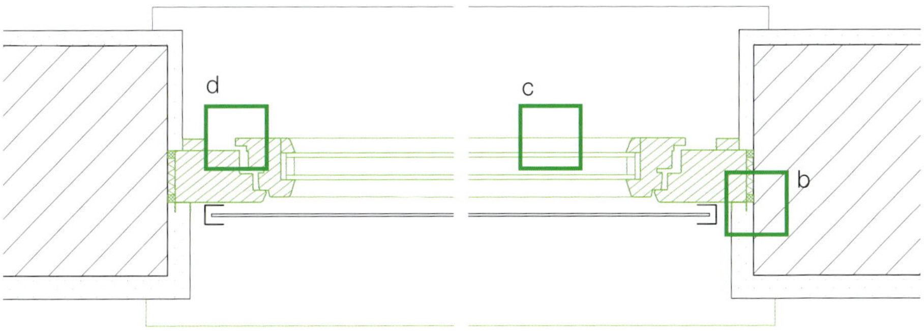

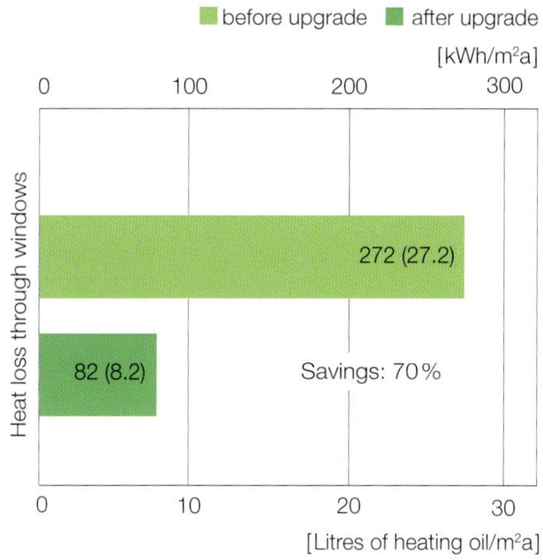

2.24

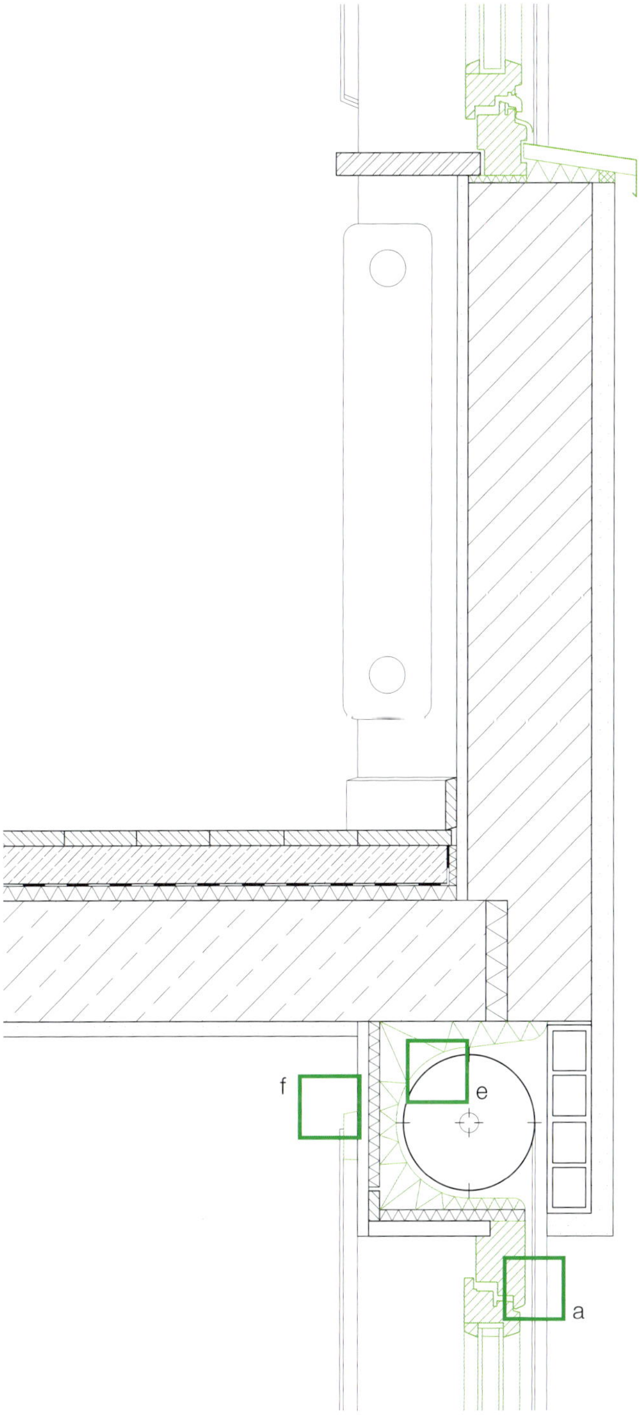

Triple insulating glazing with a U_g-value of 0.6 W/m²K is standard today. Because krypton is difficult to extract, argon is the preferred gas for filling the cavities between the panes of glass. Plastic spacers with $\psi < 0.4$ W/mK should be used between the panes.

☐ d

In terms of energy transmission and maintenance, the frame is the weakest component within the window construction. Energy optimised frames are available in all materials. They have U_f-values ranging between 0.8 and 1.0 W/m²K - this is also the case for metal frames.

In comparison to other materials, wooden frames require greater maintenance. By using a hard wood, such as oak, or an aluminium cover fixed to the outside of the frame, the maintenance work is reduced significantly.

☐ e

In order to improve the roller shutter boxes in terms of their thermal properties, various manufacturers supply prefabricated polystyrene foam inserts for insulating the existing voids within the roller shutter boxes. If the space between the rolled-up shutter and the shutter box is too narrow for the insulation, there is the possibility to exchange the existing shutter for one that rolls up in a more space-efficient way.

An airtight seal around the inspection opening is particularly important when upgrading roller shutter boxes.

☐ f

Since the opening for the roller shutter belt is a direct link between the interior and exterior, the gaps should be reduced as much as possible by installing brush seals.

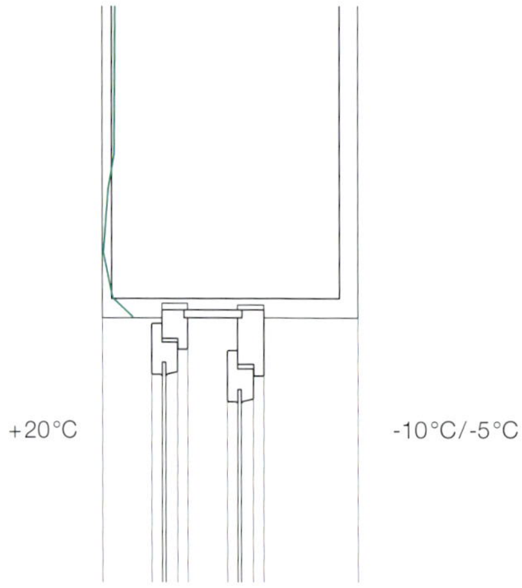

a

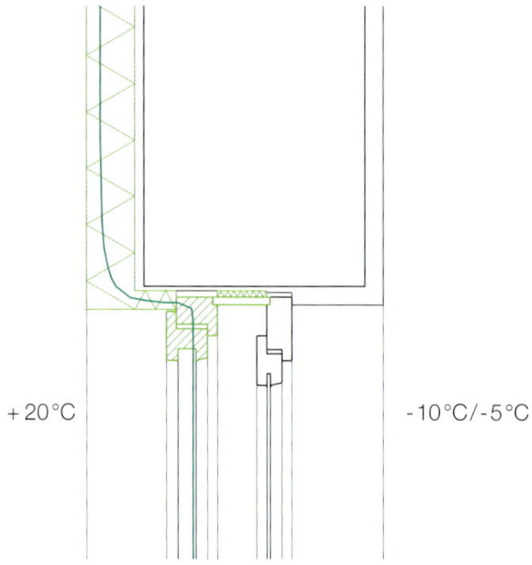

b

2.25

Requirements acc. to EnEV (enclosure 3, table 1):

Requirements for external walls which are insulated on the inside:

max. U-value 0.35 W/m²K

In comparison to wall structures that are insulated on the outside, exterior walls with interior insulation suffer from the effects of thermal bridges. In order to achieve realistic results for the associated calculations, the global thermal bridging heat loss coefficient has been doubled and therefore increased to 0.20 W/m²K in line with the tests performed by the German Energy Agency (see p. 47). The EnEV does not include such a provision.

U-value calculation of exterior wall:

	t [m]	λ [W/mK]	R [m²K/W]
R_{se}			0.040
lime cement plaster	0.025	0.870	0.029
masonry	0.300	0.610	0.492
gauged plaster	0.015	0.700	0.021
aerorock (rockwool)	0.040	0.019	2.110
R_{si}			0.130
Thermal resistance R_T			2.822
U-value [W/m²K]			**0.35**

U-values of individual windows:
Calculation according to DIN EN ISO 10077-1

U-value of inner leaf [W/m²K]	1.20
Thermal resistance [m²K/W]	0.833
U-value of outer leaf [W/m²K] (nominal value according to DIN 4108-4)	4.5
Thermal resistance [m²K/W]	0.222

U-value of double window:	R [m²K/W]
inner leaf	0.833
$-R_{se}$	-0.040
cavity > 10 cm	0.173
$-R_{si}$	-0.130
outer leaf	0.222
Thermal resistance	1.058
Total U-value (U_w) [W/m²K]	**0.95**

Interior insulation including upgrade of existing single glazing

The external appearance of the building remains unchanged if the inward facing sides of the exterior walls are insulated. This insulation method is therefore mainly used in conjunction with stucco or clinker brick facades worthy of protection.

Advantages:
- no changes to the external appearance of the building
- no issues concerning planning permission
- partial (room-by-room) exterior wall insulation is possible
- low-cost

Disadvantages:
- the heat storage capacity of the solid exterior walls can no longer be exploited for the benefit of the interior space
- danger of damage to building fabric due to the substantial drop in temperature within the load-bearing exterior walls
- thermal bridges where interior walls and floor slabs are tied into the exterior walls
- building physics problems especially in the vicinity of timber joist floors
- water and heating pipes that are incorporated into the exterior walls are in danger of freezing

When the thermal insulation is placed on the warm side of solid, heavy components, the tried-and-tested rules of construction are ignored. It is therefore necessary to perform thorough investigations of the building physics beforehand and pay special attention to good workmanship. In order to avoid high relative room air humidity during the heating period and, hence, a risk of moisture damage, it is advisable to install a controlled ventilation system at the same time as the interior insulation.

☐ a

An internal vapour barrier should always be included when adding interior insulation. Moisture that has infiltrated the insulation cannot dissipate to the outside very easily because of the relatively high diffusion resistance of the plastered and rendered masonry wall. It also takes a fairly long time for the inert masonry to heat up when the outside temperature rises, which means that the drying process occurs after considerable delay.
However, the vapour barrier can be omitted if a vapour-tight insulating material (e.g. cellular glass) is used, or if occasional condensation at the boundary between interior insulation and masonry is not expected to cause any harm.
Nevertheless, it is essential to ensure that the wall can fully dry out during periods of dry weather, and that a moisture-resistant insulating material is used (e.g. mineral foam, rigid PUR foam or calcium silicate panels).
Non-rigid insulating materials require a substructure, such as timber studs, to fasten the wall lining. All rigid insulating materials can be bonded directly to the interior surface of the exterior walls by using a suitable backing mortar or adhesive. Possible insulating materials include different sandwich panels with a core of rigid PS foam or mineral wool and a variety of linings.

☐ b

In the case of timber joist floors, special attention must be paid to the bearing points of the joists in the exterior walls. Once the interior insulation has been installed, the temperature of the load-bearing exterior wall drops dramatically. The dew point of the interior air is usually undercut at the precise depth where the floor joists are tied into the wall. Condensation at these points can raise the moisture content in the joist ends to such a

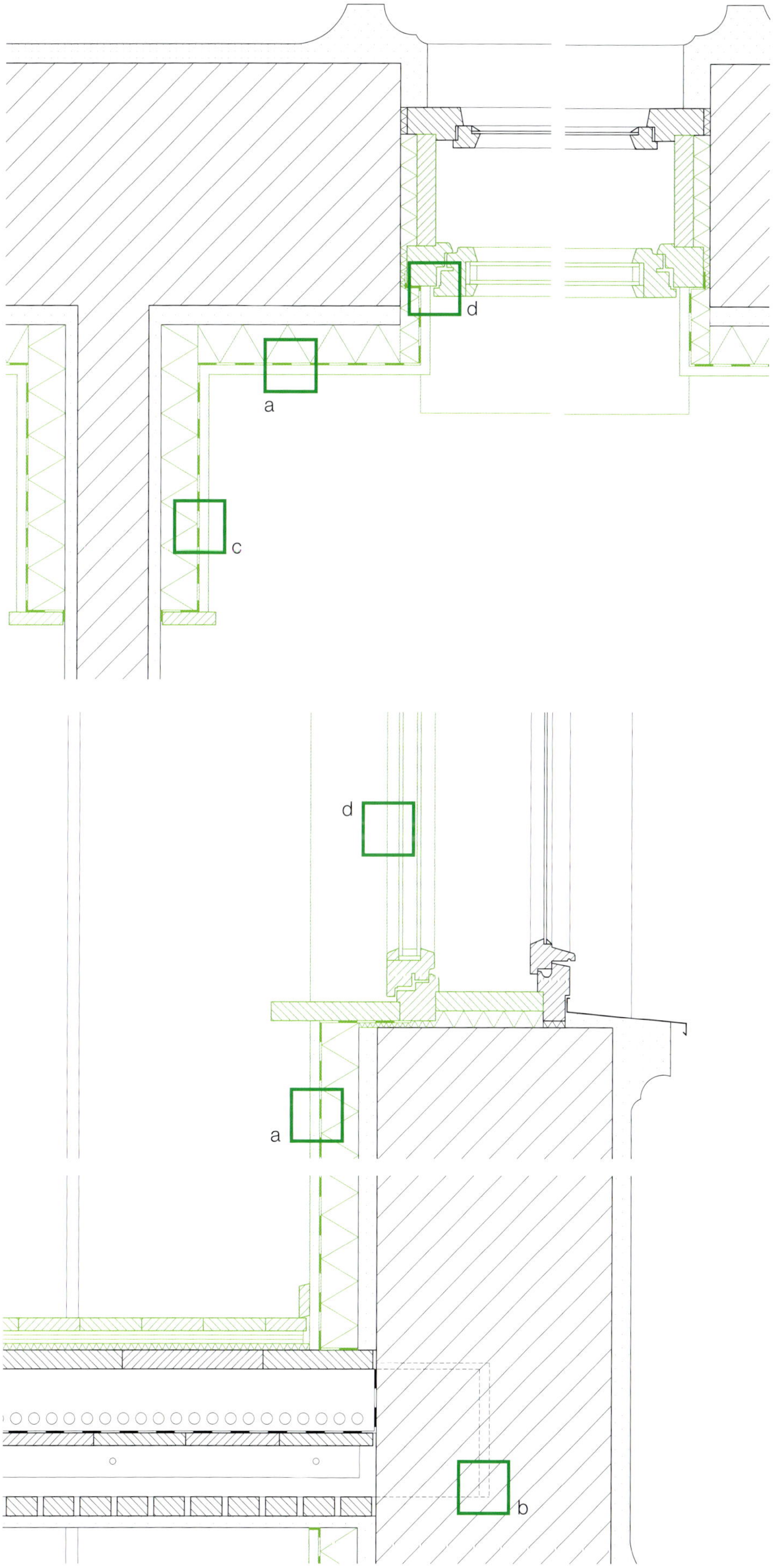

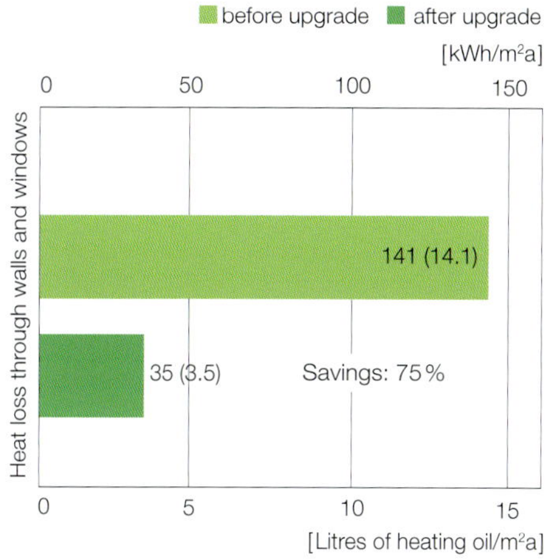

2.26

level that rotting and, finally, failure of the floor structure is inevitable. Exposing the ends of the timber joists and filling all the cavities at the support areas with PUR spray foam has proven successful in order to minimise the air convection currents between the floor cavities and the supports.

☐ c

Owing to the severe thermal bridges in areas where interior walls are tied into the exterior masonry, it is necessary to extend the insulation at these wall junctions. The insulation should continue along the interior dividing wall and reach into the room by approximately 50 cm. The same should apply to all reinforced concrete floors that are tied into the exterior walls by adding a strip of insulation to the soffits. There is no need for such insulation measures in the case of timber joist floors because the thermal bridging effect is much less significant.

☐ d

The existing original single glazing is to be retained for building conservation purposes. A new double glazed window in a wooden frame is mounted on the inside. Because the old window is not very airtight, there is no formation of condensation water on the single glazing even at low outside temperatures.

2.25 Isothermal lines before (a) and after (b) the upgrade under standard conditions according to DIN 4108-2 (outside temperature -5 °C, room temperature +20 °C, relative room air humidity 50 %)
— 12.6 °C isotherm (risk of mould growth)

2.26 Annual heat demand in kilowatt hours and in litres of heating oil per square metre of opaque wall surface in the existing building and after the upgrade (calculations performed with 84 degree days (DD), U-value of existing exterior wall 1.4 W/m²K, system loss 20 %)

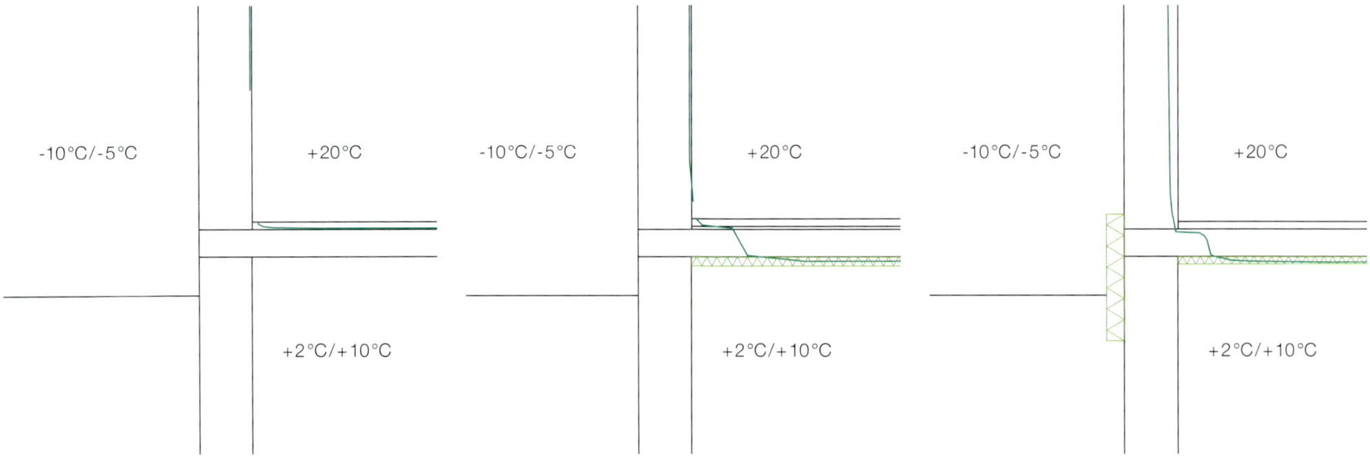

a b c 2.27

Requirements acc. to EnEV (enclosure 3, table 1):

Floor slab above unheated basement with insulation mounted to soffit:

max. U-value 0.30 W/m²K

Floor slab above unheated basement with a new floor structure on the upper side of the slab:

max. U-value 0.50 W/m²K

The requirement is also considered fulfilled when the maximum possible thickness of insulation that does not necessitate raising the door lintels is fitted on top of the floor.

U-value calculation of the ground floor slab with soffit insulation:

	t [m]	λ [W/mK]	R [m²K/W]
R_{se}			0.170
EPS insulation	0.060	0.035	1.714
concrete floor slab	0.160	2.300	0.070
EPS insulation	0.020	0.040	0.500
cement screed	0.045	1.400	0.032
ceramic tiles	0.012	1.300	0.009
R_{si}			0.170
Thermal resistance R_T			2.665
U-value [W/m²K]			**0.38**

U-value calculation of the ground floor slab with floor insulation:

	t [m]	λ [W/mK]	R [m²K/W]
R_{se}			0.170
concrete floor slab	0.160	2.300	0.070
vacuum insulation panel	0.020	0.007	2.857
UFHS mat system	0.020	0.320	0.063
natural stone (granite)	0.020	2.800	0.007
R_{si}			0.170
Thermal resistance R_T			3.337
U-value [W/m²K]			**0.30**

Ground floor insulation on the cold or warm side including plinth insulation

The majority of older buildings have a simple reinforced concrete floor slab above an unheated basement and an uninsulated plinth. Both act as extremely severe thermal bridges once a thermal insulation composite system has been installed to the exterior walls. Once the insulation work has been completed, the solid part of the exterior wall is located in the warm zone and, hence, there is a much greater heat loss through the base of the building. Plinth insulation should therefore always be added at the same time as the exterior wall insulation, even if this insulation measure is not explicitly required by the EnEV. Costs normally dictate that the floor slab over an unheated basement is insulated on the underside (cold side). The prerequisite is, however, that the ceiling height in the basement permits the installation of insulation and that the services attached to the soffit do not present any major problems. It is possible to add a strip of suspended ceiling with an insulation insert to service areas so that the room height is only reduced within the service zones.

Insulation is rarely added to the top (warm side) of the floor slab over an unheated basement. Such a solution requires a completely new floor structure throughout the ground floor. However, this approach is justified if a new underfloor heating system forms part of the refurbishment. Higher floor structures are the result, which naturally leads to a reduction of the clear heights of door and window lintels. The solution illustrated here makes do with minimal floor height thanks to modern material technology and can therefore be considered economically viable.

☐ a

Depending on the insulation material chosen, the most simple solution is to glue the insulation panels to the soffit of the floor slab. Additional mechanical fixings or a ceiling finish are not necessarily required. The soffit insulation should always be performed at the same time as the plinth insulation. Once the soffit insulation has been installed, the entire ground floor structure is on the warm side, which substantially increases the heat loss via the uninsulated plinth. Without plinth insulation, there is no appreciable increase of the surface temperature at the critical junction between the interior floor structure and the exterior wall. Part of the reduction in heat loss to the unheated basement is then simply cancelled out by the greater thermal bridging heat loss in the building's plinth zone.

☐ b

If the ground floor slab is to be insulated on its upper side, the old floor structure must first be removed down to the concrete floor slab. The new floor finish selected here consists of vacuum insulation panels (VIP) with a thermal conductivity of 0.007 W/mK. The 2-cm-thick insulation panel is the equivalent of an approximately 10-cm-thick layer of conventional insulation with a thermal conductivity of 0.035 W/mK. In single-family homes, an extremely thin, screedless underfloor heating system can be installed above the insulation. If sound proofing requirements according to DIN 4109 (increased sound proofing) are an issue, which is the case in residential multi-storey buildings, dry screed with an impact sound improvement index of at least 27 dB must at first be laid on top of the existing, smoothed floor. This can be achieved by installing a 24-mm-thick layer of mineral fibre board and a double layer of dry screed. An underfloor heating system would then further increase the height of the floor structure unnecessarily and therefore makes little sense in this case.

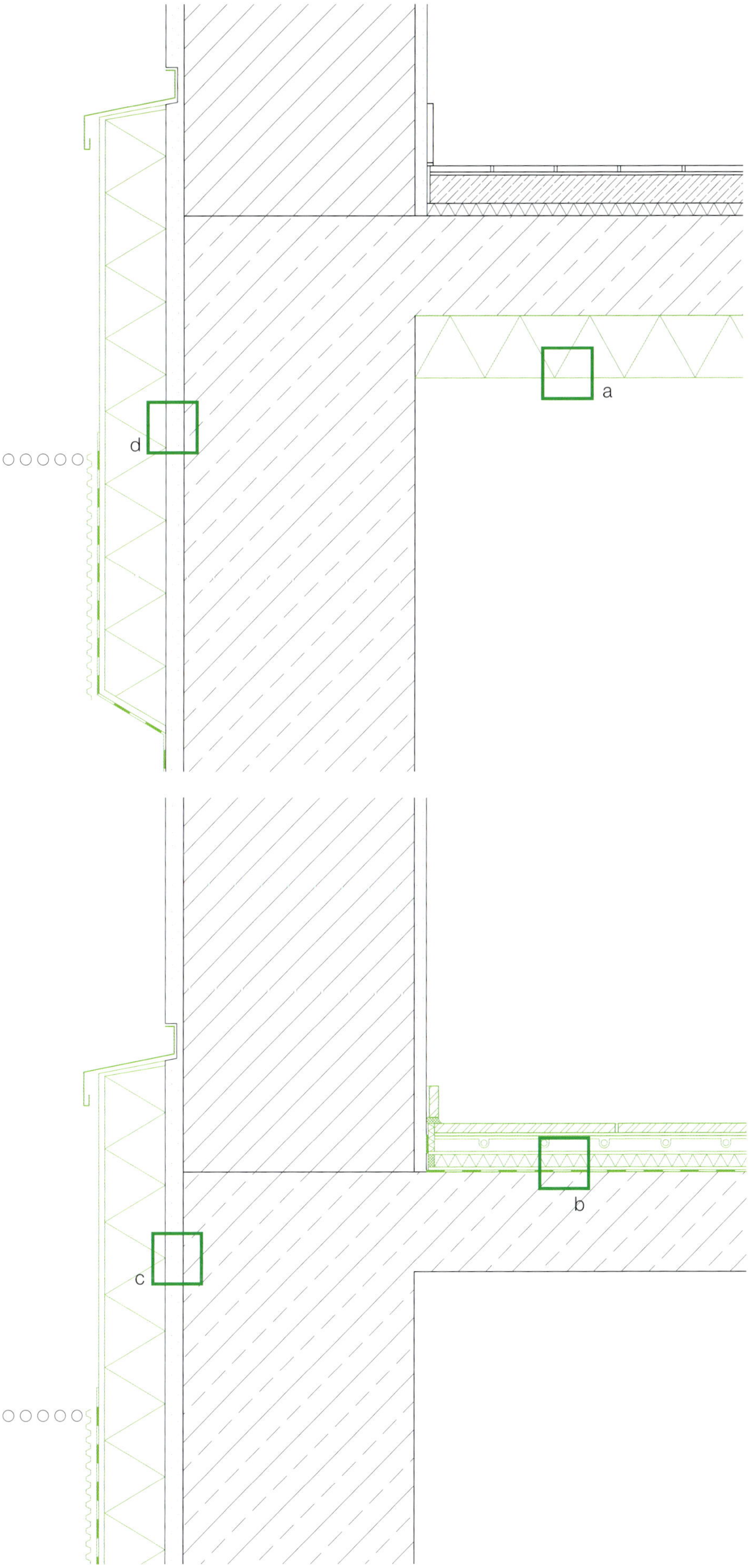

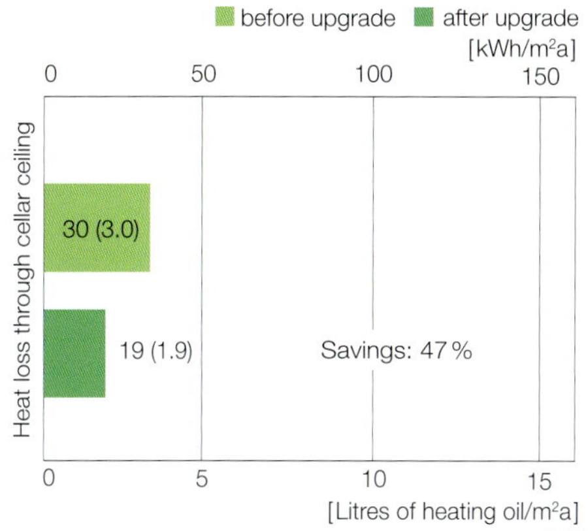

2.28

☐ c

The exterior walls of the basement are often damp due to a lack of waterproofing and must be repaired. The temperature on the interior surface is sometimes so low that there is a threat of mould growth. This can happen easily during the warmer seasons when warm air penetrates the basement rooms, due to incorrect ventilation behaviour, and the moisture content condenses on the cold exterior walls.
If a decision is made to repair the exterior basement walls, perimeter insulation should be added during the course of the necessary work processes (cleaning, rendering, waterproofing). Perimeter insulation reduces the heat loss from the basement and increases the temperature of the basement's interior wall surfaces. In the energy balance, this improvement measure is accounted for by using the correction factor F_x.

☐ d

If it is not possible or not necessary to add perimeter insulation to the basement walls, the possibility of adding plinth insulation should be considered. The insulation should be embedded into the ground (approximately 30 cm) and protected from the weather by adding render and a sheet metal coping.

2.27 Isothermal lines for a ground floor slab without (a) and with (b) insulation and (c) with soffit and plinth insulation under standard conditions according to DIN 4108-2 (outside temperature -5 °C, room temperature +20 °C, relative room air humidity 50 %)
— 12.6 °C isotherm (risk of mould growth)

2.28 Annual heat demand in kilowatt hours and in litres of heating oil per square metre of floor area in the existing building and after adding soffit insulation (calculations performed with 84 degree days (DD), U-value of the existing structure 0.6 W/m²K, temperature correction factor 0.5 and 20 % system loss)

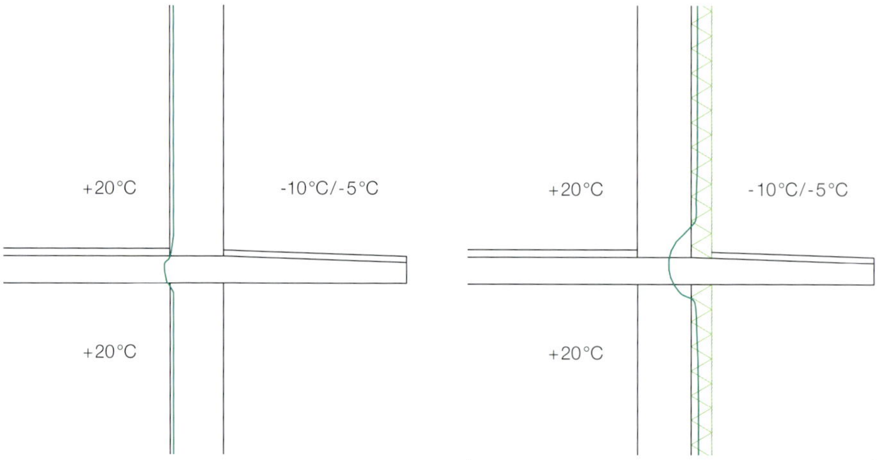

a b 2.29

Thermal bridge assessment according to EnEV:

There are basically three ways of incorporating existing thermal bridges into the energy performance calculations according to EnEV:

The global thermal bridging factor of

0.10 W/m²K

is multiplied by the total envelope area and added to the other transmission heat losses. This approach is always permissible. However, the downside of not having performed a detailed analysis is "punished" by assuming an unrealistically high factor.

The global thermal bridging factor of

0.05 W/m²K according to DIN 4108, supplement 2

is multiplied by the total envelope area and added to the other transmission heat losses. This approach is only permissible for construction designs that conform with those presented in DIN 4108, supplement 2. In the case of deviating constructions, proof of equivalence must be furnished.

Individual calculation of thermal bridges according to DIN EN ISO 10211-1

The exact calculation of the transmission heat loss deriving from all linear thermal bridges is always permissible. However, this approach is quite time-consuming: the lengths of all thermal bridges and the associated thermal transmittance values must all be determined individually.

2.29 Isothermal lines in the case of a cantilevered balcony slab made of reinforced concrete before (a) and after (b) adding a composite thermal insulation system in standard conditions according to DIN 4108-2 (outside temperature -5°C, room temperature +20°C, relative room air humidity 50%)
— 12.6°C isotherm (risk of mould growth)

2.30 Isothermal lines after upgrading the balcony slab with (a) topside insulation, (b) soffit insulation and (c) topside and soffit insulation. The position of the critical 12.6°C isotherm hardly changes in comparison to the not yet upgraded balcony slab (fig. 2.29a) by adding the various forms of insulation.

Existing balcony and thermal insulation composite system

The cantilevered reinforced concrete balcony without a thermal break is not only a major thermal bridge responsible for considerable heat loss and possible damage caused by moisture in the interior, it also poses difficulties when conducting the energy performance calculations according to EnEV. Since there is no comparable construction in DIN 4108, supplement 2, the higher global thermal bridging factor (0.10 W/m²K) has to be applied. It therefore makes sense in this case to use the individual calculation method and determine the thermal bridging factor by performing a simulation calculation. By choosing this procedure, the thermal bridge at the balcony can be compensated for by making improvements to the other thermal bridges.
The possibility of removing the balcony should also be considered. It could be removed and replaced by an independent timber or steel construction erected in front of the building or supported/suspended from point-fixed brackets on the facade.

Advantages:
- balcony design with virtually no thermal bridges
- no risk of mould growth along the edges of ceilings or floors in the adjoining rooms
- the balconies can be enlarged within the scope of the refurbishment work, which is a real gain in terms of living standards

Disadvantages:
- very expensive and elaborate solution

If mould has already formed along the edges of ceilings and/or floors in the adjoining rooms owing to the thermal bridging that results from the lack of a thermal break in the cantilevered balcony slab, the addition of insulation to the outside of the exterior wall will already help improve the situation. This solution does not require that the balcony slab be insulated, and there is no need to remove it either. As the thermal bridge simulations above show, the isotherm shifts outwards to such an extent that the interior surface temperatures, after the refurbishment, lie well above the critical temperatures. The only potential problems are at the bottom of the reveals around the balcony door, which may require some insulation to be placed on the inside.
This is done by removing the plaster at the bottom of the reveals and replacing it with, for example, calcium silicate sheets with a thickness equal to that of the original plaster. The interior surface temperatures are then higher than the critical values for mould growth. Since this internal insulation is restricted to a few very small areas, no further problems in terms of building physics are to be expected.
The impact of the thermal bridge is greater if a composite thermal insulation system is applied to the exterior walls. Insulating the balcony slab does not lead to any improvements, as is shown by the isothermal lines in figures 2.31 a–c.

☐ a

The upper connection between the insulation and the balcony slab is executed by inserting pre-compressed sealing tape into the insulation. The render at the slab is finished at an angle.

☐ b

Above the balcony slab, the composite thermal insulation system is set on a facade rail. The gap between the rail and the balcony slab is sealed using a pre-compressed sealing strip.

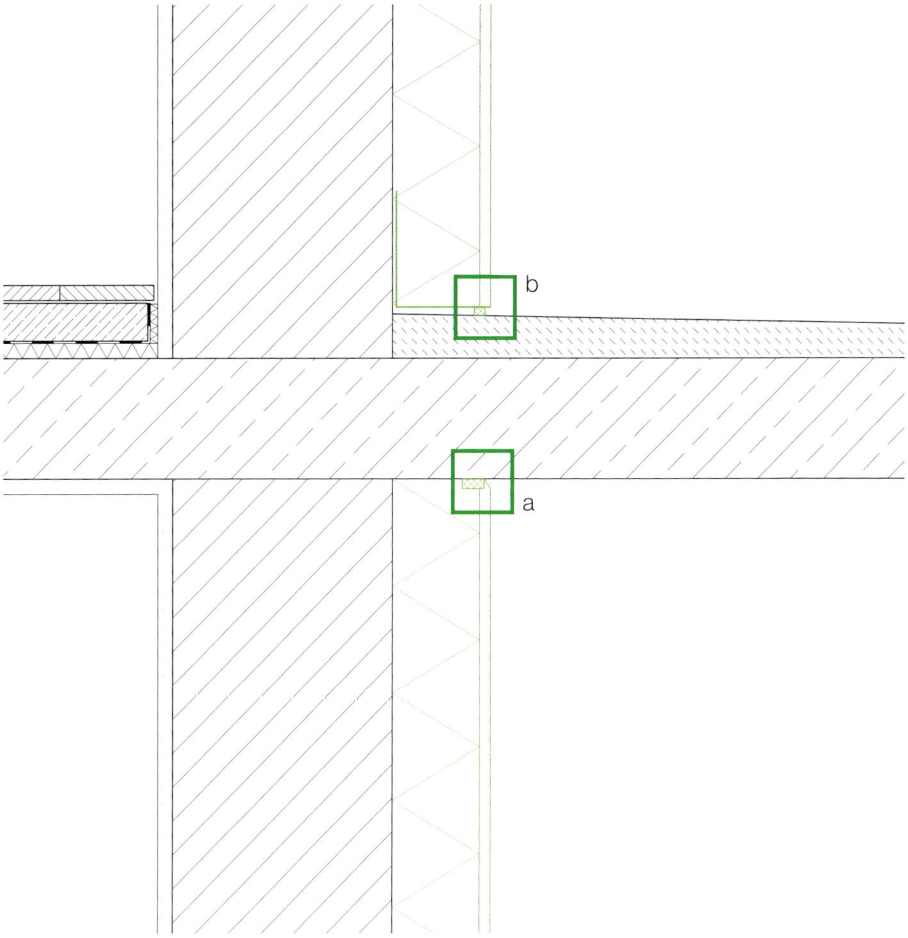

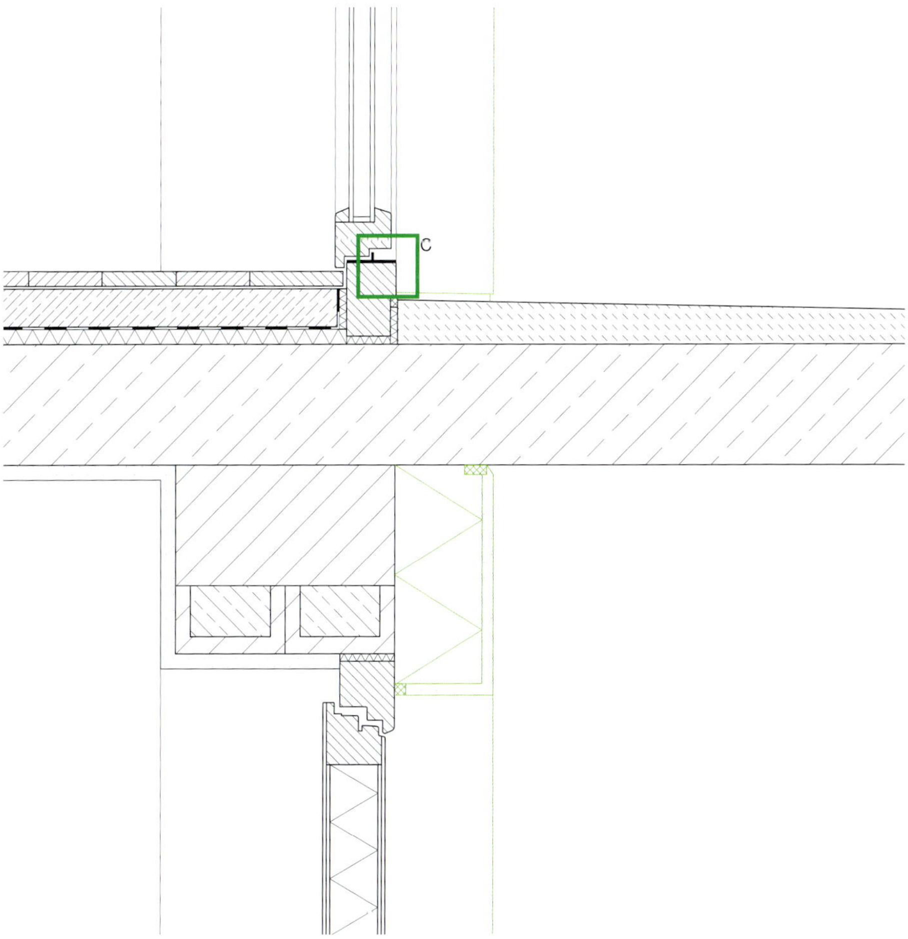

c

Most balcony doors only shut against a single L-section along their threshold. When replacing the windows, it is important to ensure that a second seal is provided by adding a door bottom seal or floor seal, which is activated automatically when closing the door (mechanical or magnetic). Barrier-free design according to DIN 18025, part 2 permits a 2-cm-high threshold. By using special sections, which are fixed to the bottom frame, the balcony door can be finished in such a way that both airtightness and barrier-free accessibility are ensured.

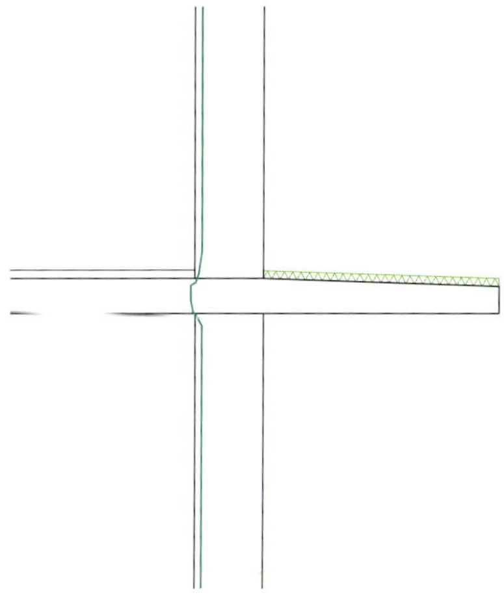

a

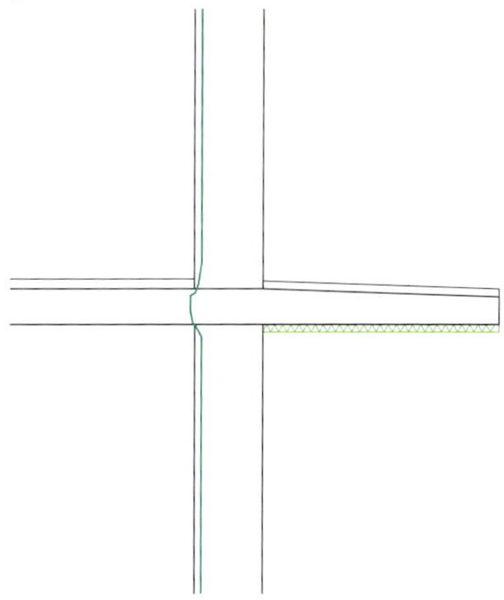

b

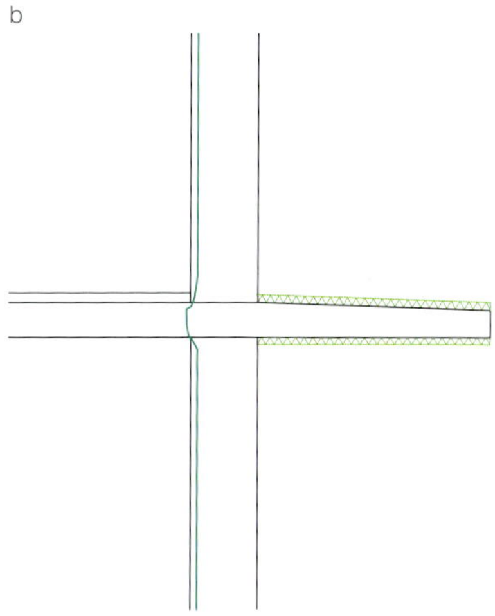

c

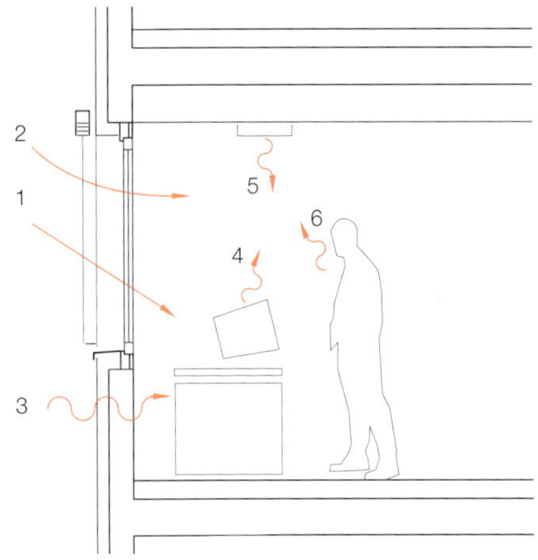

2.31

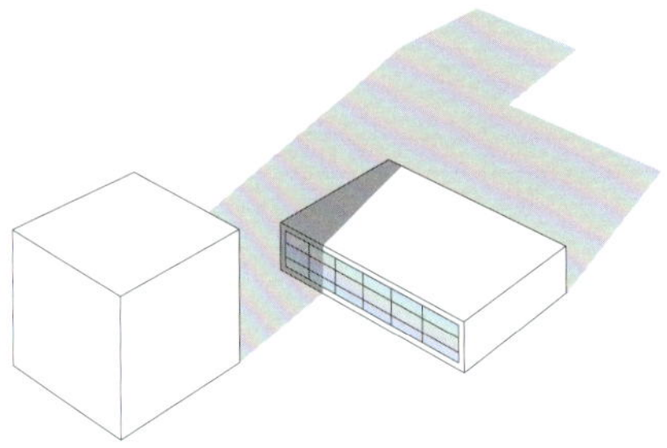

a

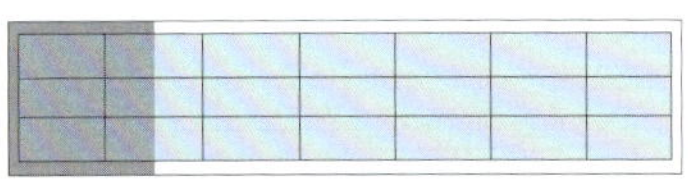

b

2.32

Classification of building types

lightweight building	C_{eff}/A_F < 50 Wh/(m²K)
medium-weight building	50 Wh/(m²K) ≤ C_{eff}/A_F ≤ 130 Wh/(m²K)
heavyweight building	C_{eff}/A_F > 130 Wh/(m²K)

C_{eff}: effective heat storage capacity of the room enclosing surfaces
A_F: net floor area of the room

2.33

2.31 Heat input
external heat loads:
1 solar radiation
2 ventilation
3 transmission
internal heat loads:
4 office equipment
5 lighting
6 occupants
2.32 Shadows cast by neighbouring buildings
a spatial configuration
b elevation of the facade with the shadow cast by the neighbouring building
2.33 Classification of building types according to DIN 4108-2
2.34 Summer climate regions for the performance of calculations for summer heat protection according to DIN 4108-2: 2003-07
a classification of summer climate regions
b threshold values of indoor temperatures
2.35 Characteristic parameters for selected glass types with optimum light transmission
2.36 Permitted values for the proportion of window surface to floor area below which a summer heat protection assessment is not required (according to DIN 4108–2, table 7)

Heat sources

All forms of heat input affecting interior space are regarded as heat sources including those penetrating the building skin from the outside and those arising from user-related processes inside. They can be used to achieve the desired heating of the indoor air, as is implied by the term "heat gains", and are therefore regarded as having a positive impact on the building's energy balance. They can, however, also be a considerable burden on the indoor climate and make measures to cool the building necessary, which is implied by term "heat loads". Whether the heat sources can be used to cover a proportion of the heat demand or whether they must be reduced by installing suitable constructions or technical systems to provide a comfortable indoor climate are facts relevant to the building's energy balance.

Minimum thermal protection according to DIN 4108

The standard minimum requirements with regard to heat protection are intended to ensure that reasonable temperature conditions are also provided inside buildings in summer by performing construction-related measures, rather than installing complicated, energy-intensive cooling systems.
The tolerable temperature conditions for the various climate regions in Germany are listed in DIN 4108-2:2003-07, table 6 (fig. 2.34). The conditions referred to in the standard may not be exceeded for more than 10% of the occupation period, which is 24 h in the case of residential buildings and 10 h in the case of non-residential buildings. For the calculations this means that the total occupation period during the so-called non-heating period is included. The already mentioned threshold value for the indoor temperature may be exceeded up to a maximum of 10% of these hours. A maximum temperature is not determined. Since this is a standardised method of analysis, which only considers the solar heat gains, it can happen that, despite having met all the requirements, some rooms can only be used when they are actively cooled. This situation can, for example, arise if certain work processes are the cause for high heat loads. The analysis method for minimum thermal protection in summer considers individual rooms. The analysis for similar rooms can be carried out by establishing a reference room.

Individual construction-related aspects
The impact that the individual factors have on the thermal input is explained in the following passages.

Proportion of window area
The sun is the most significant heat source. Its energy enters the building mostly in the form of radiation through transparent surfaces. This heat input is very much dependent on the orientation, size and inclination of the transparent surfaces; it can be influenced by using different *construction-related* measures and material properties, such as shading devices and glass quality. From a holistic viewpoint, glazing that makes up 40% of the building envelope seems a sensible choice. A fully-glazed building can never be an energy-efficient building. However, this is a subject which should never be approached solely from an energy perspective. Architects in particular should be able to justify and address the spatial quality of a large-scale or even fully-glazed facade as a "value". For they are the ones who have the skills and know-how to produce solutions that are part of a holistic, energy-based concept.
Transparent surfaces facing south, west or east should be equipped with an exterior shading system and with glass that has a reduced solar energy transmittance (g-value).

Orientation of windows
Unlike in winter, windows facing north contribute towards more comfortable room conditions in summer, simply because the solar heat gain is much lower. In some circumstances, there may even be a conflict between improving the heat input in winter by using south-facing windows and reducing the heat input in summer by having north-facing windows. In the case of non-residential buildings, summer heat input should be limited by means of *construction-related* measures because the high energy demand and considerable amount of technical plant required to provide comfortable room conditions would simply be out of proportion.
The orientation of the building, in particular the rooms prone to overheating, decides whether there will be a cumulation of heat input or whether the addition of solar heat gain and user-related heat gain can be avoided by the intelligent use of orientation. For example: from an energy efficiency viewpoint it makes no sense to place classrooms predominately used during the early part of the day

along an east-facing facade, as this would mean that the user-related heat input occurs at the same time as the solar heat input. For the same reasons conference rooms, which are used during all hours, should face north in order to prevent external heat gains being added to the internal heat loads.
In the past, it has often been very difficult to determine the influence of the surrounding buildings on the length of time during which the sun shines on a specific facade. Today, however, almost all CAD programs provide tools to illustrate the extent of shadow cast on a building surface according to the position of the sun. Hence, the amount of solar radiation reaching the transparent surfaces can be determined for each day and for each facade individually. Figure 2.32 shows a simple diagram of the shadow cast by a neighbouring building on a south-facing glass facade for a specific day, in this case 30 September at 6 p.m.

Window design
Alongside the size and orientation of the window surfaces, the design of the windows is also worth considering. In the case of transparent surfaces, the heat transmittance of the glass areas can, to a certain extent, be regulated by the g-value, the solar energy transmittance, and the design and positioning of the shading device.
Figure 2.35 includes typical parameters for insulating glass. The three parameters light transmission (τ), solar energy transmittance (g) and thermal insulation (U) cannot be improved individually since they are dependent on one another.

Thermal storage mass
The thermal storage mass of a room allows large amounts of heat to be absorbed without having a noticeable effect on the room temperature. Whereas air heats up quickly due to its low heat capacity of 0.34 Wh/m³K (a heat input of 0.34 Wh is required to increase the temperature of 1 m³ of air by 1 Kelvin), the temperature rise in solid materials is much lower. A one-square-metre piece of 10-cm-thick concrete requires a heat input of 66 Wh to increase its temperature by 1 K. In the case of a solid timber floor (lightweight construction), the same heat input would, due to the very low heat storage capacity of the construction, lead to a temperature increase of 3 K, which would probably necessitate the installation of a mechanical cooling system. In order to utilise the storage mass properly, the heat absorbed by

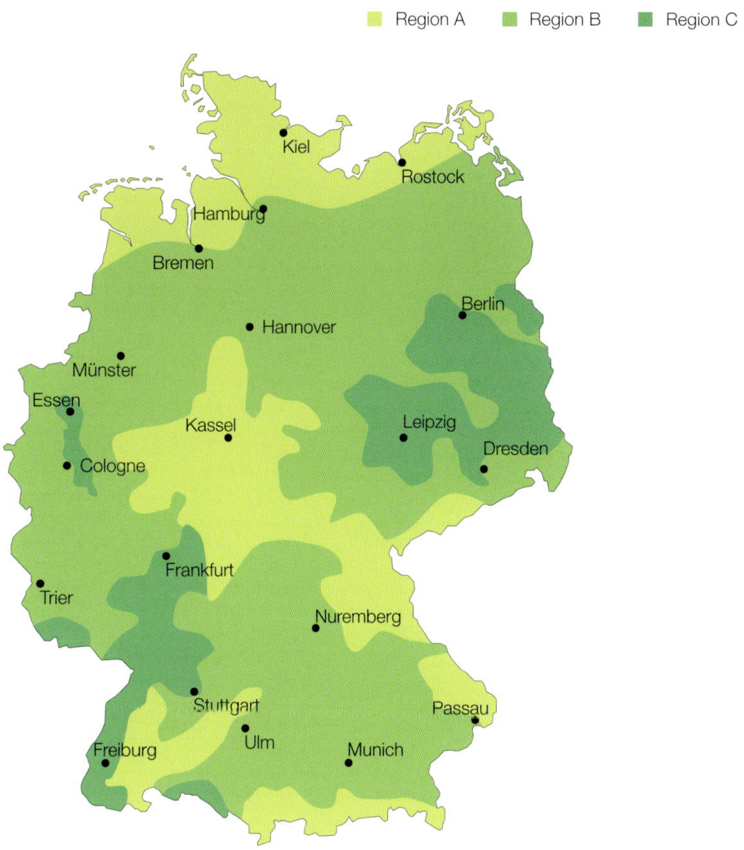

a

Threshold values for indoor temperatures in summer climate regions

Summer climate region	Characteristic of the region	Threshold value for the indoor temperature [°C]	Maximum value of the average monthly outside temperature θ [°C]
A	cool summer	25	θ ≤ 16.5
B	moderate	26	16.5 < θ < 18
C	hot summer	27	θ ≥ 18

b

2.34

Typical glass characteristics with optimised light transmission

Requirement	Structure [mm]	U-value [W/m²K]	g-value [%]	τ-value [%]
small U-value with high light transmission	4-16-4	1.1	62	80
high solar energy transmittance with high light transmission	4-14-4-14-4	0.7	64	75
low solar energy transmittance with high light transmission	6-16-4	1.0	37	70

In the case of all windows, the cavity between the panes is filled with argon.

2.35

Proportion of window surface to floor area below which a summer heat protection assessment is not required

Inclination of windows in relation to horizontal	Orientation of windows	Window-to-floor area ratio w_{FA} [%]
over 60° up to 90°	ranging from north-west, to south, to north-east	10
	all other north directions	15
from 0° to 60°	all orientations	7

2.36

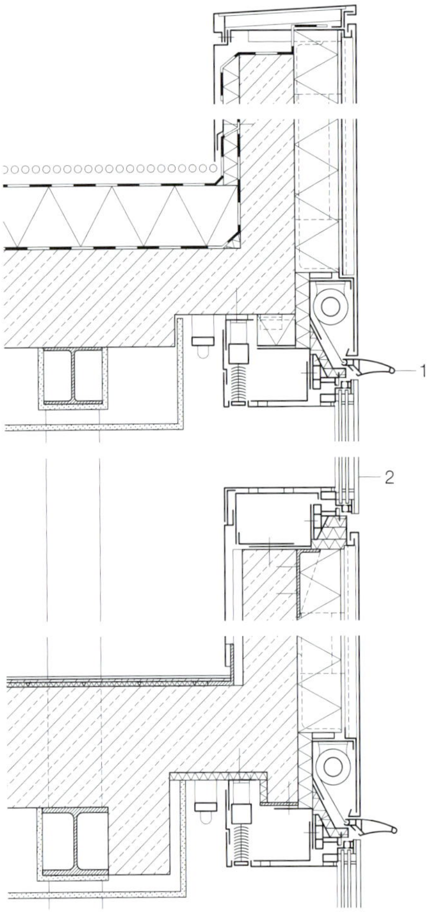

a

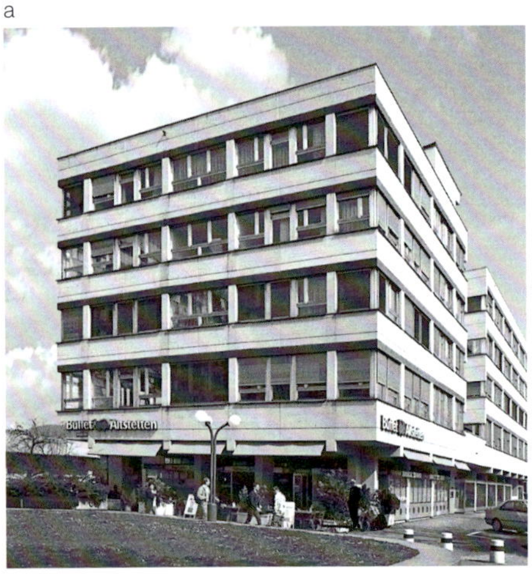

b

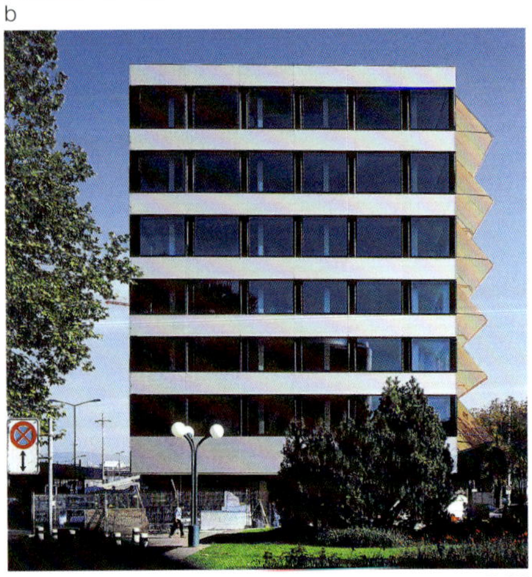

c 2.37

the building components during the day must be released again or else the temperature-regulating storage effect is lost and the temperature in the mass rises continuously. Heat can be dissipated by, for example, ventilating the respective space with cool outside air at night.
Whether a building is regarded as lightweight, medium-weight or heavyweight, can be determined by using the calculation procedure described in DIN 4108-6:2000-11. For the assessment of the critical room or rooms, all surfaces bordering indoor air are examined according to their effective heat storage capacity C_{eff} using the following formula:

$$C_{eff} = \sum (A_i \cdot d_i \cdot c_i \cdot \rho_i) \text{ Wh/K}$$

A = area of the building components that enclose the space
d = thickness
c = specific heat storage capacity
ρ = weight density
i = respective layer of the building component

For the assessment, the layers are calculated individually from inside to outside and added together for each building component. The total of all building component values C_{eff} is then divided by the room's floor area (A_f). This gives the specific heat storage capacity c_{eff} in Wh/m² K which is used to identify the relevant building type: lightweight, medium-weight or heavyweight (fig. 2.33, p. 68).
The following rules must be observed when performing the calculations:

- In the case of large room depths and windows in a single wall, the maximum depth used in calculating the room area is determined as three times the room height; where there are windows in two walls, the maximum room depth used equals six times the room height.
- The thickness of the layer considered in the calculation is a maximum of 10 cm since the charging and discharging capacity of the storage mass in a night/day rhythm is only effective up to this material depth.
- If insulation layers with a thermal conductivity of $\lambda < 0.1$ W/m K and a thermal resistance of $R > 0.25$ m² K/W are included in the material component, this layer as well as the layers positioned behind this material are not taken into account.
- If the building component is a room dividing wall and flanked by inside air on both sides, only half of the wall is considered in the case of thicknesses below 20 cm; in the case of greater wall thicknesses, the 10 cm rule applies.

Night-time ventilation
By circulating cool air through the rooms at night, the building is cooled down, which enables it to absorb more heat again the next day. This positive effect is particularly pronounced if there is sufficient thermal storage mass.
Night-time ventilation is only effective if the air exchange rate is at least 1.5/h. The issue planners have to deal with is how this air exchange rate can be achieved by opening windows in non-residential buildings without adverse weather conditions causing damage and without the property insurer restricting the insurance coverage. The air exchange can only be carried out successfully by using a cross ventilation system with weather and burglar proof facade openings. These do not necessarily have to be identical to the transparent surfaces.
If an air-handling system has been installed in the building, it could support the natural night-time ventilation by extracting the waste air.
If it is not possible to fit the kind of facade openings described, the alternative is for the air-handling system to take care of the necessary air exchange (waste air and supply air). However, this solution significantly increases the consumption of electricity to operate the fans.

Proof procedure
The procedure verifying that the minimum requirements for summer heat protection have been met is performed room by room. In the case of a building with similar rooms, a reference room, known as the critical room, is assessed instead of a large number of individual rooms.

Step 1: critical room
The critical room is identified by choosing, from a group of similar rooms, the room with the largest window-to-floor area ratio (WFA). To perform the calculation, the total window area $A_{w,i}$ (structural opening) is divided by the net floor area A_f of the room to be considered.

$$WFA = \frac{\sum A_{w,i}}{A_f}$$

According to DIN 4108-2:2003-07, table 7, all offices and communal rooms must be included in the assessment, regardless of their orientation, since proof of summer heat protection must in any case be provided for rooms with a window-to-floor area ratio of 10% and over (fig. 2.36, p. 69).

Step 2: present solar input value
Once the reference room or rooms have been identified, the solar input value is determined for these rooms according to the following formula:

$$S = \frac{\sum_i (A_{w,i} \cdot F_{c,i} \cdot g_i)}{A_f}$$

The present solar input value S is determined in relation to the window area A_w in m^2, the energy transmittance value of the glass g and the reduction factor of the solar protection device F_c and then finally related to the floor area of the room A_f (fig. 2.38). The table illustrates that the highest reduction factors and therefore the best summer heat protection is generally achieved by installing exterior-mounted shading devices.

Step 3: permitted solar input value
The permitted solar input value is a threshold value, which, if the requirements of DIN 4108-2 are to be met, may not be exceeded. The relevant construction-related factors of influence are listed and evaluated in table 9 of the standard. The more successfully the construction-related requirements for the reduction of heat input are met, the higher the total permitted solar input value (fig. 2.39).

2.37 Refurbishment of an office building in Zurich (CH), architects: Romero + Schaefle, 1970/2001
a vertical section, scale 1:20
1 fabric, straight drop awning
2 triple glazing,
g-value 0.40 (40%), τ-value 0.72 (72%)
U_g 0.80 W/m²K; $U_{w,rv}$ 1.2 W/m²K
b before the refurbishment
c after the refurbishment
The refurbishment of the office building completed in 1970 was intended not only to allow the building be operated in a more energy efficient way, but also to increase the quality of the work environment. The concrete parapets on the east, west and south sides of the building were shortened and the fake, non-load-bearing columns were removed. The new windows with fixed glazing have been optimised in terms of their heat transmittance, light transmission and solar protection. The solar protection with fabric, straight drop awnings provides optimal shading, but still enables sufficient light to enter and allows occupants to look out of the building.

2.38 Reference values for the reduction factors of fixed solar protection devices according to DIN 4108-2, table 8

2.39 Proportionate solar input values to determine the maximum permitted solar input values. It is made up of the sum of the individual values determined in lines 1 to 6.

Reference values for the diminution factors F_c of fixed solar protection devices

Solar protection device [1]	Reduction factor F_c
1 without solar protection device	1.0
2 interior mounted or mounted between glass panes [2]	
white or with a reflective, hardly transparent surface	0.75
light coloured or hardly transparent [3]	0.8
dark coloured or highly transparent	0.9
3 exterior mounted	
tiltable slats, rear ventilated	0.25
fabric roller blinds that are hardly transparent [3], rear ventilated	0.25
roller or venetian blinds, in general	0.4
roller shutters, side-hung shutters	0.3
canopies, loggias, free-standing louvres [4]	0.5
awning [4], ventilated at top and on sides	0.4
awning [4], in general	0.5

[1] The solar protection device must be permanently fixed. Decorative curtains are not considered a solar protection device.
[2] It is recommended to make a detailed assessment when planning interior-mounted solar protection device or those incorporated between glass panes since this could result in more favourable values.
[3] Hardly transparent implies that the transparency of the solar protection device is below 15%.
[4] The awning must ensure that almost no direct sunlight reaches the window surface. This is the case if:
- for south-facing windows the angle of coverage is $\beta \geq 50°$
- for east and west-facing windows the angle of coverage is $\beta \geq 85°$ or $\gamma \geq 115°$.
Each orientation is entitled to an angle range of ± 22.5°. In the case of intermediate orientations, the angle of coverage required is $\beta \geq 80°$.

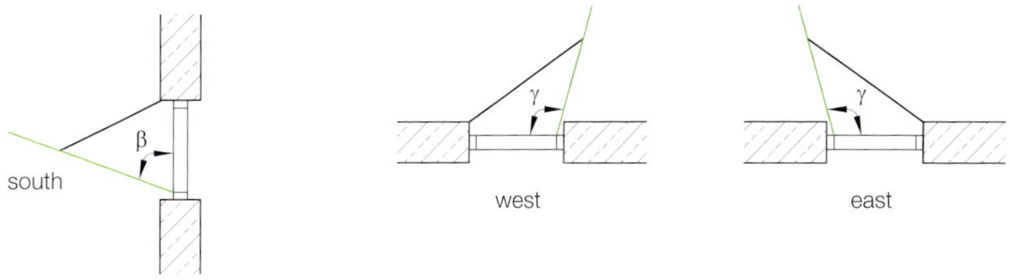

2.38

Proportionate values to determine the maximum permitted solar input value

Building position and type, window inclination and orientation	Proportionate solar input value S
1 Climate region [1]	
building in climate region A	0.04
building in climate region B	0.03
building in climate region C	0.015
2 Building type [2]	
lightweight building: no proof required for C_{eff}/A_f	0.06 f_{wei} [3]
medium-weight building: 50 Wh/(m²K) ≤ C_{eff}/A_f ≤ 130 Wh/(m²K)	0.10 f_{wei} [3]
heavyweight building: C_{eff}/A_f > 130 Wh/(m²K)	0.115 f_{wei} [3]
3 Increased night-time ventilation [4] during the second half of the night $n \geq 1.5\ h^{-1}$	
for medium-weight and lightweight buildings	+0.02
for heavyweight buildings	+0.03
4 Sun protection glazing [5] with $g \leq 0.4$	+0.03
5 Window inclination: 0° ≤ inclination ≤ 60° (in relation to the horizontal)	- 0.12 f_{incl} [6]
6 Orientation: north, north-east and north-west facing windows, so long as their inclination in relation to the horizontal is > 60° and window areas that are constantly overshadowed by the building itself.	+0.10 f_{north} [7]

[1] maximum values of the average monthly outside temperatures according to fig. 2.34, p. 69
[2] see fig. 2.33, p. 68
[3] $f_{wei} = (A_w + 0.3 \cdot A_{ew} + 0.1 \cdot A_r) / A_f$
f_{wei} weighted exterior areas in relation to the net floor area; the weighted factors consider the relation of summer heat transmittance by standard exterior building components
A_w window area (including roof lights)
A_{ew} exterior wall area (external dimensions)
A_r heat-transmitting roof and floor areas adjoining - either below or above - outside air, ground or unheated attic or basement space (external dimensions)
A_f net floor area (clear dimensions)
[4] a higher night-time ventilation can generally be assumed for single family and semi-detached houses
[5] a sun protection device that permanently reduces diffuse radiation with $g_{total} < 0.4$ is regarded as an equal measure
[6] $f_{incl} = A_{w,incl}/A_f$ with
$A_{w,incl}$ inclined window area
A_f net floor area
[7] $f_{north} = A_{w,north}/A_{w,total}$ with
$A_{w,north}$ north, north-east and north-west facing windows, so long as their inclination in relation to the horizontal is > 60° and window areas that are constantly overshadowed by the building itself
$A_{w,total}$ total window area

2.39

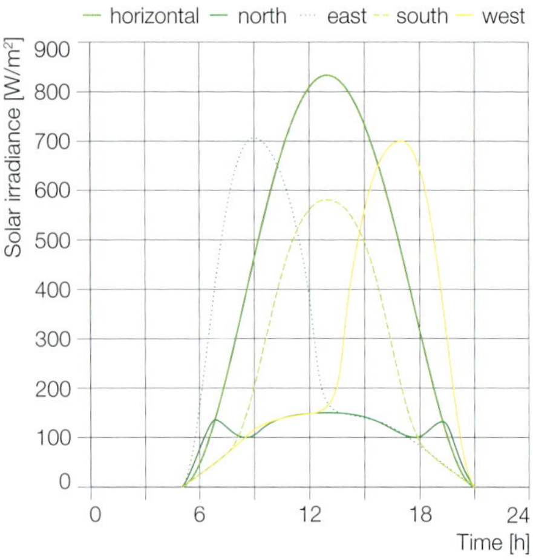

2.40

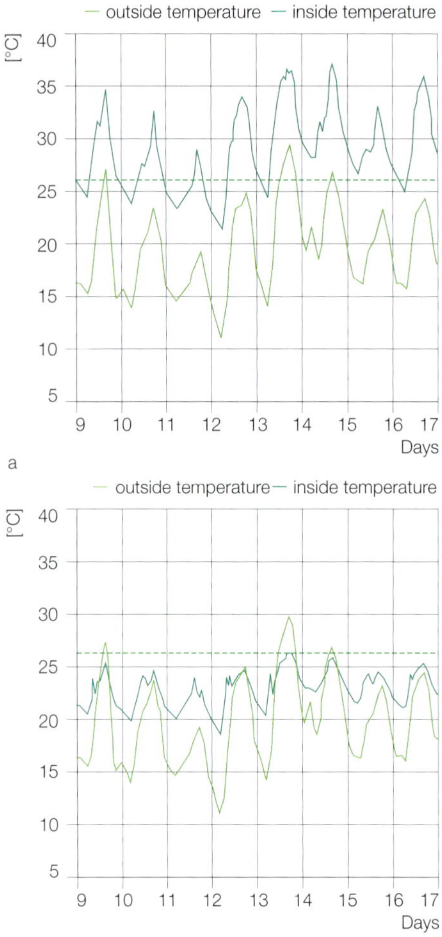

2.41

2.40 Solar irradiance on the longest day of the year in Central Europe

2.41 Simulation: temperature development in an office (corner room) with a window proportion of 70% (in relation to the floor area) from 10 to 17 July
a with interior mounted sun protection device
b with exterior mounted sun protection device

2.42 Technical solutions to reduce different heat loads (determined as thermal output according to VDI, the Association of German Engineers). The maximum heat input is related to the floor area of the room and is based on the fact that all forms of heat input occur simultaneously.

The minimum requirements for summer heat protection are met when the value $S_{permitted}$ is above the value $S_{present}$.

Upgrading existing building stock

The planning of every energy efficiency refurbishment should start by assessing whether the requirements for summer heat protection are met by the existing building. The fact that there are, for example, no exterior-mounted solar protection devices (possibly, roller shutters in residential buildings) in the case of a window-to-floor area ratio of only 20% is a sign that the minimum requirements for summer heat protection are not fulfilled. This is frequently the case for residential as well as for non-residential buildings, and improvements should be made not only concerning this issue by carrying out suitable construction-related alterations. If for a specific reason the building is to be upgraded to especially cater for summer heat conditions, the moderate threshold value $S_{permitted}$ should be reduced by 50%. This reduced value ensures that, even in the case of severe heat input through room usage, like in classrooms or in offices with a substantial amount of EDP equipment, comfortable room conditions in terms of temperature can be maintained without installing a cooling system. Nevertheless, despite elaborate improvement measures, there are still numerous uses that cannot make do without a mechanical cooling system due to the high internal heat loads (e.g. industrial productions, data processing centres, conference rooms and meeting rooms, among others).
However, especially in the case of refurbishments, it can also happen that plant for cooling the building, that was installed at an earlier date, can be removed or its performance considerably reduced thanks to significant construction-related improvements regarding the glazing, sun protection and storage mass.

Energy balance according to DIN V 18599

Every building must be assessed during its non-heating period, which according to figure 2.9 (p. 44) is defined as an excess of heat sources over heat sinks, in order to determine whether the increase of indoor temperature caused by the external heat loads is so severe that the room conditions become uncomfortable. In contrast to DIN 4108-2, all heat sources are considered in this assessment, including the heat loads caused by building use, ventilation and transmission.

The main aim of the calculations is to develop measures to restrict the temperature rise in rooms to 26°C. In various standards and industrial safety regulations, this temperature is regarded as the limit for comfortable conditions. Excess heat loads do not necessarily require that the building be cooled using plant technology because heat input is very much dependent on the weather and the building use and therefore uneven, i.e. at irregular intervals. In many cases, all possible forms of heat input occur together only for very short periods and are therefore only responsible for occasional short temperature peaks.

Influencing variables

Fundamental construction-related parameters affecting heat input are already mentioned in the section "minimum thermal protection". According to DIN V 18599, the following aspects should also be included in the overall assessment.

Solar input

The amount of solar energy $Q_{s,tr}$ penetrating the building via transparent surfaces can be determined according to DIN V 18599-2:2007-02. Based on the window area per cardinal direction (a), the solar irradiance (i_s, fig. 2.40), the effective g-value of the windows (g_{eff}) and the proportion of window frame (F_F), the following relation can be determined:

$$Q_{s,tr} = F_F \cdot A \cdot g_{eff} \cdot i_s \cdot t$$

To determine the total effective solar energy transmittance g_{eff} of the window, the g-value of the glass is reduced by a variety of factors that take into account dirt and the different ways in which the windows are fitted.

Ventilation

In the case of high air exchange rates, which are typical for non-residential buildings due to the higher occupancy rate, the temperature of the room air increases as soon as the windows are opened and the temperature of the outside air is above that of the inside air. Their are several ways to reduce this heat input, however, they all require mechanical support, i.e. some kind of plant technology. The amount of supply air can, for example, be controlled by using a simple air extraction system without elaborate ductwork. In this case, the supply air is drawn into the room via controllable exterior vents. A further possibility is to draw the supply air in through ground ducts; this cools the outside air down to approximately room temperature. The heat input via ventilation is calculated in the same way as the ventilation heat loss during the heating period (see p. 44). In this case, the temperature difference between the indoor air and the supplied outside air, the overtemperature, is used for $\Delta\theta$.

Use

The heat input from building use is independent of the time of year. Due to the consistent 24-hour use of residential buildings, a constant value of 50 Wh/m²d (d is for day) can be applied for single family and multi-family dwellings. In the case of non-residential buildings, the user-related heat input is determined according to the zone descriptions in DIN V 18599-10. The heat input includes the input through occupants, lighting, office equipment and machines, and, depending on the intensity of usage (number of occupants and computers), amounts to 60–168 Wh/m²d for an open-plan office (more than 7 work places, zone 3). The heat input through lighting, which is dependent on the supply of daylight and the selected lighting system, must be included in the total.

Calculation

The heat input can be calculated by using the cooling load calculation according to VDI (Association of German Engineers), by assessing the useful cooling demand according to DIN V 18599, or by performing thermal simulations.

Cooling load according to VDI

From a technical point of view, the worst case, i.e. the simultaneous occurrence of all possible heat inputs, should always be the basis for all decisions on whether a building requires cooling or not. The total of all simultaneously occurring loads is expressed in a specific floor-related value. For rooms with low heat loads, the value is approximately 20–30 W/m²; rooms with a special function (data processing centres) can reach values of up to 500 W/m². The values presented in figure 2.42 enable planners to make a rough estimate of the need for a permanent, or perhaps just temporary, cooling system.

Useful cooling demand according to DIN V 18599

The method to determine the heat input during the non-heating period corresponds with that to determine heat sinks (see pp. 44ff.). The non-heating period is defined as the time when the input from heat sources - relativised by the degree of utilisation - exceeds the loss caused by heat sinks. In order to determine the amount of heat which needs to be removed to maintain a room temperature below 26 °C, the loss caused by heat sinks is calculated by using the process already presented.
However, when using this standard, the statements regarding cooling are very general since they are exclusively based on the monthly average temperatures. The fact that the cooling demand is uneven, i.e. it fluctuates and depends on solar irradiance and usage (number of occupants, office equipment), is neglected. Hence, it is necessary to perform additional technical calculations or thermal simulations to accurately assess the heat input.

Simulation

The thermal behaviour of a building is best assessed by performing simulation studies. A simulation includes observing and recording the temperature development in a room or group of rooms once every hour under predetermined, detailed conditions. Based on the findings, it is then possible to determine for how many hours the threshold temperature of 26 °C

Technical possibilities to reduce heat loads according to VDI (Association of German Engineers)

Maximum heat input [W/m²]	**Plant technology**		**Comment**
	Cooling system	**Distribution**	
< 40	no plant technology required		natural cooling (night ventilation)
40 – 60	ground water pump, ground duct for ventilation purposes	surfaces (ceilings, floors) air	
60 – 80	refrigerating machine	chilled ceiling (low temperature variation), cooling sails	
> 80	refrigerating machine	chilled ceiling (high temperature variation), air (air conditioning)	danger of condensation

2.42

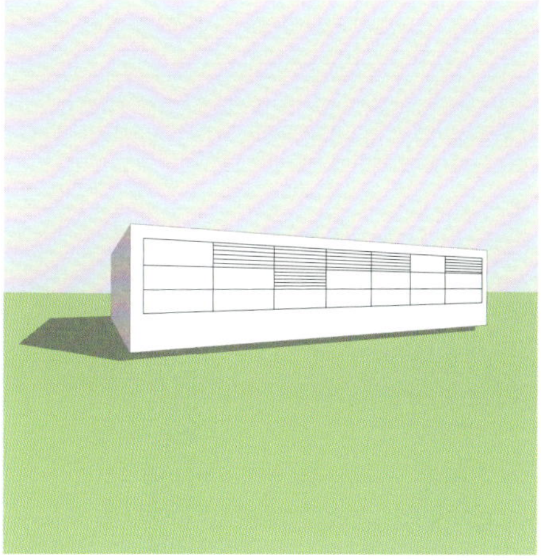

Example:
Simplified cooling load calculation acc. to VDI
The pavilion incorporates two classrooms, each with a floor area of 65 m² of which one is assessed. The building envelope has been upgraded. The east-facing window area (g-value: 65 %) is 20 m², an exterior-mounted sun protection device is installed. There is no thermal storage mass available inside. The outside temperature is 30 °C, the inside temperature 24 °C. There are 25 pupils in the room. The fresh air demand per hour is 25 m³/pupil. Firstly, the current situation is calculated and then, based on this, the upgraded condition.

External loads	**current**	**upgraded**
Solar irradiance		
max. input [W/m²]	600.00	600.00
g-value	0.65	0.37
sun protection	0.25	0.20
glass area 80 % [m²]	16.00	16.00
heat input sun [W]	1560.00	710.40
Ventilation		
air volume [m³/h]	625.00	625.00
thermal capacity air [Wh/m³K]	0.34	0.34
overtemperature [K]	6.00	-2.00
heat input [W]	1275.00	-425.00
Transmission		
This heat input can be neglected.		
Internal loads		
Persons		
number	25.00	25.00
output/person [W]	70.00	70.00
heat input [W]	1750.00	1750.00
Equipment		
number of computers	4	2
output/computer [W]	100.00	50.00
heat input [W]	400.00	100.00
Lighting		
connected load [W/m²]	15.00	15.00
active [%]	0.50	0.25
heat input [W]	487.50	243.75
Calculation		
external heat input [W]	2955.00	405.40
internal heat input [W]	2637.50	2093.75
total [W]	5200.00	2499.15
specific cooling load [W/m²]	80.00	38.45

Comment
The classroom is extremely uncomfortable in summer since the threshold value listed in figure 2.42 of 40 W/m² is clearly exceeded.
The following changes were made:
- improved g-value of the glass
- better sun protection in terms of position, colour and daylight penetration
- installation of a ventilation unit with supply air drawn in through ground duct (pre-cooling)
- reduced periods of artificial lighting
- improved technical equipment (computers, etc.)

The improvements bring about comfortable room conditions without installing an energy intensive cooling system. By activating or retrofitting thermal storage mass, providing automated control of the sun protection device and operating the ventilation unit at night, it is possible to further reduce the temperature increase in summer.

is exceeded and which maximum temperatures are reached.
The example in figure 2.41 (p. 72) illustrates how construction-related decisions can be prepared in a well-founded manner by using simulations. If, for example, an exterior-mounted sun protection device is installed, the indoor climate poses no problems and the temperature never rises above the critical value during the period investigated (10–17 July), whereas in the case of an interior-mounted sun protection device the indoor temperature frequently rises above 26 °C.

Energy efficiency refurbishments

When performing the assessment for minimum summer heat protection according to DIN 4108, the rise of the building's temperature due to user-related heat input is not considered. This however is an aspect which should be analysed carefully in the case of all construction projects and refurbishments. Among the methods described, the thermal simulation is the most effective. Simulation studies give detailed information on the temperature development in buildings and provide, based on simple variations, the opportunity to compare the efficiency of different construction-related and technical solutions.
Simple thermal simulations – the so-called single-zone model for individual rooms - can be performed by architects as an additional planning service. In terms of complexity these calculations are comparable with those for thermal bridges. Multi-zone calculations, for example the calculation of the temperature development in the fully-glazed, unheated cavity of a double skin facade and its impact on the adjoining rooms, require detailed know-how and should therefore be performed by specialists with the appropriate qualifications.
Especially in the case of refurbishments, which do not always run smoothly due to structural circumstances and other constraints, the simulation of variations contributes immensely towards finding the ideal solution to avoid overheating in summer and selecting the perfect plant technology to cater for the building's requirements.
A two-level approach is taken to identify the impact heat sources have on the building:

- The cooling load calculation according to VDI is used by engineers as a basis to dimension a cooling system, which will also work in extreme conditions.
- Whether and to what extent cooling is required, can only be identified by performing a thermal simulation, because this is the only method which illustrates the time scale of the temperature development. Thermal simulations should be a component part of every planning measure that aims at improving the building structure by reducing the negative impact of heat sources in summer.

Construction-related measures to reduce heat input

One of the biggest problems in existing non-residential buildings is the severe temperature rise inside the building during summer. These uncomfortable conditions can either simply be accepted or the rooms must be cooled with a mechanical system, which leads to an increase in energy consumption. One of the primary aims of upgrading a non-residential building is therefore to improve the summertime thermal performance (fig. 2.43).

☐ a

Retrofitting a sun protection device to the building is one of the most effective ways to reduce heat input in summer.
DIN 4108-2:2003-7, table 8, classifies sun protection devices in terms of their solar radiation permeability by using the diminution factor F_c (fig. 2.38, p. 71).
The F_c-value 1 applies to no sun protection, internal sun shading has an F_c-value between 0.75 and 0.9, external sun shading an F_c-value between 0.25 and 0.5.

☐ b

Optimising the glass itself is a very important factor. In the case of non-residential buildings, reducing the solar input, which is represented by the total energy transmission coefficient (g-value), in combination with a high use of daylight (τ-value), is more important than providing thermal protection.
The so-called selectivity of glass, which is expressed by the quotient of the τ-value (daylight penetration) and the g-value, should be around 2 for glass that has especially been improved to cater for summer conditions. These upgraded glass panes have, for example, a U-value of 1.0 W/m²K, a g-value of 37 % and a τ-value of 70 % (fig. 2.35, p. 69). The product of the F_c-value and the g-value equals g_{total}, which is the critical value for the solar energy input according to DIN EN 410. The total reduction factor of the system is determined by the type of glass and the properties of the sun shading device (type of slats, colour, distance from the window, etc.) individually for each design solution.

☐ c

Double facades with an outer skin of single glazing are a good solution, especially when the refurbishment takes into consideration structural and noise insulation problems as well as energy performance issues. The second skin protects existing concrete facades against the rigours of weather, improves the sound insulation and permits, as it is weatherproof and intruder proof, natural ventilation of the building at night. The thermal performance of the facade is also much improved by the second skin so that further insulating measures can be reduced considerably.
In the case of a double facade, it should be possible to open approximately half of the facade area to prevent the cavity from overheating. Glass louvre facades are therefore ideal for this solution. A major advantage of the second skin, which is fitted at a distance in front of the original facade, is that the sun shading device placed inside the cavity is protected from climatic conditions and can therefore be activated at any time. Without wind protection, blinds must be controlled by a wind monitor and retracted automatically as soon as the wind speed rises above 10 m/sec (wind force 6). In this situation, the glass is then exposed to the full force of solar radiation.

☐ d

In the absence of thermal storage mass, the room temperature rises very quickly, when exposed to heat input, and the interior space becomes overheated. The excess thermal energy must then be dissipated by means of a mechanical cooling system. Thermal storage mass, such as a concrete floor slab, is able to absorb a large quantity of heat without heating up very much itself. The greater the quantity of heat responsible for a temperature increase of 1 K, the more effective the storage mass is.
The lower temperature of the night-time air can be exploited to dissipate the heat stored during the day, provided reliable ventilation is possible during the night. Available thermal storage mass must therefore be uncovered during the course of refurbishment work in order to activate its effectiveness as described above.
If it is not possible to expose suitable components due to conflicting requirements (sound insulation, building services), storage mass can also be retrofit-

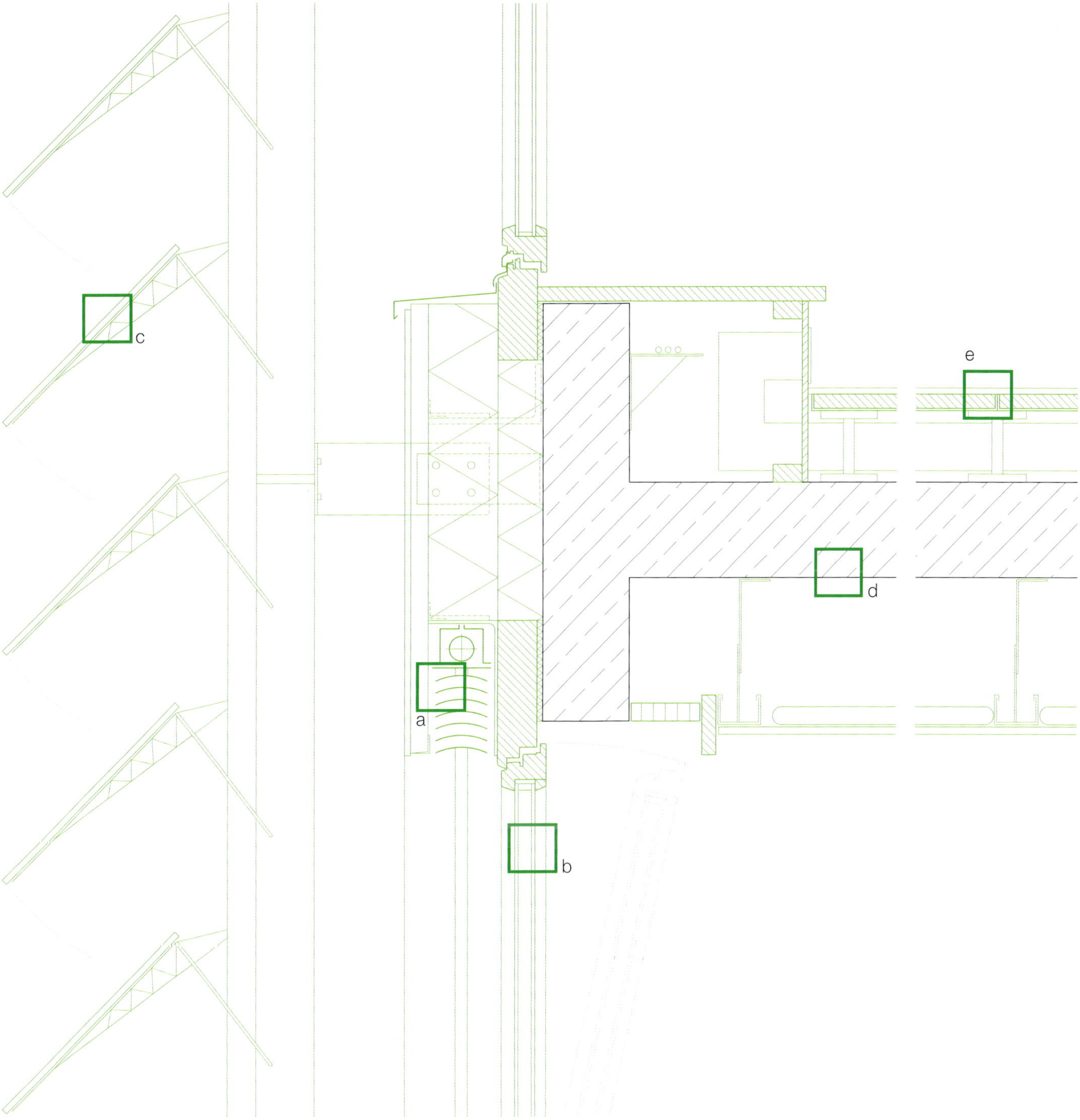

2.43

ted. By installing 2 to 4-cm-thick latent heat storage panels made of phase change materials (PCMs), which at an air temperature of 25 °C change from a solid to a liquid state, it is possible to achieve a thermal storage mass comparable to that of a concrete floor slab. In the example illustrated here, the cool night-time air is blown through the cavity between the suspended ceiling and the reinforced concrete floor slab by a fan so that the heat storage components are cooled directly.

□ e

Non-residential buildings must satisfy high demands as regards the flexibility of their floor plans. One primary requirement in this context is that it should be possible to retrofit or adapt building services with minimum difficulty. Refurbishing floors, for instance, is facilitated by installing a raised or access floor. The floor cavities can accommodate not only electric cables but also heating, cooling and ventilation ducts/pipes, while allowing access to the service systems at all times.

2.43 Construction-related measures to reduce heat input (executed project)

Ventilation

The correct ventilation of habitable rooms is extremely important in the case of new builds and energy efficiency refurbishments, since the building methods used today create an almost airtight seal between the interior and exterior space. The demand for airtightness stems from the desire to let as little outside air into the rooms as possible, which in turn has the effect of reducing the heat requirement. On the other hand, a certain amount of air exchange is necessary to ensure that the occupants are supplied with sufficient oxygen and the increasing level of moisture generated by activities taking place in the building can be removed. The CO_2 content of the air, which increases according to the number of occupants, should, by exchanging the air at regular intervals, not rise above a concentration of 1000 ppm.
§6 of the EnEV, airtightness and minimum ventilation rates, includes a description of these seemingly diverging requirements:

- 1st passage: "Buildings are to be erected in a way that their heat transmitting envelope surfaces, including all gaps and cracks, are permanently sealed according to the recognised rules of engineering."
- 2nd passage: "Buildings are to be erected in such a way that the air exchange which is required for health and heating purposes is ensured."

In order to ensure a sufficient supply of oxygen, each person requires 20–30 m³ of fresh air per hour. Thus the amount of fresh air which must, for example, be supplied to a home with four persons is approximately 120 m³/h.
Because the conditions of use in a building, such as the number of persons present, tend to vary, average air change rates are assumed for the assessment of the fresh air supply. The average air change rate n is determined in accordance with the air volume of the specified room. An air change rate of n = 2/h for a room with a net air volume of 1000 m³ means the room's air volume must be replaced with fresh air on average twice per hour when occupied. An air change rate of 2/h during a 12-hour occupancy period in actual fact equals an average daily air change rate of 1/h. In a month with 30 calender days and 22 working days, this equals an average air change rate of 0.73, and in a month with 31 calender days, an air change rate of 0.70.
Due to their consistent 24-hour use, residential buildings are easier to deal with in terms of the amount of fresh air required. Without any specific leak tests, the heat loss is accounted for on the basis of an average air change rate of 0.7/h. If the airtightness was tested on site (blower door test) and the results stated an air change rate of $n_{50} \leq 3/h$ at an overpressure of 50 pascale, the heat loss is calculated using an air change rate of 0.6/h.

To perform the blower door test, all windows are closed tightly and a powerful fan is installed into the entrance door of the building, which produces an overpressure of 50 Pa. By measuring the pressure changes in the building, it is then possible to determine the air loss through any unknown cracks and gaps (n_{50}). If the air change rate is $n_{50} \leq 3/h$, the unit is, according to the EnEV, considered airtight.

Ventilation of residential buildings

The new standard DIN 1946-6:2009-05, part 6, ventilation of residential buildings, requires a detailed analysis of all matters concerning the ventilation. It provides precise procedures to calculate whether the natural ventilation of a dwelling will suffice, or whether a mechanically supported ventilation system is required to maintain the minimum air exchange. Even though the standard refers to residential buildings, its basic approach, concerning the differentiation of air volumes in accordance with the building's use, can also be applied to other building types.
At first it is necessary to understand that there are three possible ways for air to enter a building:

- through infiltration (air leaks),
- through natural ventilation and
- mechanical ventilation.

Gaps and cracks in the building envelope inevitably lead to an exchange of air. This so-called user-independent air exchange through infiltration leads to air change rates between 0.05/h and 0.1/h in airtight buildings. In the case of older buildings with leaky windows, the air exchange through infiltration can amount to air change rates of up to 1/h.
The conscious, user-dependent supply of fresh air is usually provided by opening windows. This method, however, is subject to the presence of the residents. Depending on the type and position of the opening sash (tilted, fully open, draught by opening windows on opposite sides of room) and the duration of opening the windows (purge ventilation, permanent ventilation), it is possible to reach air change rates of up to 20/h. Due to the various parameters, it is very difficult to accurately define the actual air change rates achieved through natural ventilation. The cross ventilation of a residential space with an air volume of 100 m³ using an air change rate of 12/h, for example, means a fresh air supply of 1200 m³ per hour. If this ventilation method is applied once an hour for five minutes, the fresh air supply amounts to only 100 m³/h. If however six persons are pre-

Differential design pressure for buildings

Ventilation system	Wind zone	Differential design pressure Δp^d for	
		Single-storey units (typical in multi-storey dwellings)	Multi-storey units (typical in single family homes)
Natural ventilation (cross ventilation)	weak winds	2 Pa	5 Pa
	strong winds	4 Pa	7 Pa

2.44

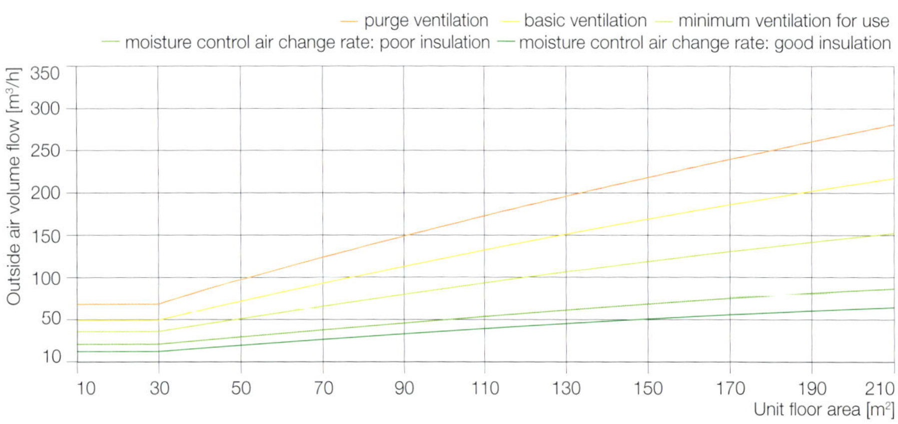

2.45

2.44 Differential design pressure Δp for buildings, extract from DIN 1946-6, table 10

2.45 Minimum outside air volume flow rates for different unit sizes according to DIN 1946-6

sent, a fresh air supply of approximately 30 m³ per person and hour is required, or in total, a fresh air supply of approximately 180 m³/h. The five-minute purge ventilation under the specified conditions is then no longer sufficient.
By using a so-called controlled ventilation system, the fresh air supply and, if required, the waste air extraction are performed mechanically. A controlled ventilation system ensures that the desired air exchange is provided independent of the occupants (see Ventilation, pp. 106ff.).

Ventilation requirements
DIN 1946-6 states the following ventilation types; they mainly differ in the air volumes supplied (fig. 2.45):

- moisture control ventilation
  The aim of this ventilation level is to remove moist air from the dwelling unit and replace it by dry fresh air. The achieved reduction of relative humidity reduces the risk of mould growth significantly. The lower the standard of insulation or the more airtight the building, the higher the demand is for ventilation to control the moisture level. This type of air exchange is the most basic requirement and must be ensured without the involvement of the residents. It can either be provided through leaks in the building envelope or by a mechanical ventilation system.
- reduced ventilation
  This ventilation level is used to provide minimum hygienic standards and general moisture protection under normal user conditions. In the short-term absence of the user, the pollutant and moisture loads may in some cases be lower. Reduced ventilation is also ensured when residents intervene, e.g. by opening the window.
- nominal ventilation
  Nominal ventilation is designed to provide minimum hygienic standards and general moisture protection in the presence of the resident/s (normal operation).
- intensive ventilation
  This type of ventilation is a rapid, necessary air exchange (purge ventilation) designed to remove large amounts of pollutants within a short period of time.

Ventilation concept
DIN 1946-6 demands a ventilation concept for new builds and refurbishments. The aim is to either prove that the necessary air exchange for moisture control can be ensured through infiltration, i.e. leaks in the building envelope, or that a controlled ventilation system must be installed to supply the fresh air mechanically. If the latter applies, the ventilation unit must be planned in detail and the mode of operation must be determined.
The first step, in this case, is to calculate the infiltration air change rate:

Infiltration air change rate
The user-independent infiltration air change rate $q_{v,inf,eff}$, i.e. the amount of air entering the dwelling unit per hour without the resident's active participation, is determined according to DIN 1946-6 as follows:

$$q_{v,inf,eff}\ [m^3/h] = 0.5 \cdot A_{du} \cdot 2.5 \cdot n_{50} \cdot (\Delta p/50)^{0.7}$$

Based on the area A_{du} of the dwelling unit and its room height (here simply set at 2.50 m), specifications concerning the airtightness (air change rate of n_{50} at a pressure difference of 50 Pa) and the wind pressure acting on the building envelope (differential design pressure Δp) are incorporated in the calculation. Depending on the property, the n_{50} value can be determined using one of the following values:

- variable input based on the measurements of a blower door test
- 4.5/h for building stock not yet refurbished
- 2.0/h for modernised single family homes
- 1.5/h for a dwelling unit in a refurbished multi-family dwelling (single-storey unit)
- 1.0/h for fan-supported ventilation

The necessary factors for Δp are stated in figure 2.44.

Air change rate for moisture control
The necessary moisture control air change rate, which is required to avoid damage caused by moisture and mould, is determined in a next step. If the determined value is higher than the infiltration air change rate, the difference between the two must at least be covered by a mechanically controlled air supply system. The necessary moisture control air change rate $q_{v,total,du,ma}$ is calculated according to DIN 1946-6 using the following equation:

$$q_{v,total,du,ma}\ [m^3/h] = f_{ti} \cdot (-0.001 \cdot A_{du}^2 + 1.15 \cdot A_{du} + 20)$$

Depending on the insulation standard of the building, f_{ti} either equals 0.3 in the case of good thermal insulation standards (at least complying with WSchVO 95) or 0.4 in the case of poor thermal insulation standards (all other buildings). The different f_{ti} values for old and new buildings show that older existing building stock

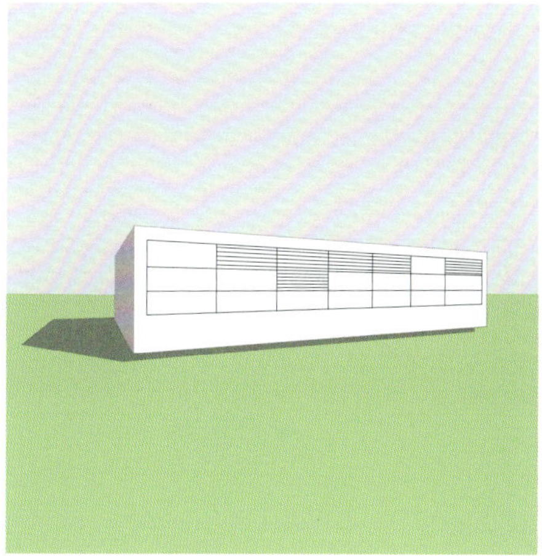

Example: air exchange
The example building, which in this case is used for residential purposes, has undergone a window replacement. Below it is examined to determine whether the infiltration air change rate is sufficiently high to cover the air change rate for moisture control.

Data

standard thermal insulation	low / f_{ti} 0.4
area of the dwelling unit (A_{du})	200 m²
Δp (fig. 2.44)	2 Pa
building type	single-storey dwelling unit
region	weak winds
n_{50} value	1.5 /h

Necessary moisture control air change rate
according to the calculation the result is
$q_{v,total,du,ma}$ 84 m³/h

Infiltration air change rate
according to the calculation the result is
$q_{v,inf,eff}$ 44 m³/h

Conclusion
The infiltration air change rate is not high enough to provide the user-independent air change rate necessary to prevent damage caused by moisture.

Measure 1
After installing window rebate vents, which enable concealed ventilation through gaps around the windows, the n_{50} value is determined by using a blower-door assessment: n_{50} = 2.9/h
The result for the infiltration air change rate is then:
$q_{v,inf,eff}$ = 88 m³/h

Conclusion
In the case of gap ventilation the airtightness is still satisfactory ($n_{50} \leq 3.0$/h) and the necessary air change rate for moisture control is met.

Measure 2
The alternative is that the moisture control air change rate is provided by an exhaust ventilation system which is operated together with controlled fresh air vents. Both the minimum air change rate as well as the nominal air change rate can be provided with this system.

Calculation of the remaining air change rate:

reduced air exchange (nominal air change rate · 0.7)	147/h
nominal air change rate (moisture control air change rate divided by 0.4)	210/h
intensive air change rate (nominal air change rate · 1.3)	273/h

Comment
Measure 1, especially in terms of a retrofit, is the most reasonable solution in terms of costs.
Measure 2 offers greater comfort and reliability, since it is a user-independent system, which can also accommodate a reduced air change rate without any difficulty.

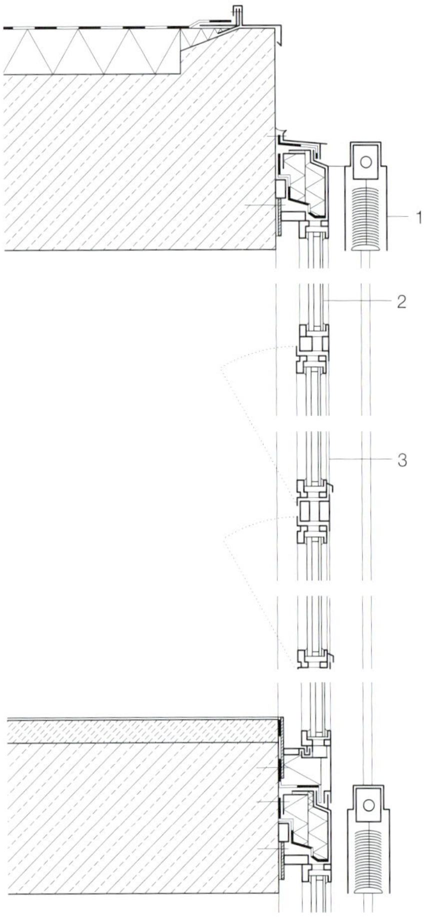

a

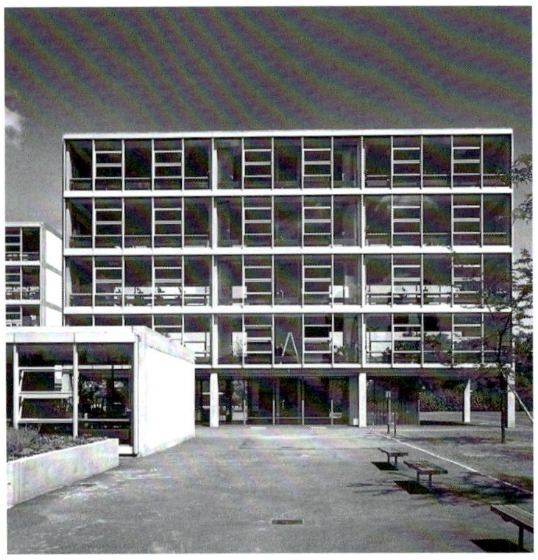

b

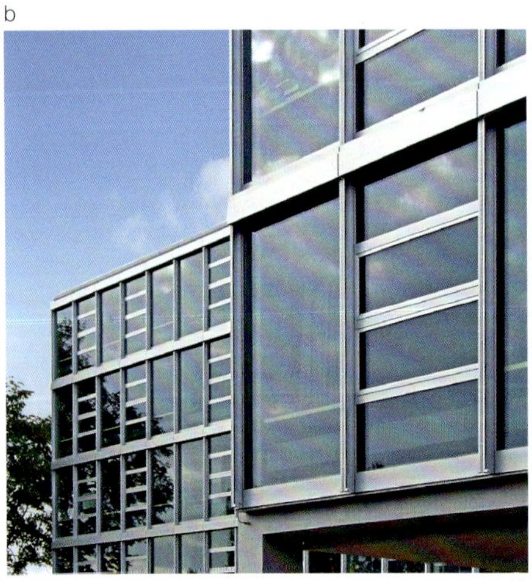

c

2.46

requires a higher air exchange rate than new builds. This is based on the fact that the inward facing surface temperature of the exterior walls and the temperature at thermal bridges is lower than that in upgraded or recently completed buildings. The danger of mould growth is therefore much greater at fairly low relative humidity levels.

Additional air change rates
If a mechanical ventilation system is needed, the equipment can also be designed to cover the remaining, higher ventilation requirements. The other air change rates are then determined according to the following factors:

- the nominal air change rate is the quotient of the air change rate for moisture control and 0.3 (in the case of high insulation standards) or 0.4 (in the case of old building stock)
- the reduced air change rate is determined by multiplying the nominal air change rate by the factor 0.7
- the intensive air change rate is calculated by multiplying the nominal air change rate by the factor 1.3

Based on the calculations above, it is now possible to decide how best to supply the necessary air volumes to the dwelling unit. If a mechanically supported ventilation system is absolutely necessary, proof must be provided of how the air volume is supplied and removed from the rooms. A building services engineer must be appointed to provide the necessary information.

Ventilation of non-residential buildings

Due to the diversity of activities taking place in non-residential buildings, the requirements concerning the ventilation vary considerably and can frequently only be fulfilled by incorporating a mechanical ventilation system.

Air change rates according to DIN V 18599
The standard DIN V 18599 states the volumes of fresh air per hour for different building uses (air change rate or factor n), which must be applied when performing the calculations for the energy balance of non-residential buildings (fig. 2.47). However, the volumes listed only apply to the predetermined utilisation periods. There is no demand for fresh air when the building is not in use.

Workplace guidelines
The German workplace ordinance (Arbeitsstättenverordnung), valid since 1975, was amended to incorporate further guidelines for workplaces established by the Federal Ministry for Labour and Social Affairs (Ar-beitsstättenrichtlinien, ASR). These German workplace guidelines (ASR) are, however, currently being replaced by the German Technical Rules for Workplaces due to changes made to the Workplace Ordinance. Although the workplace guidelines (ASR) are in fact no longer valid, they nevertheless provide some useful tools concerning numerous aspects of developing work space. ASR 5, for example, describes the circumstances which allow for a workplace to be supplied by natural ventilation methods. The different types of natural ventilation are divided into four systems:

- System I
  Single-sided ventilation with openings in one of the exterior walls for the supply of fresh air and extraction of waste air. The areas required for supply and waste air can be combined to a single opening. The wind speed assumed for the opening is 0.08 m/sec.
- System II
  Cross ventilation with openings in opposite facing exterior walls or in an exterior wall and a roof surface. The wind speed assumed for the opening is 0.14 m/sec.
- System III
  Cross ventilation with openings in an exterior wall and a ventilation shaft on the opposite side of the room. The ventilation shaft must have an opening of ≥ 80 cm², the height of the shaft must be at least 4 m. The wind speed assumed for the opening is 0.21 m/sec.
- System IV (rooms higher than 4 m)
  Like system II, except that the assumed wind speed in the opening is 0.21 m/sec.

The different activities that take place in non-residential buildings are divided into three categories:

- room category A
  workplaces with predominately seated activities (offices)
- room category B
  workplaces without predominately seated activities (sales, etc.)
- room category C
  workplaces with considerable odour nuisance and physical work.

By assigning the appropriate room category to the selected ventilation system, it is now possible to determine the size of the ventilation opening in relation to the floor area of the room concerned. Attention should be paid to the fact that the size of the opening calculated for rooms with single-sided ventilation must be doubled so that the exhaust air opening is the same size as the supply air opening (fig. 2.48).

Energy efficiency refurbishments

If in a residential refurbishment more than one third of all windows are replaced, it is, according to DIN 1946-6, necessary to develop a ventilation concept. The requirements for single-family dwellings are more stringent, since DIN 1946-6 requires a ventilation concept in the case of a window replacement, but also if more than one third of the roof surface is insulated. The greater airtightness achieved in the dwelling units of multi-family residences (according to DIN 1946-6, these are single-storey units) by upgrading the windows requires that, if the thermal conditions of the old building remain the same, a controlled ventilation system, i.e. a mechanically supported system, is installed to ensure the necessary air exchange is supplied to prevent damage caused by moisture.

If the intention is to make do without a mechanically supported ventilation system, despite the more airtight windows, ventilation gaps must be deliberately incorporated around the opening sashes. Possible solutions include the so-called rebate ventilation, which is a permanent, minimum ventilation gap, or special window handles that, when put into a 45° position by the resident, create air gaps between the sash and frame.

In both cases, it is necessary to determine the individual n_{50} value by performing blower door tests:

- if $n_{50} \leq 1.5/h$, the air change rate required for moisture control is not met
- if $1.5/h < n_{50} \leq 3.0/h$, the requirements for moisture control are fulfilled, and there is still the possibility to use the reduced average air change rate of 0.6/h in the energy balance performed according to the EnEV
- if $n_{50} > 3/h$, the dwelling unit neither meets the requirements prescribed by the EnEV for airtightness nor the requirements concerning the airtightness for minimum thermal protection in winter.

The ventilation concept should therefore also be considered carefully when carrying out refurbishment work to existing building stock. The aim is to avoid damage caused by moisture and the formation of mould due to lack of air exchange.

The installation of a controlled ventilation system with a simple air exhaust unit is a reliable and inexpensive solution. Controllable vents in the window frames and exhaust fans in the bathrooms can permanently and consistently provide the necessary air change rate to avoid damage through moisture in a dwelling unit. In order to provide the other air change rates, it is necessary to increase the fan power accordingly and, especially when catering for a high air change rate, to plan the air flow within the dwelling unit carefully to avoid draughts.

When refurbishing a school in Basel, the natural ventilation was upgraded by incorporating modern automatic control engineering (fig. 2.46). All windows were equipped with programmable electric openers. The motors are controlled according to the inside temperature, the CO_2 content of the air, wind speed, outside temperature and rain loads. The aim is, on the one hand, to ensure a sufficient amount of fresh air and, on the other hand, to avoid energy loss or extreme heat loads caused by excessive air supply. The basic setting to optimise the fresh air supply induces purge ventilation at six in the mornings and in the breaks between classes.

Night-time ventilation can be operated on demand and according to the difference between inside and outside temperature.

Size of opening for natural ventilation according to the German workplace guidelines 5 (Arbeitsstättenrichtlinie, ASR)

System	Clear room height H [m]	Maximum permitted room depth in relation to the clear room height H [m] [1]	Supply air and equally sized exhaust air opening per m² floor area [cm²/m²] [2]		
			Room category A	Room category B	Room category C
I		2.5 × H	200	350	500
II	up to 4		120	200	300
III		5.0 × H [3]	80	140	200
IV	over 4		80	140	200

[1] This column states the maximum room depth permitted for the various systems of natural ventilation in accordance with the room height.
[2] The values stated are applicable each for the size of the opening for supply and exhaust air.
[3] In the case of systems II, III and IV, the maximum permitted room depth is applicable for the distance between the exterior walls and/or the exhaust air vents in the shaft or roof.

2.48

Volume of fresh air per hour according to DIN V 18599-10, table 4

Zone	Description	Minimum outside air volume flow [m³/(m²·h)]
1	single office	4
2	group office	4
3	open plan office	6
4	meeting/conference room	15
5	ticket hall	2
6	retail store	4
7	with refrigerated units	4
8	classroom	10
9	auditorium	30
10	dormitory	4
11	hotel room	3
12	canteen	19
13	restaurant	18
14	kitchen	90
15	kitchen – adjoining room	15
16	sanitary installations	15
17	common rooms	7
18	ancillary space	0.15
19	circulation area	0
20	store, technical room	0.15
21	server, datacentre	1.3
22	workshop	20
23	audience area	40
24	foyer	25
25	stage	0.3
26	trade fair	7
27	exhibition	2
28	library – reading area	8
29	library – open area	2
30	library – stacks area	3
31	gymnasium	3
32	multi-storey carpark	8
33	if open to the public	16
34	sauna	15
35	gym	12
36	laboratory	25
37	therapy	10
38	special therapy	30
39	therapy corridors	10
40	doctor's surgery	10
41	warehouses	1

Explanation

A room in the category of Zone 1 (single office) with a useful floor area of 20 m², a clear height of 2.50 m and a volume of 50 m³, requires a fresh air supply of 80 m³ per hour (4 m³/m² · 20 m²).
In relation to the room volume, this corresponds with an air change rate of 1.6/h (80m³ / 50 m³)

2.47

2.46 Refurbishment of the Wasgenring school, Basel
Architects: Fritz und Bruno Haller
Facade (refurbishment): PPEngineering
Petignat Ingenieurbüro für Fassadentechnik
Year of completion: 1960/2004–2005
a Vertical section of south facade, scale 1:20
1 sun protection: crimped slats, 80 mm
2 fixed glazing
3 tilting sash
each with sun/heat protection glazing (double insulating glass)
g-value 42 %, τ-value 68 %; U_g = 1.1 W/m²K
b before the refurbishment
c after the refurbishment
Due to the fact that the depth of the building is not very great, the top-hinged sash in the fully glazed north and south facade already enabled a natural ventilation of the school building before the refurbishment. The natural ventilation was maintained, however upgraded by installing electric window openers operated by a state-of-the-art control system. It was not necessary to incorporate a mechanical ventilation system.

2.47 Volumes of fresh air required per hour by different building usages according to DIN V 18599-10:2007-02, table 4, which must be applied to determine the energy balance.

2.48 Size of the ventilation opening according to the German workplace guidelines 5 (Arbeitsstättenrichtlinie, ASR)

Daylight

Whether or not there is sufficient daylight to provide adequate lighting conditions inside a room depends on the size and type of the windows, the shading of windows by surrounding buildings and trees, the depth of the room and its usage (required luminosity, daytime and/or nighttime use).
Either generated by artificial light or daylight, the luminosity of a room is described by the illuminance of the individual surfaces (room-enclosing surfaces, work surfaces). The illuminance E (in lux) is the total quantity of light (in lumen) falling from all directions on a unit area of surface. It is the quotient of luminous flux incident on a surface (lm/m² = lx). Different room uses require different illuminance levels. DIN V 18599 lists the specific lighting requirements for a range of typical uses (fig. 2.54).
In order to provide comparable data on the quality of daylight conditions in interior space, the calculations to establish the amount of daylight that can be used to illuminate a room are based on the assumption of an overcast sky. On a horizontal surface outdoors, this light source generates an illuminance of 10,000 lx. The quantity of light that can actually be made use of in the room is expressed by the daylight factor D in per cent. A daylight factor of 5 % suggests, for example, that an illuminance of 500 lx is provided at a specific position in a room (usually a horizontal work surface at a height of 0.80 m). Approximately 300–500 lx are required to perform activities such as reading or writing. Alongside the size and position of window openings, the light transmission τ of the glazing and the reflective capacity of the room-enclosing surfaces are also of significance for the lighting conditions of a room. Light transmission is a characteristic feature of glass, typical values are listed in figure 2.35 (p. 69).

Requirements according to DIN 5034

From a building permission point of view, the lighting in habitable rooms is regarded as sufficient if an eighth of the floor area is provided as structural openings for windows. While obligatory, this requirement does not necessarily guarantee a sufficient supply of daylight in a room. DIN 5034-1:2010-09 takes a more differentiated approach. The standard includes a variety of minimum values, which depend on whether the window is designed to create a visual link or to pro-

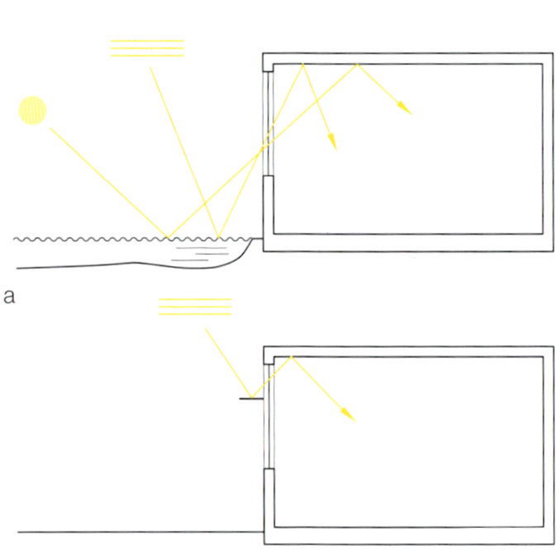

a

b

2.49

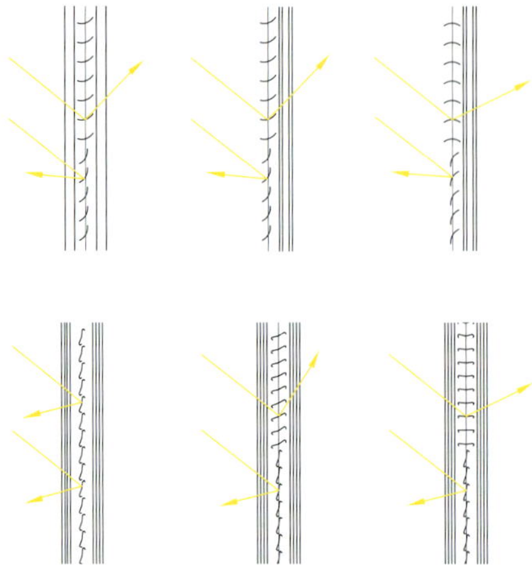

2.50

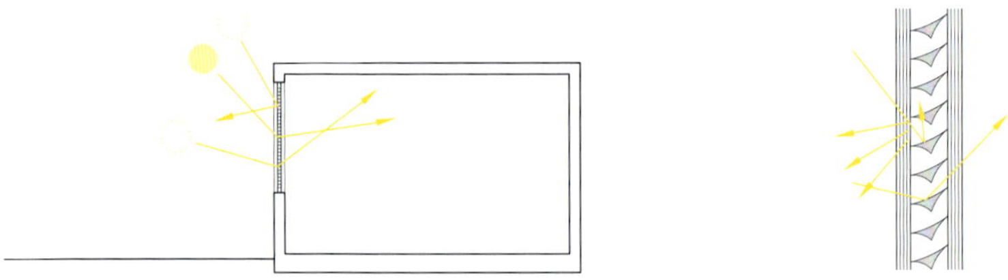

2.51

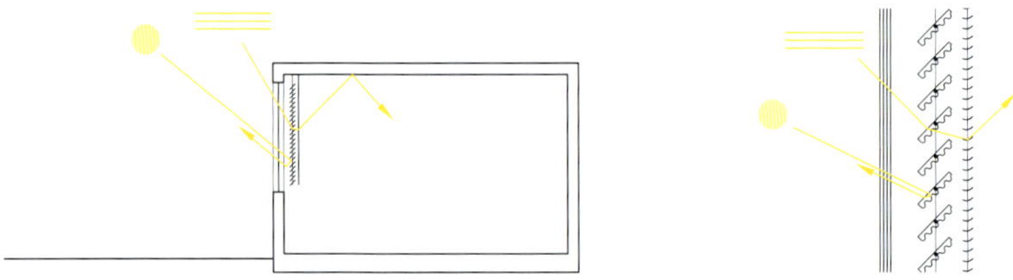

2.52

Lighting requirements for habitable rooms with windows

Requirement	**Living room**	**Work areas (only defined for room heights ≤ 3.5 m, room depths ≤ 6 m and floor areas ≤ 50 m²)**
View to the exterior	(total) width of the transparent window parts must be at least 55 % of the respective wall's width; the parapet height must be no higher than 0.9 m, the lower edge of the transparent window parts no higher than 0.95 m above finished floor level, the upper edge of the window at least 2.2 m above the finished floor level.	
		- height of the rough opening must be ≥ 1.3 m, this also applies to room heights greater than 3.5 m - width of the transparent window parts must be ≥ 1 m - minimum area of the transparent window parts for room depths smaller than 5 m: 1.25 m², for larger room depths: 1.5 m² - transparent (total) window area must be at least 30 % of the product of the room width and the room height and at least 10 % of the floor area
Sufficient luminosity	at half the room depth, 0.85 m above the floor and 1 m distance from the side walls, D should equal: - exactly between the two points at least 0.9 %; at one of the two points at least 0.75 % - in rooms with windows in two adjoining walls 1.0 %. The reflective capacity of the room enclosing walls should be as high as possible.	
Required illuminance	no requirements specified	at the most unfavourable work place, it should be at least the 0.6-fold amount of the requirements for artificial lighting as specified in DIN 5035-2.
Protection against glare, heat gain and loss	e.g. sun shading devices, roller shutters, Venetian blinds, curtains, vertical louvres	
Insolation period	insolation period of one window in each living room of a dwelling unit should be at least 1 hour on 17 January at parapet height	no requirements specified

2.53

vide daylight (fig. 2.53). Whether or not these minimum values are met can be checked during the planning phase by carrying out simulations, by performing calculations according to DIN 5034-3 or, in the case of existing buildings, by taking measurements according to DIN 5034-5.

Daylight guidance system

The use of daylight can be improved by purposely directing the light into the interior space. Specially designed light-guiding elements direct incoming light onto reflective surfaces, usually the ceiling, from where it is reflected into greater room depths. At the same time, the high daylight levels in immediate vicinity of the windows are reduced. This minimises the luminance contrast, which is often perceived as unpleasant.

For a daylight guidance system to be truly effective, the window opening should reach up to the ceiling, which should feature a white smooth surface or be fitted with light-guiding ceiling lamellae. Interruptions in the ceiling through, for example, drop beams or suspended light fittings impair the light yield and thus the effectiveness of the structural measures. Daylight guidance can be performed by using different methods and either off-the-shelf building products or individual solutions. Elements can either be fitted outside, in front of the facade, inside or in the gap between the panes of insulating glazing; they can either be rigid or moveable to track sunlight. Each type has different qualities in terms of failure susceptibility, dirt, transparency and sun protection. The most important principles are highlighted in the following sections.

Simple structural measures

A deep, smooth, white rendered window reveal directs daylight into the room; a water area in front of the facade has the same effect. As do simple horizontal elements (light shelves), which are fitted in the top third of a window, either indoors or outdoors, and reflect light onto the ceiling (fig. 2.49). These simple methods, if installed correctly, can be very effective. They compare well with technically more sophisticated systems.

Louvres and blinds

Fixed horizontal louvres and external Venetian blinds enable precise control in terms of directing light. By adding special coatings and shaping the elements accordingly, it is possible to meet specific requirements, frequently fulfilling more than one function at the same time, such as daylight guidance and sun protection. Louvres and blinds are especially effective if they can be adjusted to the differing light conditions prevailing during the course of a day, i.e. if they are equipped with pivotable fittings. Venetian blinds, of which the top third can be controlled separately, are able to direct sunlight into the interior (top third) and provide sun shading (lower area) simultaneously (fig. 2.50). Their efficiency can be further increased by adding light shelves to the interior side of the window. Further improvements in redirecting side light onto the work surface can be achieved by installing specially developed light directing ceiling elements made of concave curved lamellae with a micro-prism structured surface.

Light-directing glass

The cavity between the panes of double or triple glazing can be used to incorporate optical elements or blades which are specially designed to reflect and scatter light. They consist of, for example, parabolically shaped mirrors or acrylic sheet (total reflectance) with a prism patterned surface structure or linear grooves, which are selected to suit the particular situation (fig. 2.51).

These solutions allow diffuse light to penetrate, whereas direct sunlight is reflected either outwards or inwards. It is even possible to combine several layers of different prism patterned acrylic sheet in order to ensure perfect sun shading, glare protection and the redirection of sunlight at different angles. Movable, single prism sheets, which can be applied in a similar way to Venetian blinds, are a simple alternative (fig. 2.52). The downside is that all of these system obstruct the view out and are therefore only recommended for the top third of the window or in roof glazing.

Tubular daylight guidance systems

When it comes to making effective use of daylight, tubular daylight guidance systems, also known as light or sun tubes, with their highly reflective tubes are also an interesting alternative. Having penetrated an opening in the roof or the facade, daylight is directed through an up to 15-m-long mirrored tube into the building's interior (fig. 2.55). The light yield depends on the diameter of the tube (at least 30 cm) and the way in which it has been installed (with or without changing direction). Light is distributed in the interior by output components, which can be, in a similar way to lamps, integrated into the suspended ceiling. If required, lamps can be incorporated into the tube ends,

Required illuminance levels (maintenance values) according to DIN V 18599, table 4

Zone	Description	Maintenance value of illuminance $\bar{E}_m$ [lx]
1	single office	500
2	group office	500
3	open plan office	500
4	meeting/conference room	500
5	ticket hall	200
6	retail store	300
7	with refrigerated units	300
8	classroom	300
9	auditorium	500
10	dormitory	300
11	hotel room	200
12	canteen	200
13	restaurant	200
14	kitchen	500
15	kitchen – adjoining room	300
16	sanitary installations	200
17	common rooms	300
18	ancillary space	100
19	circulation area	100
20	store, technical room	100
21	server, data centre	500
22	workshop	500
23	audience area	200
24	foyer	300
25	stage	1000
26	trade fair	300
27	exhibition	200
28	library – reading area	500
29	library – open area	200
30	library – stacks area	100
31	gymnasium	300
32	multi-storey carpark	75
33	public multi-storey carpark	75
34	sauna	200
35	gym	300
36	laboratory	500
37	treatment room	500
38	special care room	300
39	corridor in care facility	125
40	doctor's surgery	500
41	warehouse	150

Explanation

The maintenance value of the mean illuminance ($\bar{E}_m$) on the surface to be illuminated may not be undercut. It is for this reason that the requested illuminance value is increased by a maintenance factor so that, even in the case of the lamps soiling, the predetermined value is met (see Step 5: maintenance factor MF, p. 115).

2.54

2.49 Simple methods to redirect daylight
a water area in front of the facade
b exterior light shelf

2.50 Light-directing louvres and blinds can be fitted outside, inside or in the cavity between the panes. Fitting the sun shading device in the cavity has the advantage that it is not only wind protected, but it also prevents the interior from becoming overheated. However, this solution requires the components to be durable and low maintenance.

2.51 Mirror elements in the cavity of double glazing

2.52 Prism patterned acrylic sheet as movable blades in combination with an interior blind

2.53 Requirements for habitable rooms with windows from a light-technical point of view according to DIN 5034-1:1999-10, appendix A

2.54 Required illuminance levels (maintenance values) for different room uses according to DIN V 18599 10:2007-02, table 4 which switch on as soon as daylight ceases. This inexpensive way of directing daylight can be combined with heliostats in order to provide direct sunlight into dark areas throughout the day. Heliostats are mirrors that track the sun so as to keep reflecting sunlight towards a fixed target, for example from the roof of a building through a full height atrium into lower storeys (fig. 2.56). Tubular daylight guidance systems and heliostats are particularly suited to sunny regions.

Skylights

Skylights are able to improve the lighting conditions indoors immensely. However, they can also, depending on their shape and orientation, increase the heat load to such a degree that all the energy saving achieved through lighting is eaten up by the greater need for cooling, or at least ventilation.
The disadvantages can be avoided by choosing an appropriate orientation of the transparent surfaces or by providing effective shading devices.
A study performed on behalf of the German association for daylight and smoke protection (Fachverband Tageslicht und Rauchschutz e.V.) states that continuous rooflights with openings taking up approximately 15 to 20 % of the roof area are sufficient to cover the required supply of daylight [1]. By increasing the surface area of the skylights to approximately 20 to 25 %, the useful floor area can be lit evenly without any disturbing luminance contrasts. Furthermore, to prevent glare, it is recommended to use light scattering or patterned glass.
An even distribution of light and a sensible arrangement of the rooflights from an energy perspective is achieved if the distance between the continuous rooflights is roughly equal to the room height and the width of the openings corresponds to half the room height (fig. 2.59).

2.55 Daylight spotlight with a flexible light tube
2.56 Heliostat to redirect sunlight
2.57 Circular skylight: area provided with daylight according to DIN V 18599
2.58 Continuous rooflight: area provided with daylight according to DIN V 18599
2.59 Arrangement and size (in % of the roof area) of skylights (dimensions in m)
a for a sufficient supply of daylight
b optimised for an even distribution of light
2.60 Steps to calculate the supply of daylight in a room measuring 6 × 8.4 × 3.2 m. The grey area defines the height of the work surface (dimensions in cm).
a daylit area A_{dl}
b transparency index I_{tr}
c room depth index I_{rd}
d daylight obstruction index I_{do}: angle γ for opposite, linear buildings ($I_{do,lb}$)

Energy balance according to DIN V 18599

In the case of non-residential buildings, the energy demand for artificial lighting plays a significant role when calculating the overall energy efficiency of a building. The energy consumption of a lighting system is dependent on the structural parameters, which in turn affect the supply of daylight to the interior space.
By using DIN V 18599-4, it is possible to determine a room or room zone's supply of daylight in a few simple steps.

Step 1: daylit areas

The area of a room supplied with daylight is determined independently from the actual window size by multiplying the lintel height (h_{li}) less the specified height of the work surface (h_{ws} - usually a table height of 80 cm) by the factor 2.5. The result expresses the maximum depth of the room supplied with daylight.

$$a_{dl,max} = 2.5 \cdot (h_{li} - h_{ws}) \text{ [m]}$$

If, in the case of ribbon windows, the windows are as wide as the room, the space supplied with daylight is as wide as the room; in the case of a punctuated facade, the width of the area supplied with daylight is the width of the window plus a

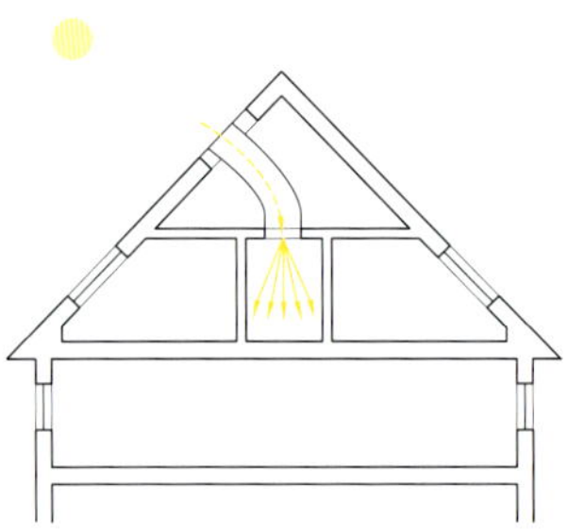

2.55

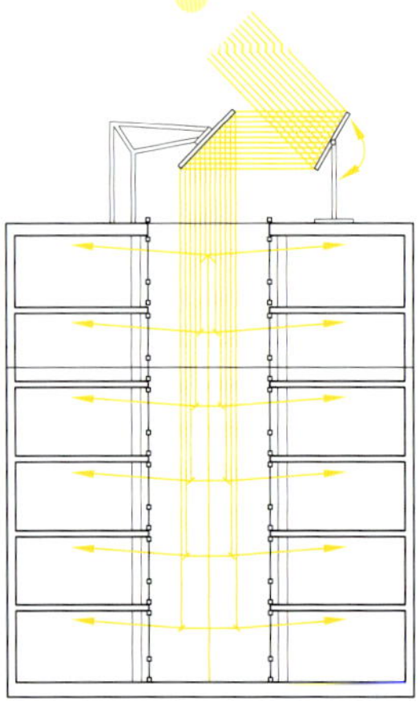

2.56

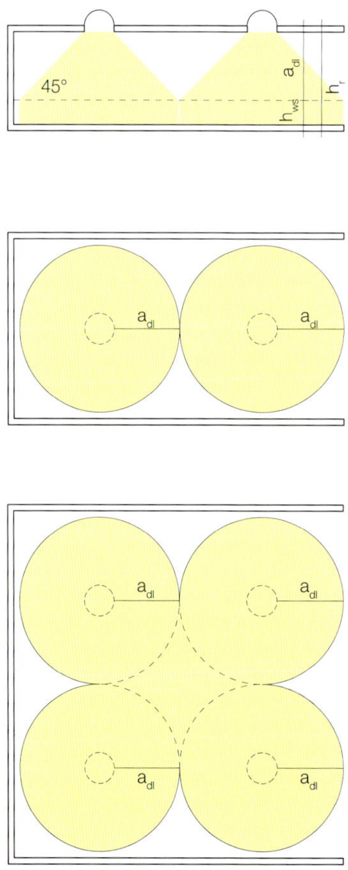

2.57

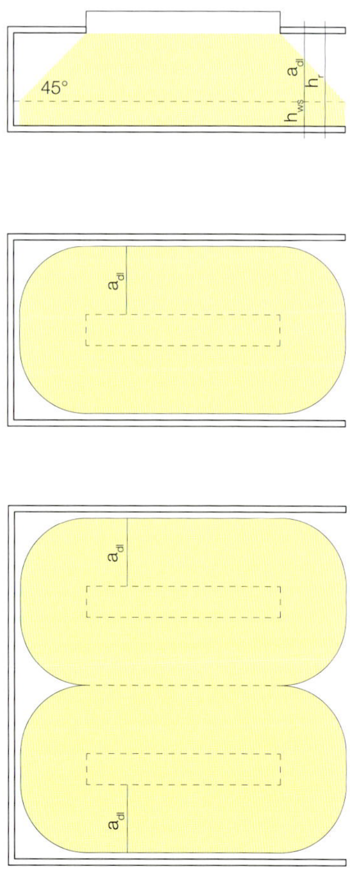

2.58

quarter of the daylit depth ($a_{dl,max}$) for either side of the window (fig. 2.60a). This information allows the planner to divide the total floor area of a room into an area supplied with daylight (A_{dl}) and an area not supplied with daylight (A_{ndl}). The greater the proportion of area supplied with daylight in relation to the total floor area, the better the building is, in fact, naturally lit (see Example: artificial light, p. 117).

In the case of skylights, the area provided with daylight is determined by drawing a line around the opening at a distance which equals the height of the skylight (lower edge above the finished floor level) minus the height of the work surface. The area defined, projected onto the floor, defines the space supplied with daylight (fig. 2.57 and 2.58).

Step 2: daylight quality

The quantity of light actually penetrating into the daylit areas is dependent on the true window size (transparency index), the room depth (room depth index) and obstructions caused by surrounding buildings (daylight obstruction index).

The transparency index (I_{tr}) expresses the ratio of window area (rough opening) to daylit area within the room (fig. 2.60b):

$$I_{tr} = \frac{A_{rough}}{A_{dl}}$$

The room depth index (I_{rd}) is used to establish a relation between the depth of the daylit area (a_{dl}) and the height of the window (fig. 2.60c):

$$I_{rd} = \frac{a_{dl}}{h_{li} - h_{ws}}$$

The effect the height of the lintel above finished floor level has on the amount of daylight penetrating the room is incorporated into this equation.

The quantity of daylight entering a room is also dependent on the surrounding buildings. Particularly in dense urban environments, the incidence of daylight is often, in some way or other, obstructed by neighbouring buildings. In these situations, the upper levels of a highrise have a much greater daylight penetration factor during the building's operating hours than the lower storeys. In order to consider these circumstances in the calculations of the energy balance, the daylight obstruction index (I_{do}) must be determined. If the building is not overshadowed by neighbouring structures or protruding components the index equals 1; the greater the obstructions caused by surrounding buildings, the smaller the reduction factor I_{do}. An individual room with uniform fenestration can therefore be separated into areas with different light conditions in accordance with the obstruction angle γ (2.60d).

The quality of daylight conditions in the considered room zones must then be calculated individually according to the daylight factor relevant to the specific area of the window's rough opening (D_{ro}).

$$D_{ro} = (4.13 + 20 \cdot I_{tr} - 1.36 \cdot I_{rd}) \cdot I_{do} \; [\%]$$

This equation shows that not only the size of the window (I_{tr}), but also its position (I_{rd}) and the obstructions caused by neighbouring buildings (I_{do}) are incorporated. The daylight factor D_{ro} is therefore a realistic value to describe the daylight quality of a specific room. DIN V 18599-4 determines four categories to differentiate the quality of daylight (fig. 2.62).

Step 3: effective light transmission factor of the glazing

The efficient use of daylight is very much dependent on the size and design of the window.

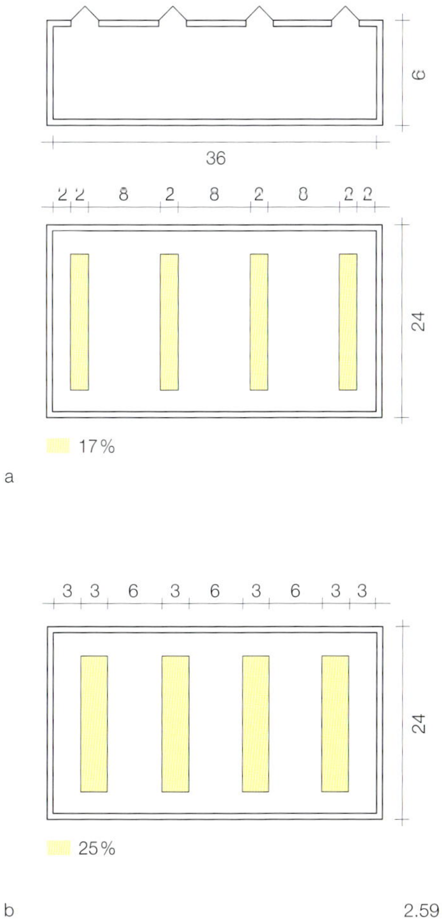

2.59

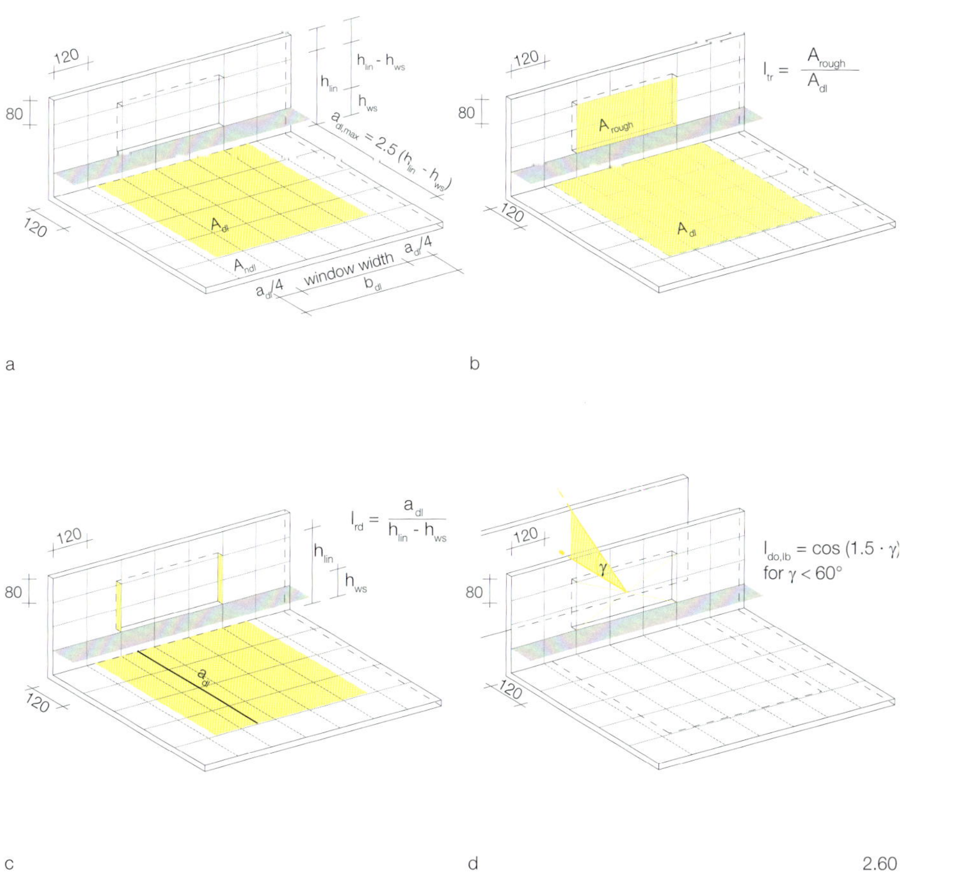

2.60

The effective light transmission value of the glass τ_{eff} is determined according to properties of the glass affecting the light transmission τ and the reduction factors for the proportion of frame (k_1), dust and dirt (k_2) as well as for the non-vertical light incidence (k_3).

$$\tau_{eff} = \tau \cdot k_1 \cdot k_2 \cdot k_3$$

Step 4: daylight penetration factor
With the help of the effective light transmission factor (step 3) and the daylight factor (step 2), DIN V 18599-4, table 9 can be used to determine the daylight penetration factor $C_{dl,pen}$ (fig. 2.63). This is performed with due regard to the required illumination intensity level (illuminance maintenance value as in fig. 2.54, p. 81) and the orientation of the facade. In the case of $\tau_{eff,\,sna}$ (sna: shading not activated) equalling 60 %, an "average" daylight penetration category and a requested illuminance of 500 lx, the daylight penetration factor of a room facing north is 0.75.
The energy-saving effect of a light guidance system is incorporated, through the assessment of the sun shading device, in the daylight penetration factor. It is therefore also an integral part of the building's energy balance.

Step 5: effective use of daylight
In a final step, calculations are performed, in accordance with the specific orientation of the facade (here: north-facing), to determine the period of time $t_{eff,day,al}$ during which the space can only be used with artificial light despite being provided with daylight. On the supposition that the north facade does not feature

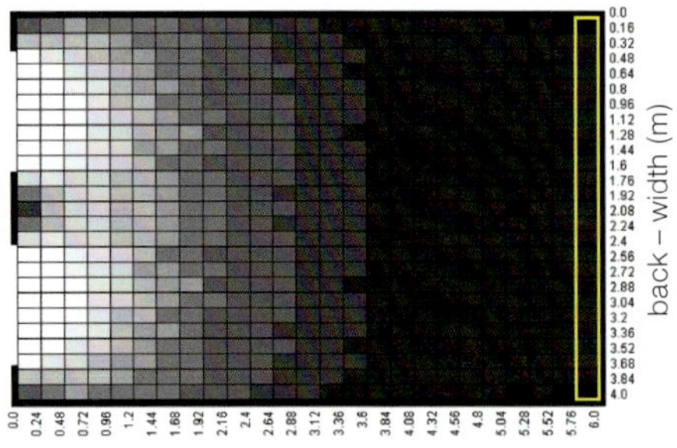

2.61

Classification of daylight provision as a function of the daylight factor for the respective rough opening $D_{ro,s}$ according to DIN V 18599-4

Daylight factor $D_{ro,s}$	Classification of daylight provision
$D_{ro,s} \geq 6\%$	good
$6\% > D_{ro,s} \geq 4\%$	average
$4\% > D_{ro,s} \geq 2\%$	low
$D_{ro,s} < 2\%$	none

2.62

Daylight factor $C_{dl,pen,sna,s}$ for rooms facing north

$\tau_{eff,sna}$ [%]	Daylight factor of the rough opening D_{ro} low (3 %)			average (5 %)			good (8 %)		
	300 lx	500 lx	750 lx	300 lx	500 lx	750 lx	300 lx	500 lx	750 lx
< 10	0	0	0	0	0	0	0	0	0
10	0.15	0.12	0.08	0.25	0.20	0.14	0.39	0.30	0.21
20	0.30	0.23	0.16	0.47	0.37	0.26	0.67	0.53	0.39
30	0.43	0.34	0.24	0.64	0.51	0.37	0.82	0.68	0.52
40	0.55	0.43	0.31	0.76	0.61	0.46	0.90	0.76	0.62
50	0.64	0.51	0.37	0.83	0.69	0.54	0.93	0.82	0.70
60	0.71	0.57	0.43	0.88	0.75	0.60	0.95	0.86	0.76
70	0.77	0.63	0.48	0.91	0.79	0.65	0.96	0.89	0.81
80	0.82	0.68	0.52	0.93	0.82	0.70	0.97	0.91	0.84

2.63

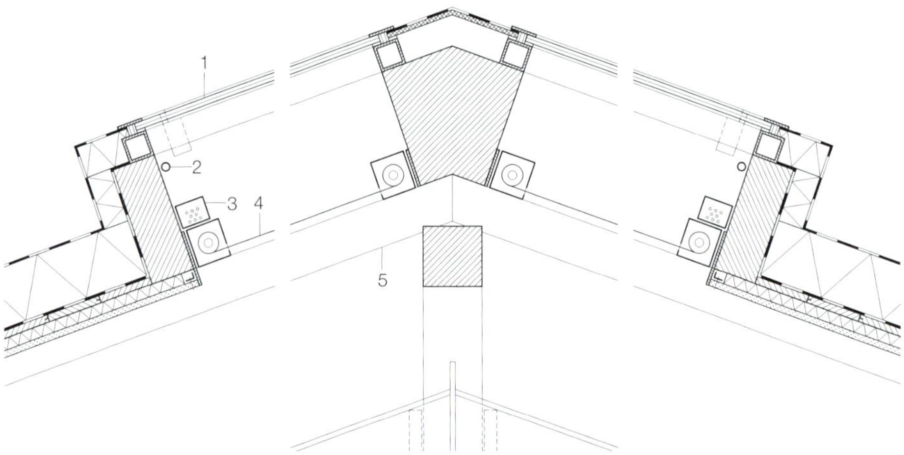

a

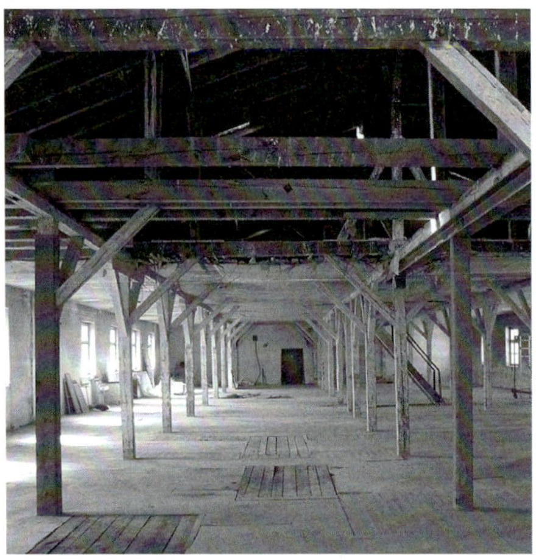

b

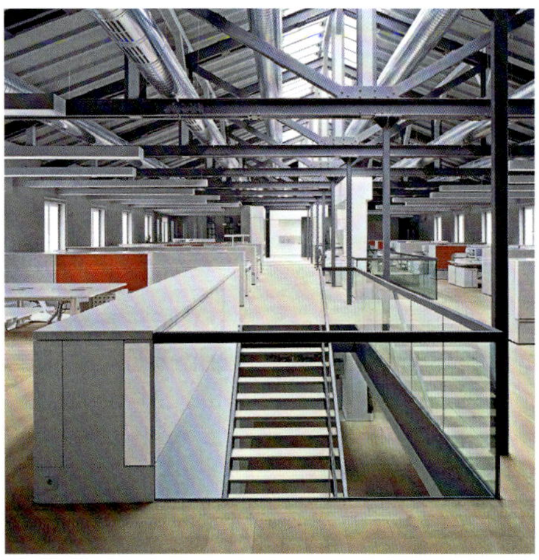

c

2.64

2.61 Simulation of an office (4 × 6 × 2.7 m with individual windows)
2.62 Quality of daylight provision according to DIN V 18599-4, table 6
2.63 Daylight penetration factor for north-facing rooms according to DIN V 18599-4, table 9 (excerpt)
2.64 Conversion of an old factory building in Rehau Architects: weber + würschinger, 1920/2005
a detail section, 3rd floor, scale 1:20
1 laminated safety glass on light metal frame
2 heating pipe
3 cable duct
4 polyester antiglare protection device
5 rafter (existing) 130/150 mm
b hall before the conversion
c hall with skylight after the conversion
In the conversion of this former porcelain factory into office space, the large depths of approximately 20 metres and the demand to retain the small windows in the facades posed the biggest challenge for the planning of daylight zones. It was for this reason that the architects added two continuous skylights, one to each side of the ridge, to minimise the use of artificial light in the centre of the room.

sun shading, the usage period (determined when describing the use zones) is now multiplied by the daylight penetration factor $C_{dl,pen}$. In the case of a daily usage period of 10 hours, the area provided with daylight can be used for 7.5 hours without artificial light ($t_{eff,day,dl}$ = 7.5 h). For 2.5 hours a day, artificial lighting is required. This is considered accordingly in the energy balance.
Further detailed calculations can be made to determine the impact the selected sun shading system and its control mechanism have on the operating time of the lighting plant.
The most important structural parameters that contribute towards minimising the operating time of artificial lighting and therefore towards a reduction of lighting energy consumption are:

- the size of the daylit area:
  - size and arrangement of windows
  - height of lintel above finished floor level (if possible without a visible lintel)
  - uniform fenestration (ideally ribbon windows)
- quality of daylight:
  - type of glass
  - minimum proportion of frame
  - perfect sun shading device
  - orientation of the facade

Simulation
Naturally, the light distribution in a room is not homogenous as is allegedly expressed in the daylight factor D_{ro}, which is calculated according to DIN V 18599 for the total area. Nevertheless, the simple method is a useful tool to easily estimate the provision of daylight and perform improvements to individual parameters to achieve better results. More precise and differentiated values concerning the daylight conditions in interior space can be determined by using computer aided simulations (fig. 2.61). The programs take into account all structural boundary conditions (overshadowing caused by neighbouring buildings, flexible sun shading devices, room shape, reflective capacity of room-enclosing surfaces) even if these overlap, as is frequently the case.

Energy efficiency refurbishments
It is most important, particularly when carrying out an energy efficiency refurbishment to a non-residential building, to improve the daylighting conditions. First evaluations of the building stock should therefore always consider simultaneously construction-related and technical aspects.
Possible improvements concerning the use of daylight should, however, never be performed independently from other aspects affecting the energy demand, but rather with a view to upgrade the total energy demand of the building. It might, for example, be wise to accept a deterioration of daylight incidence by replacing the old, uncoated insulating glass (τ-value 0.8) by triple low-e glass (τ-value 0.7). The higher energy demand for artificial lighting would be compensated by significant savings in heating and, depending on the orientation of the facade, also in cooling.
In the context of refurbishment work, it is always worthwhile checking whether the size of daylit areas can be increased by performing structural alterations, such as the creation of atria and lightwells. Improvements to windows, for example reducing the heights of lintels and parapets, minimising the proportion of frame or tapering the reveals, can also contribute towards increasing the size of daylit zones. Further details worth considering include the quality of glass and the position and type of sun shading devices. Possibilities to direct more daylight into the depths of the building should not go unmentioned.
Architects who take a holistic approach should always see the provision of daylight as part of the design concept and should therefore not reduce these issues to merely energy-related aspects. Especially in the case of existing building stock, measures to improve the incidence of daylight can often increase spatial qualities considerably and give areas which were formerly unused a new life and a higher quality of use (fig. 2.64).

Note
[1] Study on the impact of skylights in halls regarding their heat and light-related energy demand based on DIN V 18599. Report VK 43218, Centre for Sustainable Building (Zentrum für Umweltbewusstes Bauen, ZUB). Kassel 2008, unpublished.

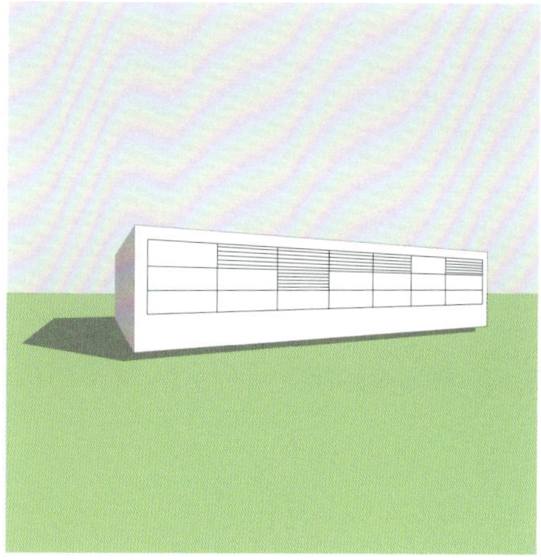

Example: provision of daylight
The pavilion's open plan office with windows on one side is examined in terms of its daylight conditions. The procedure corresponds with the steps 1–5 explained in the section "Energy balance".

Data:

office floor area (16.8 × 7.2 m)	121 m²
rough opening of window (16.8 × 2.25 m)	37.8 m²
underside of lintel above FFL (h_{li})	3 m
height of work surface above FFL (h_{ws})	0.8 m
facing north, no sun shading device	

Use profile according to DIN V 18599-10: profile 3 – open plan office (fig. 1.27, p. 24)	
working hours per day per year	2543 h
working hours per night per year	207 h
requested illuminance	500 lx

Step 1: determination of daylit area A_{dl}
$a_{dl,max}$ = 2.50 · (3.0 - 0.8) = 5.50 m

b_{dl} = 16.80 m (total width)

A_{dl} = 5.50 m · 16.80 m = 92.40 m²

A_{ndl} = 121 m² - 92.40 m² = 28.60 m²

Step 2: daylight quality
transparency index I_{tr} = 37.80 m² ÷ 92.40 m² = 0.41

room depth index I_{rd} = 5.50 m ÷ 2.20 m = 2.5

daylight obstruction index I_{do} = 1
(there are no buildings overshadowing the pavilion)

Daylight factor
D_{ro} = (4.13 + 20 · 0.41 - 1.36 · 2.5) · 1 = 8.9 %
assessment (fig. 2.62): "good"

Step 3: effective light transmission factor
τ_{glass} = 0.7 (70 % according to product data sheet)
k_1 = 0.7 (30 % proportion of frame)
k_2 = 0.9 (standard value)
k_3 = 0.9 (standard value)

τ_{eff} = 0.7 · 0.7 · 0.9 · 0.9 = 0.4

Step 4: daylight penetration factor
$C_{dl,pen}$ = 0.76 (taken from fig. 2.63)

Step 5: effective use of daylight
$t_{eff,day,dl}$ = 2543 h · 0.76 = 1933 h/year
$t_{eff,day,ndl}$ = 2543 h - 1933 h = 610 h/year

Comment
Due to the considerable depth of the room, only 24 % of the relevant floor area is provided with daylight. By adding a skylight or extra windows to the sides of the pavilion, the room would be supplied with sufficient daylight. By reducing the proportion of frame to 10 % (factor k_1 = 0.9), the light transmission factor τ_{eff} would increase to 0.5, the daylight penetration factor to 0.82 and the effective use of daylight to 2085 h/year.

- Heating
- Domestic hot water
- Cooling
- Ventilation
- Artificial light
- Photovoltaics

The function of building services is to provide the appropriate climate conditions in buildings so that the space can be used in accordance with its designated purpose. This is to be performed independent of the building envelope's quality, the local climate and the individual demands of the occupants concerning thermal comfort. The use of technical equipment is usually associated with the consumption of resources, unless renewable energy sources are applied. But even if this is the case, the equipment still requires a considerable amount of energy for its production, operation and maintenance. The main aim of an energy efficient concept is therefore to restrict the use of technical plant to the absolutely essential by optimising the structural features and carefully analysing the standard aimed at in order to provide the desired thermal conditions. The demand for building services determined according to this approach must be efficient, i.e. it must be possible to operate the plant with a low consumption of resources. It is the planners' task to incorporate the components of the building services into the structural and architectural concept of the building.

Building entrance room

Every building is connected to the public utilities for the supply and discharge of services. The following grid infrastructure is usually provided by the public utilities:

- domestic water grid
- natural gas grid (not comprehensive)
- district heating grid (not comprehensive)
- sewer network
- mains grid

In order to define clearly the transition point from the public grid to the private network, each building should be provided with a so-called building entrance room close to the street. Up to the building entrance room, the services are taken care of by the public utilities. At the transition point, they are then fitted with appropriate shut-off valves. In the case of work being performed either in the building or on the public grid, the building can be disconnected from the infrastructure. This is also the interface for responsibilities in terms of maintenance, upkeep and repair work. The building entrance room should be accessible from the hallway and fitted according to figure 3.1.

Boiler room

Due to the danger deriving from the burning of resources, the rooms for furnaces must comply with the appropriate requirements. The German fire safety regulations (Muster-Feuerungsverordnung, MFeuVO) lists differentiated guidelines for the equipment of boiler rooms.
The floor area required in boiler rooms is frequently underestimated. At least one wall in the boiler room is taken up by the heating distribution lines; alongside the actual heat generator, there is usually a domestic hot water storage tank and possibly a buffer storage tank for the heating as well as an additional storage tank for the use of solar thermal energy. All pieces of equipment must be accessible on three sides for maintenance purposes. In order to provide sufficient space for fixing pipes to the underside of the ceiling, the clear room height of the boiler room should be at least 2.50 m.

Ventilation plant room

The space required for ventilation equipment is very much dependent on the system selected for handling the air and the volume of air being moved. The plant can either consist of a centralised system to supply the entire building, or decentralised units to cater for separate zones. The latter solution is especially suited to buildings with a variety of zones and thus different ventilation requirements.

Refrigeration plant room

The refrigeration machine should be in immediate vicinity of the recooling unit, which removes the heat from the waste air extracted from the interior space and discharges it into the outside air.
The refrigeration machine and the recooling system can be placed side by side on the building's roof. However, it is also possible to place the refrigeration machine on a basement level next to other mechanical installations or, in the case of decentralised units, only on certain levels within the building.

Sanitary plant room

In the course of discussions about sustainable building, water supply and discharge has taken on an increasingly important role. It is necessary, in this context, to mention plants for water purification and the recovery of heat from waste water. These changes have led to the establishment of separate sanitary plant rooms. In the case of rainwater harvesting, sanitary plant rooms are the ideal place to feed the collected rainwater into the water distribution system.

Electricity plant room

A reliable supply of electricity requires a primary and secondary distribution system and a metering device. These pieces of equipment are also ideally accommodated in their own separate room.

Low voltage plant room

The comprehensive fit out of buildings with a variety of different communication networks requires a separate space for these installations.

Space requirements for mechanical plant rooms

The size of mechanical plant rooms in buildings can be estimated roughly by using the standard VDI 2050 (fig. 3.2).

Mechanical plant rooms can, depending on the property and the requirements concerning the technical fit-out, either be arranged in the building as a centralised or decentralised system, or even a mixture of both.
The basement location has the advantage that it is very easy to connect to the public infrastructure. Furthermore, heavy plant and storage tanks can, if required, be set on their own separate foundations. Necessary sound insulation measures can also be carried out very easily on basement levels (fig. 3.3a).

Location of mechanical plant rooms
Placing mechanical plant rooms on the roof of a building has the advantage that it is very easy to organise the ventilation equipment in terms of exhaust air outlet and fresh air intake. In the case of an air conditioning unit, the direct connection between the refrigeration machine and the recooling system is also extremely beneficial (fig. 3.3b). When using solar thermal energy, the distance to the buffer tank's feed-in point is very short. However, in these circumstances, it is necessary to support and distribute the considerable weight of the equipment positioned on the topmost floor slab. Furthermore, the volume of such rooftop units is often underestimated and must therefore be considered carefully. The appearance of the building, in particular from a distance, can be harmed severely if the units are not planned and designed with utmost care.
In multi-storey buildings, the mechanical plant rooms can also be placed half way up on a separate mechanical equipment floor. The shorter lengths of the distribution lines, especially in the case of ventilation units, lead to much smaller duct dimensions.
It is also possible to organise the heating system in a centralised fashion and, depending on the requirements, use decentralised units to cool or mechanically ventilate individual building zones.
Especially in the case of room ventilation, decentralised systems are increasingly being favoured over centralised units. In some cases, a building may have between six to eight ventilation units. Classrooms in school buildings or rooms with similar functions are frequently fitted with individual room units to supply fresh air and/or extract waste air. Similar to those in single-family dwellings or large apartments, these units are able to provide an air exchange of approximately 400 m³/h. The advantage of these decentralised ventilation units is the ability to adapt the settings to suit the respective use of the room.

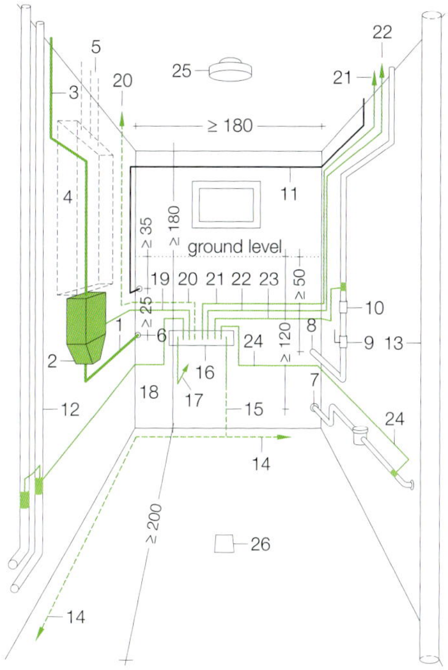

3.1

1 high current service entrance
2 high current fuse box
3 high current power cable
4 meter cabinet (if required)
5 high current lines from metering device to individual circuit distributors
6 protective cable cover
7 water service entrance with meter
8 natural gas utility line
9 gas shut-off device
10 insulation section
11 telecommunications utility line
12 heating pipes
13 sewage pipe
14 foundation earthing
15 connection lug of foundation earthing
16 equipotential bonding
17 lightning protection system
18 earthing conductor for heating pipes
19 trunk line as a protective measure in TN system
20 earthing conductor as a protective measure in TT system
21 earthing conductor for telecommunications system
22 earthing conductor for aerial system
23 earthing conductor for gas pipes
24 earthing conductor for water pipes
25 ceiling mounted light fixture
26 floor drain (if required)

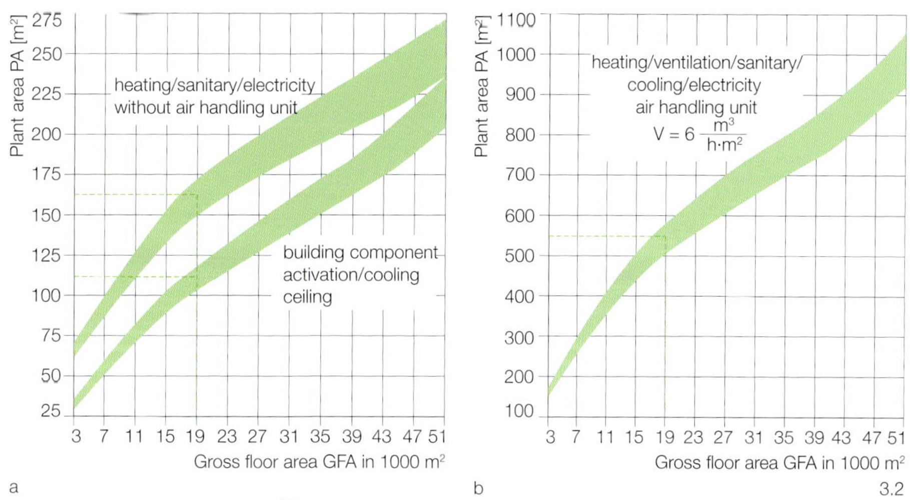

3.2

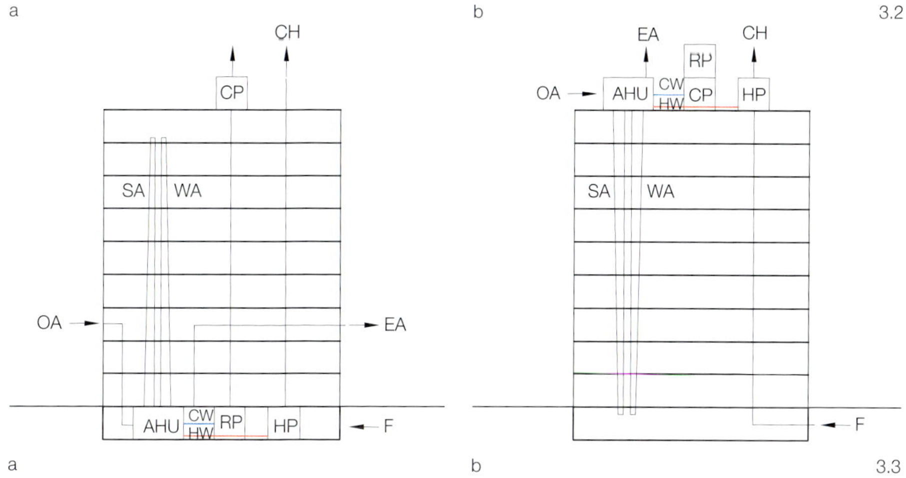

3.3

3.1 Building entrance room in a residential building (electric lines - green, all dimensions in cm)
3.2 Space requirements of mechanical plant in administration buildings according to VDI 2050
a without air handling unit
b with air handling unit (AHU)
3.3 Mechanical plant room
a in the basement
b on the roof
RP refrigeration plant

CP cooling plant
HP heating plant
CW cold water
EA exhaust air
OA outdoor air
WA waste air
SA supply air
AHU air handling unit (heating and cooling)
HW hot water
CH chimney
F fuel

Heating

Buildings either have their own system for providing heat or are connected to a local or district heating network. In the case of a local or district heating network, a heat exchanger is installed in the building in place of a heat generator. In terms of energy efficiency, the use of district heat is a sensible solution, but owing to the high heat loss through distribution, is only viable within a radius of 5–10 km of the respective power station. The norm is therefore to install the heat generator in the building itself.
Heating plant can either be arranged in a centralised or, as is usually the case in multi-family dwellings, in a decentralised fashion. When a centralised system is opted for, a plant installed in the roof space has a number of advantages over a plant in the basement: a chimney is not required and a solar energy system can readily be connected without the need for lengthy pipework. In the case of a decentralised system, each apartment or dwelling unit has its own separate heating plant. The integration of renewable energy sources is not feasible in these cases and often only possible at great expense.

System

A heating system is a closed circuit. Water is usually used as the heat transfer and storage medium. The water is heated in a boiler and, with the help of a circulation pump, transported to the heating surfaces through pipes. In systems requiring different temperatures, a distribution manifold in combination with three-way mixing valves ensures that the necessary quantity of hot water is fed into the respective heating circuit at the right temperature. The heating surfaces emit the heat into the rooms. This has the effect that the hot water cools down; it then flows back to the boiler via a return circuit where it is once again heated. Control engineering is responsible for the automatic link-up of components in accordance with the corresponding programmed specifications. The heat output in the rooms can either be controlled by thermostatic valves on radiators or by room temperature sensors. External temperature sensors ensure that the forward flow temperature is adjusted to suit the respective weather conditions and that the water is only heated to peak values at very low outdoor temperatures.
The water flow temperature differs depending on the type of heating system chosen. In the case of a typical oil or gas-fired system with ribbed radiators, the forward flow temperature is usually set at 70 °C; the return flow temperature at 50 °C. In the case of surface heating, which is fed by a heat pump, the temperatures are 35 °C and 28 °C respectively. In principle, heating plants with low system temperatures are more efficient and create more comfortable thermal environments.
Water tanks are frequently incorporated into the system to provide domestic hot water and to act as a buffer tank for the heating. Buffer tanks reduce the cycle frequency of cold start-ups in non-modulating systems and hence improve the efficiency of the heating plant (fig. 3.6 a).
In non-residential buildings and residential buildings with high energy efficiency standards, such as passive houses, the heat is distributed mainly or even entirely through air. This principle makes use of the incorporated ventilation system to transfer the heat and therefore eliminates the need for a water-based distribution system (fig. 3.6 b).
In order to operate such a system in an energy efficient way, it is essential to coordinate the heat demand for providing the determined temperature with the air volume needed to meet fresh air requirements. Hence, the quantity of necessary heat on the coldest day should be met by supplying warm air at a maximum temperature of 40 °C without having to increase the air change rate, i.e. the amount of air which is, in relation to the net volume of the room, supplied per hour, above the level required for the supply of fresh air. If the air volume at 45 °C is not sufficiently high to provide the necessary room temperature, the expenditure only for heating, for larger ducts, air outlets and drive systems, rises inexorably. The requirements mentioned can be met by buildings that have a low heat demand; in other words, buildings that are well insulated and have a moderate window to wall ratio of approximately 50 % throughout the entire building envelope.

Pumps

Pumps are responsible for keeping the heating circuit running and form an integral part of every heating system. For greater energy efficiency of the heating system, the pump capacity should be selected in accordance with the demand. An uncontrolled pump can merely be set at a maximum of four different levels. The settings are guided by the maximum amount of water output in m³/h on the coldest day. If only a low flow of water is required, the pump capacity reduces only marginally and the excess power is released by the pump as heat. Speed controlled pumps are, on the other hand, able to adapt their speed according to demand. The electricity consumed by the pump is thus reduced by almost 90 % (fig. 3.4).

Control

Whereas older boilers, so called constant temperature boilers, only operate in an on / off mode, the application of electronic components in more up-to-date boilers allows for the temperature of the heating water to be controlled in accordance with the outdoor temperature. Preconfigured control commands determine, by changing the heating curve of the boiler, the water temperature according to the indoor and outdoor temperatures. If the heating curve is, for example, set at 0.8, the temperature of the forward flow will be 50 °C at an outdoor temperature of -15 °C. The forward flow temperature reduces continuously in accordance with the rising outdoor temperature (fig. 3.5).

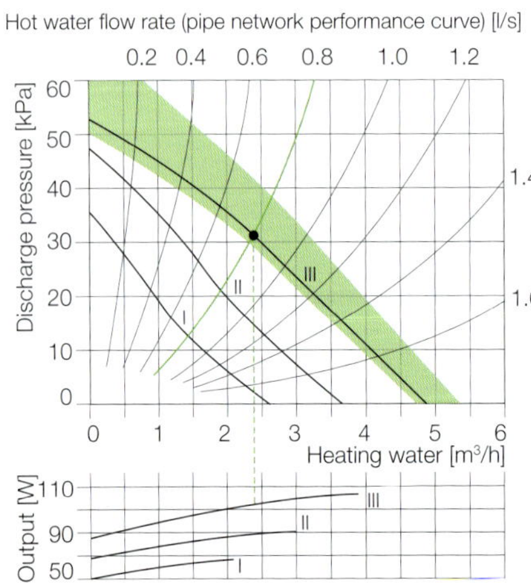

3.4

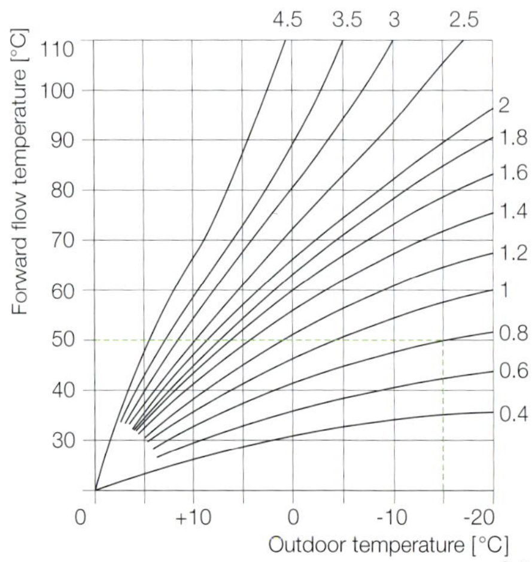

3.5

The heating curve must be set by a building services specialist according to the heating load and the available or planned heating surfaces. Due to their innovative control mechanisms, these new boilers, which were introduced at the beginning of the nineties, are referred to as low temperature boilers.
In larger systems that work with a distributor, control is exercised by mixing hot and cold water at the respective circuit. Depending on the availability, the hot water is either provided by the boiler or by the solar hot water storage tank. In smaller systems, the water temperature is controlled directly by the boiler – as is the case with a modulating gas burner – or indirectly by adding the appropriate amount of water from the buffer storage tank.

Heat generation
Heat is normally generated by the direct combustion of raw materials. The raw materials available include gas, oil and biomass. Modern heat generators that are operated with oil or gas modulate their heat output to suit the heat demand in such a way that it is possible to generate heat with minimal loss. The modulating operation method is limited when it comes to using biomass (wood) for the combustion. It is for this reason that a means of storing heat is required.
In the case of local heat supply systems, one boiler supplies several buildings with heat; in the case of district heating, whole communities or districts are supplied by local utility companies with heat from combined heat and power (CHP) plants. The waste heat produced in electricity generation or waste incineration systems is frequently fed into local or district heating networks (fig. 3.10).

Combustion – low temperature boiler
Low-temperature boilers adjust their heat generation automatically to suit the circumstances. The water is only heated to the temperature level required to suit the outdoor temperature or the hot water production. This is one of the main reasons for the higher efficiency of modern low-temperature boilers in comparison to conventional older models. In addition, these boilers are better insulated to reduce the loss through cooling to an absolute minimum.

Combustion – condensing boiler
Condensing boilers are more efficient than low temperature boilers since they also make use of the heat contained in

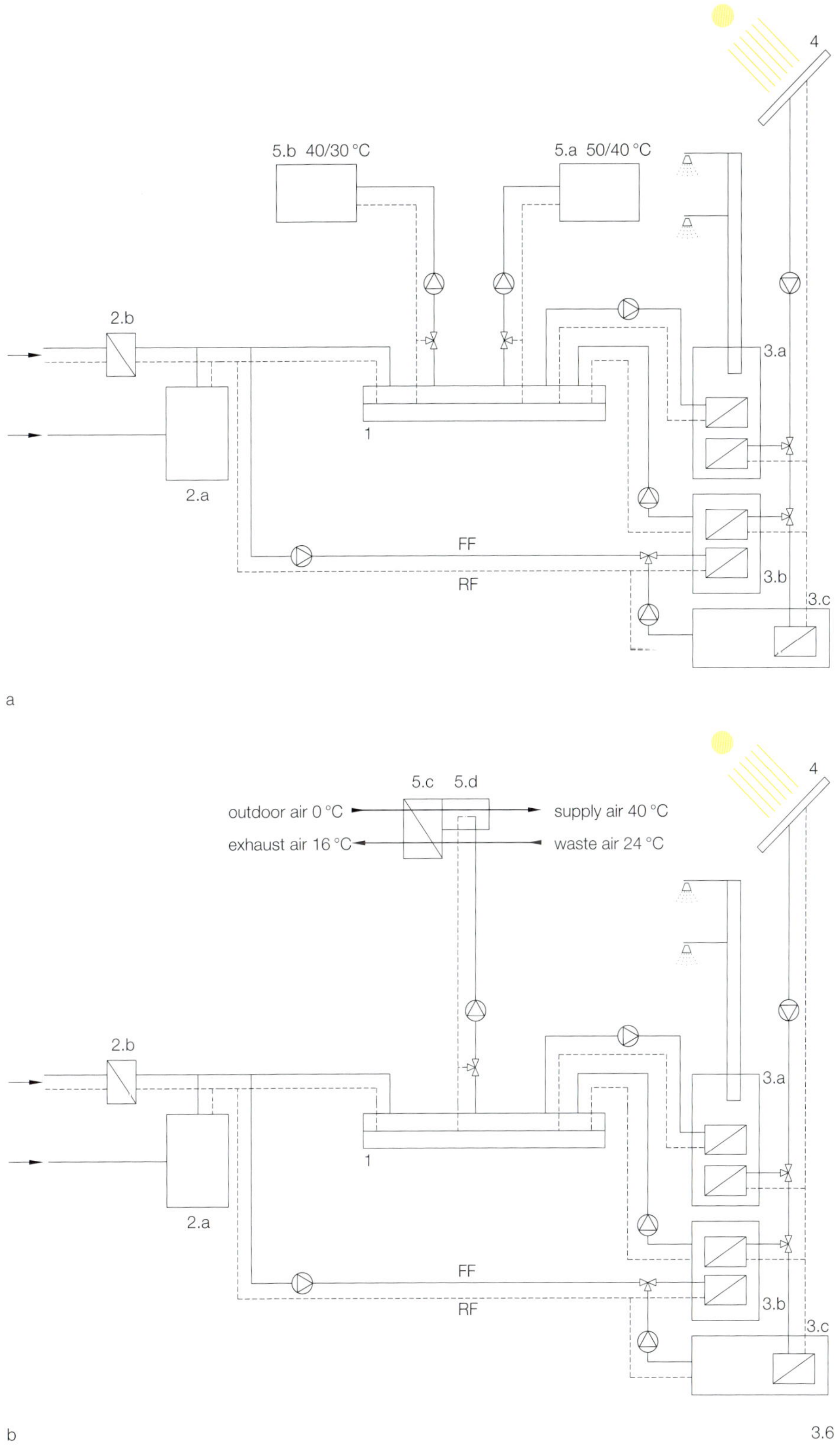

3.6

3.4 Performance diagram of an uncontrolled pump
The pump is set according to the maximum output on the coldest day (level III).
As soon as the demand for heat decreases, the pressure in the pipe drops. In order to save electricity, the pump is reset to provide a lower output
(level II or I).

3.5 Heating curve for a room temperature of 22 °C and a forward/return flow temperature of 50/40 °C

3.6 Schematic diagram of a heat distribution system
a through water
b through air

1 heat distributor
Heat production:
2.a heat generator
2.b district heat
Storage:
3.a domestic hot water storage tank
3.b buffer storage tank
3.c seasonal storage tank
4 solar collectors
Heat output:
5.a high temperature circuit (radiator)
5.b low temperature circuit (surface heating)
5.c warm air heating: heat exchanger
5.d warm air heating: reheater

the flue gas. This is performed by passing the fairly cold water from the return circuit through a heat exchanger where it is pre-heated by the hot flue gas. When cooled below the dew point, the flue gas also transfers the condensation heat to the return flow water.
The downside of a condensing boiler is that the flue gas no longer has sufficient thermal buoyancy to rise through the chimney and must be supported by a fan. Owing to the condensation and overpressure in the flue gas system, existing chimneys must be upgraded for the application of this boiler type. This is achieved by inserting a plastic or stainless steel chimney liner into the existing chimney.

Combustion – new concepts
There is great potential to increase the energy efficiency when it comes to the combustion of primary resources. Two different strategies for the combustion of gas are presented here:

- In a gas motor operated heat pump, the medium circulating in the heat pump is, through a heat source, raised to a higher temperature level according to the principle of absorption. When using environmental heat, e.g. air, as the heat source, a cubic metre of gas allows, on an annual mean basis, the extraction of approximately 14–16 kWh of heat from the environment. (In comparison: in the case of direct combustion, this value is only 10–11 kWh.)
- A further strategy to increase the efficiency of gas burners, is the use of recently introduced adsorption heat pumps. By loading and unloading a zeolite heat store with solar thermal energy, environmental heat is used. This heating method allows for approximately 15 kWh of heat to be extracted from the combustion of one cubic metre of gas.

Electricity – direct heating
In the case of electric heating, electricity is directly converted into heat or, when using an electric storage heating system, stored in heat retaining clay bricks. Owing to the high energy loss in the production and distribution of electricity, the total degree of efficiency is very unfavourable. The use of electric heaters only makes sense in circumstances where disproportionate cost and effort are required to generate heat with other systems. One situation in which it would be worthwhile is when using a solar thermal plant to generate domestic hot water. Rather than switching on the boiler for a few hours in summer, the production of hot water could, for example, be supported by an electric water heating system.

Electricity – heat pump
By making use of the thermodynamic cycle (the process that is also used in a refrigerator), it is possible to use a small amount of electrical energy to drive a compressor and then produce a considerably greater amount of heat (fig. 3.7). The outside air, or better even the soil or groundwater, are a suitable heat source. Considering the high primary energy input to generate electricity, the coefficient of performance (COP), the ratio of electricity input to heat output, should be at least three on an annual mean basis. The smaller the rise in temperature between the heat source and the temperature of the heating forward flow, the higher the COP of the heat pump (fig. 3.8). The COP is the status quo under fixed laboratory conditions, for example the electrical input and heat output with an initial temperature of 0 °C (temperature of the heat source) and a set final temperature of 35 °C (forward flow temperature of the heating system). The COP was created to provide a basis for comparing the energy efficiency of different heat pump systems available on the market. It is important to consider the electrical input

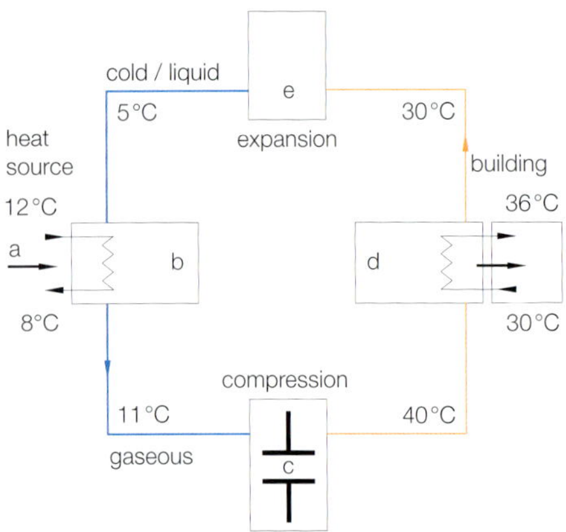

3.7

Coefficient of performance (COP) of heat pumps

Forward flow temperature of heating water	Heat sources								
	Air							Ground-water	Ground
°C	-15	-10	-5	±0	+5	+10	+15	~+10	~±0
30	–	–	–	3.10	3.20	3.30	3.50	5.80	5.00
35	2.50	2.60	2.70	2.90	3.00	3.20	3.40	5.20	4.40
40	2.20	2.30	2.40	2.60	2.80	3.00	3.20	4.70	3.90
45	2.00	2.10	2.30	2.40	2.60	2.80	3.00	4.30	3.50
50	1.90	2.00	2.10	2.30	2.50	2.70	2.90	4.00	3.30
55	1.80	1.90	2.00	2.20	2.40	2.60	2.80	3.80	3.10

3.8

3.7 Schematic diagram of a heat pump used for heating purposes
a heat source (e.g. earth pile, ground collector, groundwater or outdoor air)
b heat exchange through water or brine/refrigerant (refrigerant evaporates)
c compressor: refrigerant is heated by compression (energy supply)
d heat exchange through refrigerant/water (heat output into heating circuit)
e expansion valve: expansion of the refrigerant (liquefaction and cooling)
3.8 Efficiency of heat pumps: approximate mean coefficient of performance (COP) for different heat sources and temperatures
3.9 Global warming potential of different refrigerants for heat pumps in relation to carbon dioxide (see figure 3.27, p. 101)

Global warming potential (GWP) of refrigerants in comparison to CO_2

Abbreviation	Name	Substance group	GWP
R 12	dichlorodifluoromethane	CFC	6640
R 134 a	1, 1, 1.2 tetrafluoroethane	HCFC	1300
R 404 A	mixture of various HCFCs	HCFC	3260
R 407 C	mixture of various HCFCs	HCFC	1530
R 410 A	mixture of various HCFCs	HCFC	1730
R 290	propane	CFC-free	3
R 600 a	butane	CFC-free	3
R 744	carbon dioxide	CFC-free	1
R 717	ammonia	CFC-free	0
R 1070	propene	CFC-free	3

3.9

of the entire system including the electricity consumption of fans for air-to-water heat pumps or the circulating pumps for groundwater or ground heat pumps.
The so called annual performance factor β is the better value to perform energy assessments. The APF compares the amount of electricity required to operate the respective system per month with the thermal energy supplied by the system. Naturally, the annual performance factor fluctuates in accordance with the temperature of the heat source and the desired temperature of the heating system.
Whereas, in the case of heat pumps that use ground water (10 °C) or ground heat (2 °C) as their heat source, the temperature remains fairly constant, the situation when using fresh air as the heat source is very different. The temperature of the heat source, air, fluctuates between -12 and + 30 °C degrees. Hence, the different performance factors of air heat pumps independent of the desired target temperature.
No matter which heat source is selected, a low forward flow temperature in the heating system is a fundamental prerequisite for the efficiency of a heat pump. A low forward flow temperature will only suffice if the heat demand is low and the heat distribution is performed via surface systems (floor or wall heating). In some circumstances, it may be possible to install either very large conventional radiators or innovative low energy fan heaters as an alternative to a surface heating system.
As regards the supply of domestic hot water, it is equally inefficient, from an energy viewpoint, to heat water to 60 °C only by using a heat pump. In this case, it makes sense to install solar thermal collectors to support the generation of heat. The period during which the heat pump is then used to heat water is reduced significantly. A further way of achieving higher energy efficiency, is to use decentralised fresh water modules in the hot water system. These reduce the target temperature and thus increase the annual performance factor of the heat pump (see Domestic hot water, pp. 97 ff.).
The means used to keep the circulation process in the heat pump going are however fraught with problems from an environmental perspective. If the gases leak, there is a high risk of causing damage to the environment. It is therefore important, when selecting a heat pump, to check whether environmentally friendly operating resources can be used (fig. 3.27, p. 101).

Combustion-based combined heat and power plant
The heart of a combined heat and power (CHP) plant is the engine or a gas turbine that drives an electric generator. The electricity generated is either used at the place of its origin, i.e. in the building, or fed into the mains network; the waste heat is used for heating the building. Normally, a CHP plant provides the base loads of the heating, whereas an additional boiler covers the peak loads. For economic reasons, CHP plants should only be used if heat is also required, and not only for generating electricity. Furthermore, they should, if at all possible, be used throughout the year. CHP plants are best suited and work to their full capacity in local heating systems where the heat demand is fairly even.

Hydrogen-based combined heat and power plant
In fuel cells, hydrogen and oxygen, which are separated by an electrolyte, are combined electrochemically to form water. The reaction induces an electron flux and the fuel cell generates electricity. In the same way as in a CHP plant, the waste heat developed is used to heat the building (combined heat and power). The required hydrogen can either be produced from water, by way of electrolysis, or by extracting it from natural gas. If the electrolysis process is performed using electricity from solar panels, hydrogen can be generated without using finite resources. Nowadays, it is possible to store hydrogen generated by solar power. Nevertheless, fuel cell technology is still in the infancy stage.

Renewable energy – biomass
The combustion of biomass, generally wood, achieves a very positive rating in the relevant standards since a regrowable resource is being used and therefore little primary energy is consumed (merely for production and transportation). In terms of CO_2 emissions, the combustion of biomass almost works out emission-free, because, for its growth, biomass extracts the same amount of CO_2 from the environment as is then released during its combustion. Nevertheless, the combustion of wood must be questioned very carefully for a number of reasons. Wood is an excellent CO_2 store; it is not at all clear why this store is dismantled during combustion. It would in fact be better to retain the CO_2 store for as long as possible and find other uses for wood, for example as a construction material

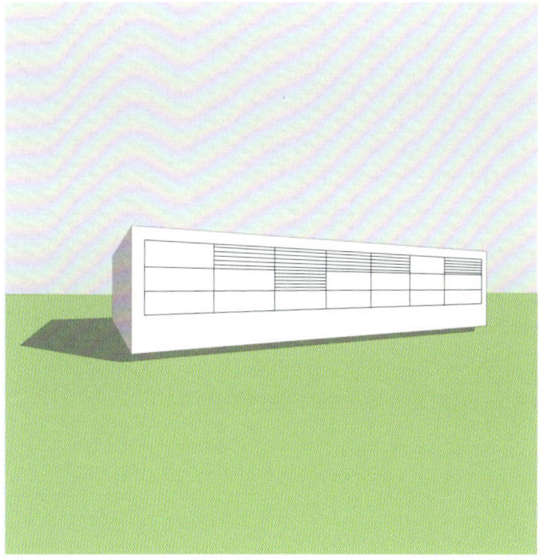

Example: heat pump
The entire energy supply (heating and hot water) in this building is to be performed with an air-to-water heat pump. The annual performance factor (APF) of the heat pump is calculated according to VDI 4650.

Input data
Coefficient of performance (COP) of the heat pump (as specified by the manufacturer; S = source/ T = target temperature)

ε_{N1} at S–7/T35	3.00
ε_{N2} at S2/T35	3.50
ε_{N3} at S10/T35	4.40

Correction factor for forward flow temperature of 55 °C and outdoor temperature of –16 °C (see table below) 1)
$F_{\vartheta 1}$ 0.180
$F_{\vartheta 2}$ 0.564
$F_{\vartheta 3}$ 0.147

Consideration of the temperature difference ($F_{\Delta\vartheta}$) in the liquefier (VDI 4650, table 1)

$\Delta\vartheta_m$ at the test stand	5 or 10 K
$\Delta\vartheta_b$ on the specific project	calculated
$F_{\Delta\vartheta}$ (simplified)	1.000

α	proportion of coverage achieved by the heat pump in terms of heat generation for heating and hot water (in a mono energy, monovalent or bivalent operation mode with different energy media)	1.00
x	proportion of heat demand	0.82
y	proportion of the heat demand for hot water	0.18

Annual performance factor for heating mode

Equation:
$\beta_h = (\varepsilon_{N1} \cdot F_{\vartheta 1} + \varepsilon_{N2} \cdot F_{\vartheta 2} + \varepsilon_{N3} \cdot F_{\vartheta 3}) \cdot F_{\Delta\vartheta}$

β_h = APF for heating mode = 3.16

Annual performance factor: hot water production

Equation:
$\beta_w = (\varepsilon_{N1} \cdot 0.103 + \varepsilon_{N2} \cdot 0.903 + \varepsilon_{N3} \cdot 0.061) \cdot F_{\Delta\vartheta}$

β_w = APF for hot water production = 3.74

Total annual performance factor

Equation:

$$\beta_{hp} = \frac{1}{x \cdot \frac{\alpha}{\beta_h} + y \cdot \frac{\alpha}{\beta_{hp}} + 1 - \alpha}$$

β_{hp} = total APF= 3.25

Correction factors according to VDI 4650, table 4 a
(for forward flow temperature of 55 °C, TL = 15 °C)

Operation mode	Standard outdoor temperature [°C]			
	-10	-12	-14	-16
$F_{\vartheta 1}$ (S–7, T35)	0.051	0.081	0.109	0.180
$F_{\vartheta 2}$ (S2, T35)	0.635	0.635	0.626	0.564
$F_{\vartheta 3}$ (S10, T35)	0.217	0.179	0.160	0.147

Furthermore, the combustion of wood, especially in the case of small sized plants, is responsible for a higher fine particulate air pollution. The use of wood as a fuel should therefore be restricted to larger plants in local heating systems, which are equipped with efficient fine particulate air filters. This is the only effective means of emission control. Finally, the combustion of biomass can only be justified if the full process is based on a sustainable concept. Uncontrolled deforestation is just as unreasonable as the use of valuable ground to produce biomass.

Renewable energy – environmental heat
Heat that regenerates itself can be extracted from the environment with the help of heat pumps. However, the heat in groundwater, in the ground or air can only be used for heating purposes by investing a considerable amount of auxiliary energy. In order to utilise the heat extracted from the environment to heat a building or produce domestic hot water, the temperature has to at first be increased to approximately 30–60 °C by installing electricity powered heat pumps. Solutions to reduce the required amount of auxiliary energy are therefore of prime importance for an energy efficient design in particular when this type of heat generation is selected.

Renewable energy – solar thermal plants
The sun provides a lot more energy than we need. South-facing windows make passive use of solar radiation to heat the building interior. By installing a solar thermal plant in conjunction with storage media, the solar energy can also be used actively, i.e. irrespective of the structural concept. Here, the solar energy is absorbed by collectors and transported via a water circuit to a storage tank.
Solar thermal plants are primarily used for providing hot water. At today's energy prices, it is recommended, from an economic viewpoint, to cover up to 70 % of the hot-water demand with solar thermal energy. Owing to the high requirements in terms of storage and control mechanisms, the more intensive use of solar thermal energy in conjunction with a seasonal thermal storage system to provide space heating is not yet economically feasible and therefore less frequently encountered in practice. However, backup for the heating system can be achieved fairly cheaply by simply increasing the size of the buffer storage tank for the solar-powered hot water system. Solar thermal plants can also be retrofitted fairly easily in the case of existing heating systems.
The solar gain from collectors in accordance with the orientation and inclination of the collectors can be calculated by using the table of solar radiation included in DIN 4108-6 for 35 regions in Germany (fig. 3.11). Due to the extreme plant loss concerning the distribution and storage of

Comparative analysis of heating in a residential building

	System efficiency [-]	Final energy [kWh]	Resource	Primary energy [kWh]	Global warming potential [kg CO_2 equiv.]	Costs/ year [€]	Costs in 20 years [€]	Investment 2) [€]	Annuity [€]	Total [€]	Comment
Heat generation: combustion											
Constant temperature boiler											
manufactured before 1985	1.3	13,000	gas	14,300	3159	1040	38,257				
Low temperature boiler											
manufactured before 1995	1.2	12,000	gas	13,200	2916	960	35,314				
manufactured after 1996	1.15	11,500	gas	12,650	2795	920	32,371				
manufactured 2010	1.1	11,000	gas	12,100	2673	880	32,371	6000	8890	41,261	
Condensing boiler	0.14										
manufactured before 1995	1.05	10,500	gas	11,550	2551	840	30,900				
manufactured after 1996	1	10,000	gas	11,000	2430	800	29,428				
manufactured 2010	0.95	9500	gas	10,450	2309	760	27,917	8000	11,773	39,690	
Wood pellet boiler											with store and buffer
manufactured 2010	1.2	12,000	biomass	2400	480	500	18,329	16,000	23,546	41,875	storage tank
Gas motor heat pump											with buffer storage
manufactured 2010	0.75	7500	gas/renew. 1)	8250	1822	600	22,071	12,000	17,659	39,730	tank
Zeolite heat accumulator											with 10 m² of solar
manufactured 2010	0.6	6000	gas/solar	6600	1458	480	17,657	20,000	29,434	47,091	thermal collectors
Heat generation: electricity											
Direct electricity											electric radiators
manufactured 2010	1	10,000	electricity	26,000	6830	2200	80,928			80,928	
Air heat pump											with buffer storage
manufactured before 1995	0.55	5500	elec./renew. 1)	14,300	3756	770	28,324				tank
manufactured after 1996	0.5	5000	elec./renew. 1)	13,000	3415	700	25,749				
manufactured 2010	0.45	4500	elec./renew. 1)	11,700	3074	630	23,174	14,000	20,602	43,776	
Ground heat pump											200-m-deep earth
manufactured before 1995	0.45	4500	elec./renew. 1)	11,700	3074	630	23,174				piles
manufactured after 1996	0.4	4000	elec./renew. 1)	10,400	2732	560	20,599				
manufactured 2010	0.35	3500	elec./renew. 1)	9100	2390	490	18,024	22,000	32,357	50,381	
Groundwater heat pump											production and
manufactured before 1995	0.35	3500	elec./renew. 1)	9100	2390	490	18,024				re-injection well
manufactured after	0.3	3000	elec./renew. 1)	7800	2049	420	15,449				
manufactured 2010	0.25	2500	elec./renew. 1)	6500	1708	350	12,874	22,000	32,357	45,231	
Combined heat and power											
Motor											with buffer storage
heat generation		20,000	gas	6400	762	940	34,578	22,000	32,357	66,935	tank and peak load
excess electricity credit		6000									boiler
Fuel cell											non-renewable
heat generation		18,000	gas	4200	276	780	28,693	40,000	58,865	87,558	resource
excess electricity credit		6000									

1) The renewable energy is generated with the help of electricity/gas (auxiliary energy). 2) for heat generator

3.10

solar energy, only approximately 50 % of the radiation energy reaching the collector can actually be used for the generation of heat. If, for example, 113 kWh of solar energy reach a south-facing collector panel measuring 1 m², mounted at a 30° angle, approximately 55 kWh only remain for the generation of heat. The dilemma as regards the use of collector heat for heating purposes results from the time lag between heat generation and heat demand.

Flat plate and tube collectors

The collectors are the centrepiece of a solar thermal plant. Black absorbers collect the solar radiation and heat up the water. In the case of flat plate collectors, the absorber sheets are fitted in a flat, rectangular housing and covered with a sheet of glass (fig. 3.21, p. 98). Vacuum tube collectors are composed of multiple evacuated glass tubes each containing an absorber plate (fig. 3.22, p. 98). The vacuum ensures better insulation and hence improves the efficiency of the collectors. In both cases, the absorber plates can be adjusted within the collector housing to achieve optimum heat input even if the collectors are not facing south.

Air collectors

Air collectors work only in conjunction with a ventilation system. The outside air is drawn into a solar collector consisting of an array of aluminium tubes fitted beneath a sheet of glass. Once the air is heated, it is either supplied to the rooms directly or, if required, heated further in a heat exchanger incorporated in the ventilation system by using the heat extracted from the waste air.
If the heat is not required for ventilation, it can also be used to produce hot water with an air-to-water heat pump. In conjunction with the principle of adsorption, air collectors can also be used to provide heating and cooling (see Cooling, pp. 100ff.).

Heat distribution

When it comes to the distribution of heat, a distinction is made between horizontal distribution pipes, vertical distribution pipes and connection pipes. The connection pipes are the pieces between the vertical distribution pipes and the radiator or the distributor of the underfloor heating system.
To avoid energy loss, all of these pipes must be insulated in accordance with the requirements of the EnEV, enclosure 5

Solar insolation for Würzburg (reference climate for Germany)

Orientation		**Average monthly radiation intensity in [W/m²] and monthly amount of energy in [kWh/m²]**												**Annual available solar radiation [kWh/m²]**	**Value for heating period [kWh/m²]**
		Jan	Feb	Mar	Apr	May	Jun	Jul	Aug	Sep	Oct	Nov	Dec	Jan–Dec	Oct–Mar
Horizontal	0°	33 24.6	52 34.9	82 61.0	190 136.8	211 157.0	256 184.3	255 189.7	179 133.2	135 97.2	75 55.8	39 28.1	22 16.4	1120.0	225.0
South	30°	51 37.9	67 45.0	99 73.7	210 151.2	213 158.5	250 180.0	252 187.5	186 138.4	157 113.0	93 69.2	55 39.6	31 23.1	1216.0	295.0
	45°	57 42.4	71 47.7	101 75.1	205 147.6	200 148.8	231 166.3	235 174.8	178 132.4	157 113.0	97 72.2	59 42.5	34 25.3	1187.0	310.0
	60°	60 44.6	71 47.7	98 72.9	190 136.8	179 133.2	203 146.2	208 154.8	162 120.5	150 108.0	95 70.7	60 43.2	35 26.0	1104.0	310.0
	90°	56 41.7	61 41.0	80 59.5	137 98.6	119 88.5	130 93.6	135 100.4	112 83.3	115 82.8	81 60.3	54 38.9	33 24.6	810.0	270.0

3.11

Insulation of hot water and heat distribution pipes

Line	**Type of pipe/fitting**	**Minimum thickness of insulation in relation to a thermal conductivity of 0.035 W/(mK)**
1	internal diameter up to 22 mm	20 mm
2	internal diameter between 22 mm and 35 mm	30 mm
3	internal diameter between 35 mm and 100 mm	equal to the internal diameter
4	internal diameter above100 mm	100 mm
5	pipes and fittings as in lines 1 to 4 in wall and ceiling penetrations, at intersections, at connection points, in the case of central pipe network distribution systems	½ the requirements of lines 1 to 4
6	pipes in central heating systems as in lines 1 to 4 that were installed after 31 January 2002 in building components between heated rooms with different users	½ the requirements of lines 1 to 4
7	pipes as in line 6 incorporated in floor structure	6 mm
8	cold distribution pipes, cold water pipes and fittings in air handling and air conditioning systems	6 mm

3.12

Annual heat loss through distribution [kWh]

Forward flow/return flow [°C]	**Without insulation**	**Insulation according to EnEV, enclosure 5**	**Double insulation thickness**
90/70	73,108	16,409	10,970
70/55	53,742	12,020	8045
55/45	39,877	8872	5978
35/28	19,398	4229	2926

Energy balance of building (living area 2800 m²):
energy demand for heating according to EnEV standard: 150,000 kWh
energy demand for heating in the existing building (completion year 1965): 450,000 kWh

3.13

3.10 Comparison of different plant technologies to cover the heat demand (10,000 kWh/year) in a building. The energy demand incorporates the useful energy demand (heating) as well as the energy loss caused by distribution and transmission. The system is operated at a temperature of 50/40 °C.
Assumptions: Costs for gas 0.08 ct/kWh, electricity 0.22 ct/kWh, power for heat pump 0.14 ct/kWh. Life expectancy of the system is 20 years, financing at a 4 % interest rate, price increase of 5 %

3.11 Solar insolation for Würzburg (EnEV's reference climate for Germany) according to DIN 4108-6:2003-06. The top numbers in each line show the average monthly solar radiation for a number of different inclinations. If this value is multiplied by the number of hours per month (e.g. 31 days × 24 h/day), the result is the amount of energy provided by the sun per month (bottom value in each line).

3.12 Insulation of heating pipes, hot water pipes, cooling pipes, cold water pipes as well as all the fittings according to EnEV, enclosure 5

3.13 Annual heat loss through distribution in a nine-storey residential building with 36 dwelling units (living area: 2800 m²) using different insulation standards of the distribution lines (200 m of horizontal distribution pipes in the basement, 200 m of vertical distribution pipes and 200 m of horizontal connection pipes to the individual radiators)

Energy loss through heat generation, distribution and storage

Description	Heating circuit temperature	Age of building	System performance factor – Useful area acc. to EnEV [m²] 150	500	2500
1. Heat generation for heating					
a) Central heating system					
Constant temperature boiler	70/55 °C	before 1986	1.47	1.36	1.28
		1987–1994	1.34	1.26	1.19
		after 1995	1.33	1.23	1.16
Low temperature boiler		before 1986	1.24	1.21	1.18
		1987–1994	1.19	1.15	1.13
		after 1995	1.14	1.11	1.09
Condensing boiler		before 1986	1.11	1.09	1.07
		1987–1994	1.09	1.06	1.04
		after 1995	1.07	1.05	1.04
Improved condensing boiler	55/45 °C	after 1999	0.99	0.98	0.97
District heat transition system	all	all	1.02	1.02	1.02
Electric heat pump, outdoor air	55/45 °C	1979–1994	0.41	0.41	0.41
		after 1995	0.39	0.39	0.39
Electric heat pump, ground		1979–1994	0.33	0.33	0.33
		after 1995	0.27	0.27	0.27
Electric heat pump, groundwater		1979–1994	0.27	0.27	0.27
		after 1995	0.23	0.23	0.23
Central electric heating system (block storage heater)	all	all	1.02	1.02	–
b) Decentralised heating system					
Water heater (circulation heating system)	all	before 1994	1.24	1.24	1.24
		after 1995	1.14	1.14	1.14
Gas condensing water heater		after 1995	1.07	1.07	1.07

3.14

Description	Heating circuit temperature	Age of building	Heat loss [kWh/(m² · a)] – Useful area acc. to EnEV [m²] 150	500	2500
2. Heat output					
Central heating system, thermostat controlled	all	all	3.3	3.3	3.3
Single furnace	–	all	0	0	0
3. Heat distribution					
Central distribution	70/55 °C	before 1978	75.1	43.5	32.7
		before 1978, insulation retrofitted	40.9	28.2	23.9
		1979–1994	20.2	13.8	11.6
		after 1995	9.3	5.4	4.1
Central distribution	55/45 °C	before 1978	57.4	32.9	24.4
		before 1978, insulation retrofitted	30.8	21.0	17.6
		1979–1994	15.3	10.3	8.5
		after 1995	9.3	3.9	2.9
Distribution per unit	all	before 1978	8.4	8.4	8.4
		1979–1994	5.4	5.4	5.4
		after 1995	1.3	1.3	1.3
Decentralised system (without distribution)	–	all	0	0	0
4. Heat storage					
Electric central storage	70/55 °C	before 1994	3.2	1.3	0.4
		after 1995	2.5	1.0	0.3
Buffer storage tank for electric heat pump	55/45 °C	before 1994	2.5	1.0	0.3
		after 1995	1.9	0.7	0.2
Buffer storage tank for wood burner	70/55 °C	before 1994	4.4	1.8	1.1
		after 1995	3.4	1.4	0.8

3.15

3.14 Lump sum approach to energy loss in the case of heat generation systems
3.15 Lump sum approach to the energy loss in the case of heat output, distribution and storage
3.16 Approximate determination of storage tank sizes for the use of solar thermal energy

(fig. 3.12, p. 93). The heat loss caused by lack of insulation in the distribution system is considerable, particularly in existing building stock (fig. 3.13, p. 93).

Heat storage

The objective of heat storage is to separate the time of heat generation from the time of heat extraction and utilisation.
A distinction is made between short-term and seasonal storage, and between the type of storage media. Sensible heat storage media change their temperature when absorbing or releasing thermal energy; in the case of latent heat storage media, on the other hand, the aggregate state of the material changes. Water in its frozen state, for example, is a latent heat storage medium since it absorbs thermal energy upon melting without changing its temperature. Once the ice has melted, the water represents a sensible heat storage medium until it reaches its boiling point. Water is commonly used as a short-term, sensible heat storage medium in the space heating and hot water systems of buildings.
Thermal stratification sets in automatically in storage tanks. This phenomenon should be taken into account when introducing thermal energy, i.e. hot water should be fed into the top of the tank, cold water into the bottom.
The reference values listed in figure 3.16 can be used to determine the size of tanks for different storage types. The final size should however be calculated by a building services engineer in accordance with the selected system (hot water production, forward flow temperature of the heating system, size of solar collectors, solar coverage rate).
It is generally possible, in a well-insulated single-family home, to fully cover the demand for heating and hot water by using exclusively solar thermal energy. A 100 % coverage requires a collector area of approximately 80 m², a buffer storage tank with a capacity of 200 litres, a domestic hot water storage tank, also with a capacity of 200 litres, as well as a seasonal thermal storage tank with a capacity of 40,000 litres.

Domestic hot water storage tank

Instantaneous water heaters are only suitable for heating small amounts of water. Furthermore, the use of solar thermal energy for such local heating installations is not feasible. The tanks therefore hold only enough hot water ready to suit the number of residents/users and their demand for hot water.

Buffer storage tank

In non-modulating heating systems, buffer storage tanks have the effect of reducing the cycle frequency of the heat generator. Buffer storage tanks are therefore always required when wood or other types of biomass are used to generate heat. It also makes sense to install a buffer storage tank when using a heat pump.

In the case of all heat generators, the use of solar thermal energy to boost the heating system is only possible if a buffer storage tank has been incorporated. Both the heating system and the solar thermal collectors feed hot water into the buffer storage tank.

Seasonal thermal store

Seasonal thermal stores provide the facilities to store energy for longer periods, normally a whole season. The different types available include well-insulated water tanks, latent heat stores and the ground. Near surface geothermal energy, which can be tapped, for example, by using ground piles that reach down to a depth of 100 m, should, depending on the usage and the regeneration ability of the ground, not be considered as an energy source, but rather as an energy store. With the aid of a seasonal thermal store, the solar thermal energy generated in summer can be used in winter to provide substantial support to the heating.

Heat output

The aim of an energy efficient design must always be to achieve, in accordance with the structural standard of the building, a low forward flow temperature in regard of the heat output. The heat emission in the respective rooms can be achieved by means of radiators, a surface heating or warm-air system.

Radiators

Convectors, ribbed radiators and flat radiant panels are the most common radiators in existing buildings. They emit heat partly by way of radiation and partly by way of convection. Their efficiency depends on the size and the temperature of the hot water. If the refurbishment is used only as an opportunity to replace the heat generator and reduce the water flow temperature, the radiators must either be enlarged or replaced by radiators with a better heat output capacity. If however the heat load is reduced during the building upgrade by adding insulation to the building exterior, the water flow temperature can be decreased whilst retaining the original radiators.

Surface heating

This type of heat output expands across large areas. It is operated with low water temperatures and can be fitted in floors, walls, and, in rooms with sufficient height, in ceilings, too. Since the heated surfaces emit warmth evenly through gentle radiation, a very pleasant interior climate is generated. However, due to the time lag between receiving and emitting the heat, the system is considered sluggish and difficult to control. Surface heating systems are therefore primarily suited to permanently heated rooms in which the conditions remain constant. When using surface heating systems, the extent to which alterations to the interior layout or usage can be accommodated is limited. Radiators, on the other hand, can adapt better to changes, either by being replaced or relocated, or by resetting the water temperature accordingly.

Since surface heating systems can only be operated with low forward flow temperatures, their use in refurbishment projects is only recommended if the heat demand is reduced at the same time.

Warm air heating

The temperature of air that is blown into a room must not exceed 45°C. Because air has only a limited heat absorption capacity, high air change rates and large duct cross-sections are required to supply large quantities of heat. Owing to the

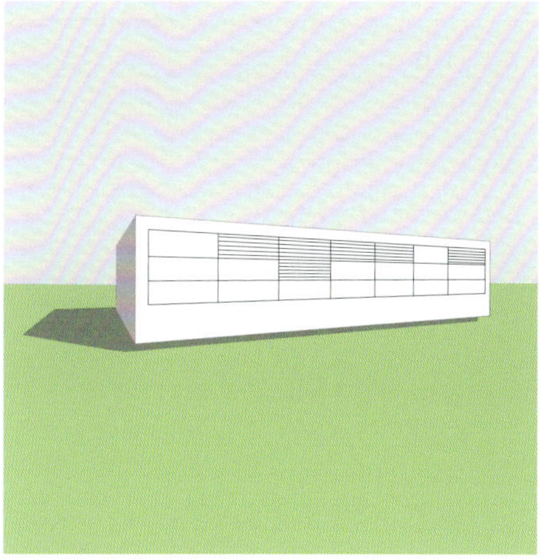

Example: heat output

The heat output can be performed using three different methods:

- waterborne with radiators
- waterborne with a surface heating system
- airborne using the ventilation system

The heat load is determined for each room individually according to the coldest day (here one room is used as an example). The exact dimensioning of the heat emitting components is performed by a building services engineer.

Initial situation

- coldest day in Munich -12°C
- desired room temperature 20°C
- floor area of the examined room 120 m^2
- opaque envelope area 357 m^2
- transparent envelope area 38 m^2
- $\bar{U}_{opaque}$ 0.3 W/m^{2}K
- $\bar{U}_{transparent}$ 1.4 W/m^{2}K
- average air change rate 0.7/h
- objective: low forward flow temperature

Heat load

- transmission heat loss
  0.3 W/m^{2}K · 357 m^2 · 32 K = 3427 W
  1.4 W/m^{2}K · 38 m^2 · 32 K = 1702 W
  Total: 5129 W
- ventilation heat loss
  360 m^3 · 0.7/h · 0.34 Wh/m^{3}K · 32 K = 2741 W
- total heat loss: 7870 W
- specific heat loss: 65 W/m^2

Radiators

- installation below windows
- 7 radiators each w 200 cm x h 50 cm
- FF/RF 55/45°C
- three columns with 43 sections (d = 6.2 cm)
- output per radiator 1161 W (27 Watt/section)
- 7 radiators 8127 W

The required heat output of 7870 W can be achieved by these radiators.

Surface heating (underfloor heating system)

- forward flow/return flow 45/35°C
- floor covering parquet flooring
- pipe distance 20 cm
- heating surface 18 m^2

Air heating

- maximum air supply temperature 40°C
- required air change rate in order to balance the transmission heat loss with the overtemperature of 20°C:
  5129 W = 360 m^3 · x/h · 0.34 Wh/m^{3}K · 20 K

Result: x = 2 (double air change rate)

Comment

The necessary high air change rate is responsible for a considerable increase of the ventilation unit's size and a much higher electricity demand for the fans. Not until the heating load is reduced to approximately 2500 Watt by insulating the building, is it worthwhile, from an energy viewpoint, to distribute the heat via air.

Comparison of storage systems

Store/ store size	Reference value	Costs (delivery and assembly)	Comment
Domestic hot water			
Domestic hot water (without solar)	60 l/person	€2/l	specification for 5 persons
Domestic hot water (with solar)	80 l/person	€3/l	
Buffer store (heating)			the m^2 specification of the reference value refers to the size of the collector area
Buffer store (without solar)	10 l/kW	€2/l	
Buffer store (with solar)	50 l/m^2	€3/l	
Seasonal store			
20 kWh/m^{2}a heating backup	300 l/m^2	€0.20/l	the m^2 specification of the reference value refers to the size of the collector area
40 kWh/m^{2}a heating backup	400 l/m^2		
60 kWh/m^{2}a heating backup	500 l/m^2		

3.16

amount of plant and energy that is required to move air, the air volume should be limited to the amount absolutely necessary in terms of hygiene.
For all of these reasons mentioned, warm-air systems should only be used in well-insulated buildings with very low heat demand, no higher than 20 W/m². Warm air systems have a clear advantage in low energy buildings:
a single system is able to provide ventilation, which is recommended or even necessary in these airtight buildings, and heating simultaneously.

Energy efficiency refurbishments

The energy that must be generated by a boiler is a conglomeration of the useful energy demand of the building itself and the energy lost through storage, transmission and distribution. Independent of the amount of energy resulting from this equation, the efficiency of the system used to provide the required amount of energy is a fundamental assessment criterion when selecting the heat generation system.
Depending on the boiler's construction, the amount of energy wasted during the heat generation process differs. It can be related to flue gas loss or heat loss caused by the poor insulation of the boiler. Frequent cycling (cold starts), which always involves a reduction of the boiler temperature and then the need to reheat it, is a further cause for unnecessary heat loss.
The efficiency of the heat generator is therefore always a primary issue to contend with when it comes to energy efficiency refurbishments. The system performance factor expresses the ratio of energy input (final energy consumption of the heat generator) to the thermal energy produced (heat demand).
If, for example, an old low temperature boiler with a system performance factor of 1.19 is replaced by a condensing boiler with a factor of 0.99, the amount of energy required by the building is reduced by approximately 20% simply by introducing this one measure (fig. 3.14, p. 94).
When exchanging a boiler without upgrading the heat distribution and output system, it is important to make sure that the boiler settings are adapted to the reduced heat load as soon as the requirements change, for example due to the installation of insulation. The energy efficiency of old heating plant that is still in working order and is to remain in place until the end of its life cycle can be improved by adding a solar thermal collector system. When eventually the boiler is replaced, the solar thermal collectors can be retained.

Heat pumps

The retrofitting of a heat pump is always doomed to failure in existing buildings if the aforementioned requirements of a low forward flow temperature are not achieved due to bad building fabric. It only makes sense to install a heat pump in these circumstances if the pump is supplemented by a peak load boiler, which is able to heat water quickly to the required heating forward flow temperature, or if the size of the heating surfaces is increased to allow a significant reduction of the water temperature.

Energy Saving Ordinance

According to the German Energy Saving Ordinance (Energieeinsparverordnung, EnEV), paragraph 10 and 10a, concerning heating plant, the following retrofitting measures or modifications are obligatory:

- All water distribution pipes without insulation must be insulated in accordance with the requirements listed in enclosure 5 of the EnEV (fig. 3.12, p. 93). The transition period for this upgrade measure has expired.
- All electric storage heating systems installed before 1990 (electric direct heating) may only be operated until 2019. All appliances installed thereafter may be used for a maximum of 30 years.

Hydraulic balancing

Heating systems can only be operated in an energy efficient manner if the separate components, i.e. the heat generator, heat distribution and heat output system, are set according to calculations performed by a heating specialist and adjusted to achieve a sensible balance between the individual parts. The following aspects should be considered:

- accurate and precise determination of the forward and return flow temperature with the objective of lowering the overall temperature level
- exact calculations to identify the amount of circulating water. These are necessary to establish accurately the pump performance and the heat output at the radiator.
- coordination of the pipe sizes to match the amount of water and avoid flow generated noise

Whereas these calculations are common everyday practice in the case of new heating systems (DIN 18380, DIN EN 12828), their application to older plant is often limited due to a lack of knowledge concerning all heating components. This is particularly the case if the system is only partially upgraded, for example only the boiler is replaced, or the heat load is reduced after installing insulation. The hydraulic balancing of existing plant should, however, incorporate technical calculations concerning the entire heating system, since the objective is to operate the system efficiently upon the newly established basis of having performed construction-related and technical upgrading work. The measures undertaken should include the identification of new forward and return flow temperatures, in consideration of the existing or replaced radiators, as well as improvements to the distribution system by selecting a demand-oriented, highly efficient circulation pump. Once the system is operating under balanced pressure conditions, fine adjustments can be made to the heat flow settings at the thermostatic valves, and the predetermined temperature can thus be obtained in the respective space.
The improvements that can be made to an existing heating system through hydraulic balancing can bring about energy savings ranging between 10 and 20%. The fees for planning these measures can be determined according to the HOAI (German Official Scale of Fees for Services by Architects and Engineers). Since the calculations concern an existing plant, the chargeable costs relevant for the fee calculation are based on the costs that would be incurred for the installation of a new plant of similar size. According to §53 and 54 of the HOAI in fee band II, the calculations for a heating system would lead to planning costs amounting to approximately 5 to 7% of the installation costs (chargeable costs).

Plant energy loss

Plant energy loss in the form of an inefficient heat generator or exposed and uninsulated water pipes, storage tank heat loss or energy loss due to a lack of possibilities concerning the control of heat output in the rooms are the cause for considerable inefficiency. The objective of an energy efficiency refurbishment should be to reduce these different forms of energy loss as much as possible (fig. 3.15, p. 94).
The upgrade of technical installations is frequently the most effective refurbishment measure, in fact sometimes even more so than replacing windows and insulating the building envelope.

Domestic hot water

The basis for calculating the heat demand for the hot water production is the volume of hot water required per person and day. The estimated amount for residential buildings is approximately 20–25 litres of hot water (60 °C) per person and day. In the case of non-residential buildings, the hot water demand is very much dependent on the use of the building (fig. 3.17). The amount of energy required to heat one litre of water from 10 °C (temperature of the water in the supply pipe) to 60 °C (temperature of the water in the storage tank) can be calculated by considering a few details. The specific heat capacity of water is 4.19 kJ/kg or converted into Wh/lK 1.16. Hence, 58 Wh of energy are required to heat a litre of water by 50 Kelvin. In a month with 30 days and an average daily hot water demand of 25 litres per person, the heat demand to produce domestic hot water therefore amounts to 44 kWh. Depending on the selected type of water heater, the hot water production involves different forms of energy loss, which have a considerable effect on the energy demand of the overall system. In the case of all hot water distribution networks, it is a clear advantage from an energy efficiency point of view to provide a compact arrangement of all sanitary rooms or hot water outlets (fig. 3.18).

Central hot water system

In buildings with central heating, the generator for the provision of heat is usually also responsible for the production of hot water (fig. 3.20 b, p. 98). The advantage of this system is the ease of incorporating solar thermal energy into the heating process of the water. However, this configuration also leads to greater energy loss, in particular distribution heat loss during the circulation (long distances) and, depending on the quality of the heat generator, boiler energy loss. Estimates have shown that the plant energy loss when using a central hot water system is often as high as the energy actually required to heat the water.
The high energy loss caused by the distribution (circulation) and storage could be reduced significantly by lowering the water temperature to 40 °C. However, because this temperature is not high enough to avoid the growth of Legionella, further heating of the water should be performed in a decentralised fashion by heat exchangers, so called fresh hot water modules, that are fitted in

Water demand in different buildings

Type of building	Total water demand [l/day] 1)	Cold water demand [l/day] 1)	Hot water demand [l/day] 1)
simple residential building	90/140	65/80	25/60
average residential building	100/170	70/100	30/70
single family home	110/180	70/110	40/70
sophisticated residential building	140/230	90/150	50/80
mansion with large garden	180/280 and more	120/180 and more	60/100
hotel etc. per bed			
simple inn	60/100	20/30	40/70
hotel	90/130	30/40	60/90
upscale hotel	110/200	40/70	70/130
luxury hotel	150/350	70/200	80/150
hospital per bed			
150–300 beds	250/450	200/340	50/110
300–600 beds	300/500	240/380	60/120
600–1000 beds	400/600	320/470	80/150
old people's home per resident	100/150	70/90	30/60
children's home per child	100/130	60/80	40/50
school (without pools) per pupil	5/10	5/10	–
barrack per person	100/150	70/90	30/60
office building per employee	25/35	15/20	10/15
department store per employee	25/50	20/40	5/10
restaurant			
per seat	30/50	15/20	15/30
very busy	50/80	20/30	30/50
swimming pool per visitor			
swimming pool	140/200	70/120	50/80
washing facilities/shower bath	40/90	15/40	25/50
washing facilities/public bath with tubs	280/370	150/180	130/190

1) the first figure specifies the average, the second figure the highest water demand

3.17

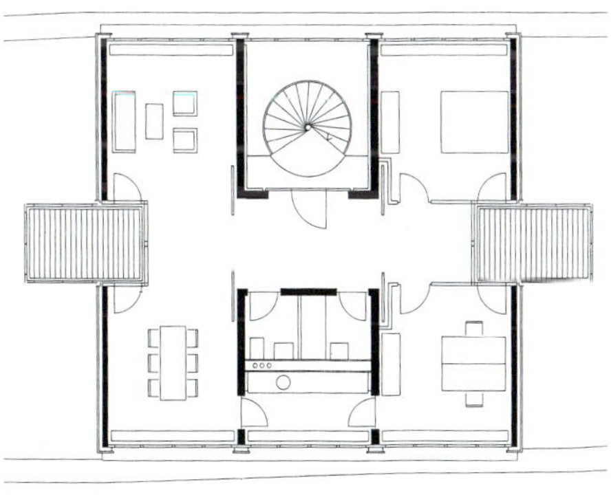

a

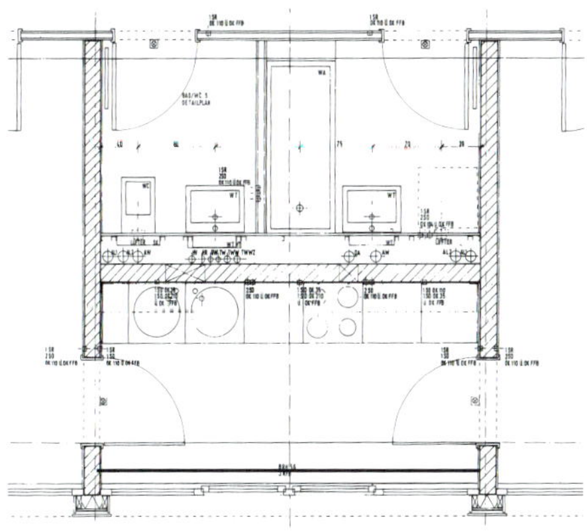

b

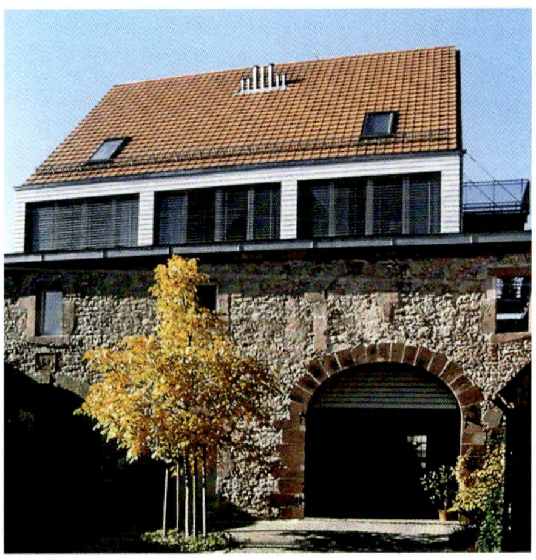

c

3.18

3.17 Water demand for buildings with different uses, per person and day
3.18 Reduction of distribution heat loss by arranging the water outlets in a compact core
Construction of a multi-family dwelling within the walls of a former tithe barn in Weil der Stadt
Year of completion: 1995
Architects: Richarz Schulz, Munich
a floor plan, scale 1:300
b floor plan of the sanitary core, scale 1:100
c elevation

Comparative analysis of hot water production systems in a multi-family dwelling

	Central hot water system, condensing boiler (gas) with domestic hot water storage tank	Decentralised hot water system, one electric instantaneous water heater per dwelling unit
Distribution / generation		
distribution pipes in the basement	200 m	–
vertical distribution pipes	200 m	–
connection pipes	360 m	360 m
Energy demand		
useful energy	40,000 kWh	40,000 kWh
distribution	25,000 kWh	–
storage	5000 kWh	–
boiler energy loss	5000 kWh	–
Final energy demand	75,000 kWh	40,000 kWh
Environment		
primary energy for hot water production	82,500 kWh	104,000 kWh
primary energy for production /year	2000 kWh	500 kWh
total primary energy	84,500 kWh	104,500 kWh
CO_2 from hot water production	18,225 kg	25,720 kg
CO_2 from production	300 kg	70 kg
total CO_2	18,525 kg	25,790 kg
Costs		
costs for distribution (incl. fire protection)	€40,000	€10,000
proportion of costs for central water supply	€20,000	–
cost of individual modules	–	€36,000
cost of construction work	€300,000	€70,000
cost of rent reduction	€20,000	€5000
Total costs of investment	€380,000	€121,000
Costs for the provision of water	€6000	€8000
Renewable energies		
partial coverage	70 % coverage of the final energy demand by installing 120 m² of solar thermal collectors	70 % coverage of the final energy demand by installing 300 m² of photovoltaic panels
100 % coverage	100 % by using biogas	100 % by using green power
Others		simple accounting; refurbishment of individual apartments is possible; most reasonable solution for the investor

Basis: nine-storey residential building with 36 dwelling units and a total living area of 2900 m², four apartments per level, water distribution using four vertical, stainless steel pipes
Gas costs €0.08/kWh, electricity costs €0.20/kWh, life expectancy of the systems/observation period: 20 years

3.19

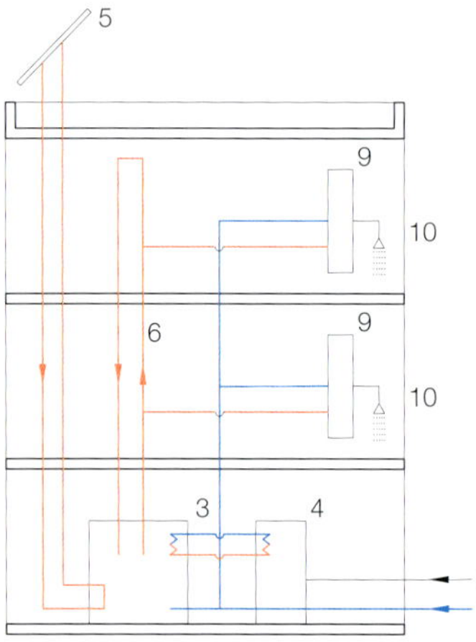

a

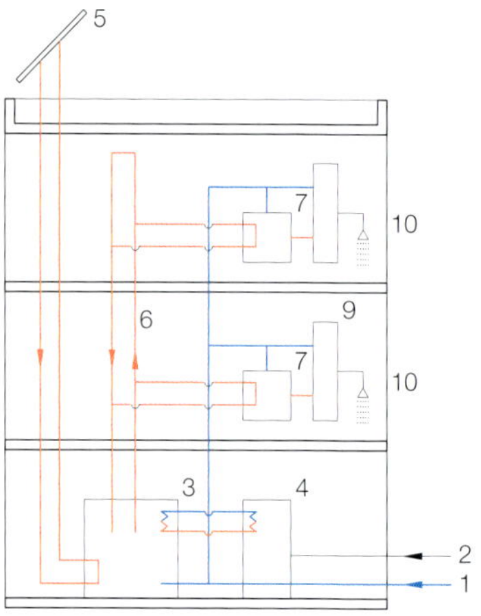

b

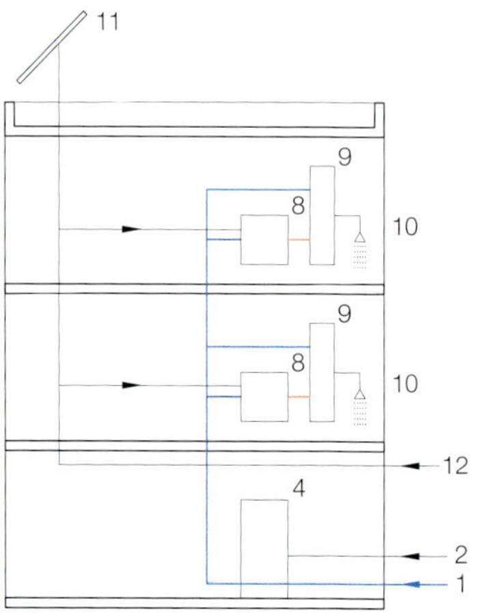

c

3.20

3.19 Comparative analysis of centralised and decentralised hot water production systems
3.20 Schematic diagrams of water supply systems
a centralised system with hot water recirculation
b centralised system with hot water recirculation and one fresh water module per dwelling unit
c decentralised hot water supply system
1 water
2 gas
3 storage tank
4 heating
5 solar collector
6 recirculation
7 fresh water module
8 instant. water heater
9 fitting
10 hot water
11 photovoltaic panel
12 electricity
3.21 Flat plate collector
3.22 Vacuum tube collector

the individual dwelling units. The lower temperature of the hot water allows much more efficient use of solar thermal energy as well as a reduction of electricity consumption when using a heat pump.
By making use of innovative concepts in larger apartment blocks, it is also possible to extract the remaining heat from waste water. In these circumstances, it is necessary to separate water with fecal matter from water without. The heat from the fecal-free, 25 to 30 °C-warm water is extracted by using a heat pump and reintroduced into the domestic hot water storage tank. If the hot water production system is operating at a low temperature level (e.g. 45 °C), this type of heat recovery is particularly effective, and there is no need to install a solar thermal collector plant.
The separation of waste water also has a beneficial effect on the reduction of fresh water and waste water volumes, since the fecal-free water, after having been purified, can either be discharged into a seepage pit or reused for toilet flushing.

Decentralised hot water production

Decentralised hot water production is either performed by using electricity or by making use of a single-unit gas heating system (fig. 3.20 c). Modern instantaneous water heaters with an output of up to 20 litres of 60-degree-hot water per minute enable two persons to have a shower at the same time.
The energy demand, in comparison to a central hot water system, is reduced by 50 % simply because there is no need for an elaborate distribution and storage system. By connecting the decentralised heat generator to the electricity grid with the pay-as-you-go principle, the time and effort required to meter and calculate the energy costs for the hot water production are also no longer an issue.
The disadvantage of gas-operated instantaneous water heaters is the lack of opportunity to incorporate renewable energy sources. In the case of electricity-powered instantaneous water heaters, it is possible to cover the total energy demand for the production of domestic hot water without the use of finite resources, instead using energy generated from renewable resources.
If this electricity is produced at the building itself, the amount can be considered positively in the energy efficiency assessment performed according to EnEV.

Furthermore, in contrast to average systems, approximately 4 kg of steel per m² of useful living area are saved since there is no need for a distribution network and storage tank. This in turn leads to a reduction of the primary energy demand in the production and the associated CO_2 emissions of approximately 4000 kg. This reduction, converted according to the life expectancy of the plant technology, should be considered when comparing the differences between centralised and decentralised hot water production systems (fig. 3.19).

Energy efficiency refurbishments

The view still taken frequently by technical experts, including the EnEV, that it makes little sense from an energy point of view to heat water with electricity should indeed be judged very critically. Especially in the case of refurbishments, the installation of a new central hot water production system often involves high expenditures, in particular if the work is to be performed whilst the building is in use. Furthermore, the full replacement of the heating system can only be performed for the whole building in a single operation, whereas a decentralised concept allows individual apartments to be upgraded one at a time. If a decentralised system is chosen, it may be necessary to only replace the decrepit cold water pipes. Since Legionella growth is not an issue in decentralised systems due to the short pipe runs, the water no longer has to be heated above 40 °C. If buildings are not in use during the refurbishment, it makes sense to install a central hot water production system. The water outlets should, in this case, be arranged so that each apartment can be connected to a single vertical supply pipe.
Furthermore, when using a central supply system – irrespective of the building's structural quality – the integration of a solar thermal collector plant for the production of hot water is recommended. Between May and September, the necessary energy for the hot water production can be covered entirely by renewable resources so long as the area of the collectors amounts to approximately one square metre per resident.
Generally speaking, it is always advisable to use solar thermal energy for the generation of heat, if the heat is being used on the premises. This is the case for hot water production. If it is not possible to carry out a comprehensive energy efficiency refurbishment due to the age of the building and possible restrictions imposed by building authorities (listed building) or the building structure, solar thermal plants are an inexpensive way of contributing towards the reduction of resource consumption.
The additional heat required is then provided by the heat generator that is also responsible for the heating.
The installation of fresh hot water modules should be considered as an option to avoid plant energy loss. When it comes to detail planning, it is also necessary to decide whether it might make sense to install hot water connections for washing machines and possibly also dishwashers. A further aspect worth considering is the recovery of heat from waste water.
Even in the case of non-residential buildings with a low hot water demand, the decentralised arrangement of hot water production systems should always be the first choice. After all, it is also possible to integrate renewable energies in decentralised, electricity-powered systems by using so-called green power or by generating the electricity on the premises through the use of photovoltaic panels mounted on either the roof or the facade.

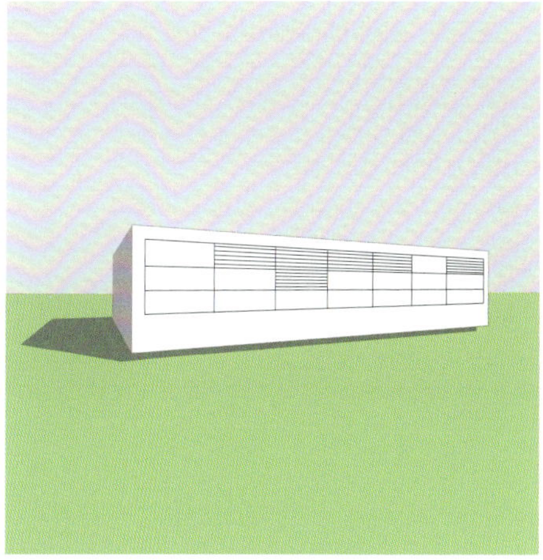

Example: use of solar thermal energy

The intention is, by using solar thermal energy, to support the hot water production system in the pavilion, which is inhabited by five persons. The existing gas heating system is, for this reason, supplemented by solar thermal collectors and a new water storage tank.

Data

- location: Würzburg
- number of persons: 5
- hot water demand HW (60 °C): 125 litres/day (5 · 25 litres/day)
- heat demand HW: 7.27 kWh/day (heated from 10 to 60 °C)
- final energy demand HW: 15.63 kWh/m² day
- size of the collector plant: 6 m² (1–1.5 m²/person for 70–80 % coverage)
- type of collector: vacuum tube collector
- orientation: 60° south (optimised for heating period)
- storage tank: 600 l
- distance between storage tank and collector: 6 m

Yield (annual balance)

- heat demand HW: 2654 kWh
- final energy demand HW: 5705 kWh
- solar insolation (fig. 3.11, p. 93): 6368 kWh
- solar yield: 2163 kWh
- efficiency (solar thermal plant): 34 % (2654/6368)
- coverage rate HW: 81 % (2163/2654)
- coverage rate of final energy: 34 % (2043/5705)
- specific collector yield: 361 kWh/m²

Ecology

- annual savings in resources (taking account of energy loss): 260 m³ gas
- annual CO_2 savings: 632 kg
- annual savings in primary energy: 286 kWh

Economy

- annual savings in resources: €182 (gas €0.70/m³)
- investment: €7000
- annuity 25 years (3 % interest rate): €10,049
- savings in resources in 25 years (8 % price increase): €12,937

Comment

In the case of rented properties, the investment for solar thermal collectors can be written off as operating costs over a 10-year period, which further increases the economic advantage.
By extending the plant by 4 m² and installing a buffer storage tank, the efficiency of the system can be raised further. The additional solar gain can then also be used to support the heating system during the between-season periods. The additional expense of approximately €2000 for the investment leads to additional savings of around €70 per year.

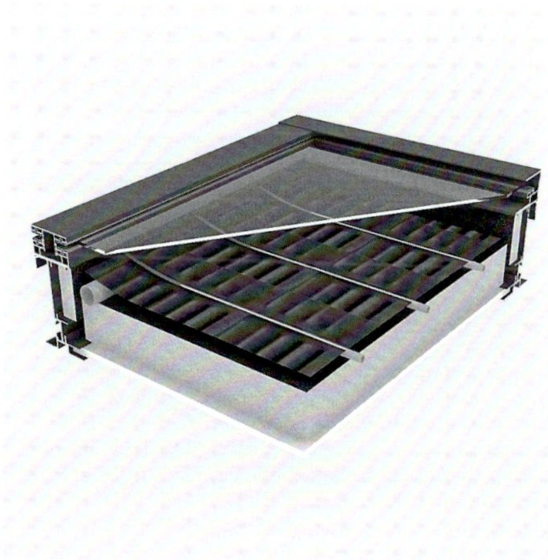

3.21

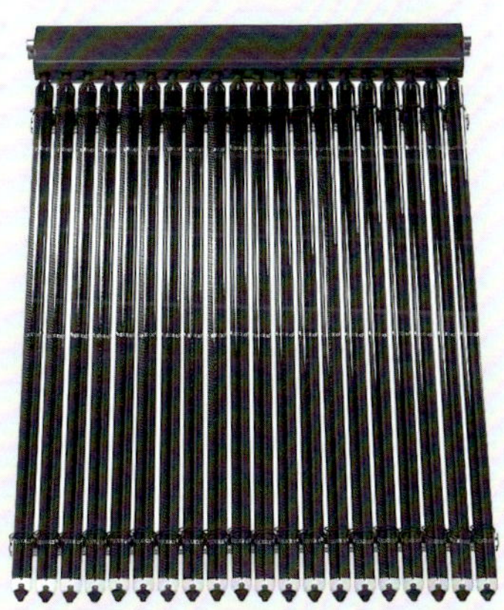

3.22

Cooling

When excess heat can no longer be discharged from the room to the exterior through ventilation, it is necessary to install mechanical systems to cool the building. Depending on whether water or air is used as the heat transfer medium, a cooling system includes the cooling surfaces or air outlets in the interior space to distribute the cold, a duct or pipe system to transport the cool air or water, a cold generator and a recooling unit. The heat is removed from the room either by exchanging the air or by circulating cold water through a pipe system. In this process, the transfer medium absorbs the heat from the interior space and becomes warmer. This heat, in turn, is transferred to the refrigerant circulating in the refrigeration unit (fig. 3.27 and 3.9, p. 90). Following this process, the refrigerant is heated further (above the temperature of the outside air) so that the heat absorbed can be released into the outside air. Depending on the working principle of the refrigeration machine, the heating of the refrigerant is performed by means of either mechanical or thermal compression. In the case of thermal compression, solar thermal energy or district heat can be used. If the required water temperature is to range between 10 and 15 °C, the heat can be dissipated directly to the ground or fed into the groundwater.

Cold generation

The generation of cold water or air can either be performed in a centralised fashion for the whole building or decentralised using so-called split or multi-split units for individual building zones (fig. 3.23). The cooling of individual rooms with separate units always leads to problems concerning the best location of the device since they can spoil the appearance of the building when placed in the facade.

When choosing the type of cold generation, energy-related aspects should be considered alongside the functional criteria since these have a significant influence on the efficiency of the installation (fig. 3.26).

Compression refrigeration system
The cold vapour compression process performed by a heat pump is to date the most common form of cold generation. It uses a low pressure refrigerant in the evaporator, i.e. on the cold side, that vaporises even at a low temperature. In doing so, it absorbs heat. A motor-driven compressor is used to compress the vapour, which raises the temperature of the refrigerant.
The high temperature enables the refrigerant to reject the heat with the help of a condenser, which is dissipated to the environment using a recooling unit. Subsequently, the expansion valve enables the cooled refrigerant to expand in the evaporator and cool down even further. Then the cycle begins anew (fig. 3.24).

Refrigeration system with thermal compression (absorption)
The heat from the room is transferred to the refrigerant ammonia, which vaporises with increasing temperature and is subsequently recovered once it meets water in the absorber. Part of the heat is released during this process, which can be used to perform further operations. In the generator, the refrigerant is boiled off the water using a small amount of heat and guided to the condenser, where it transfers its heat to the recooling circuit. Then the water returns to the absorber (fig. 3.25). In comparison to the compression refrigeration machine, the absorption refrigeration machine requires considerably less power. In contrast, however, the generator requires additional thermal energy.

When cooling with thermal compression, i.e. using heat, it is also possible to use solar thermal energy as an alternative to traditional energy sources, such as gas, oil or district heat.

Adsorption refrigeration machine
The adsorption refrigeration system is especially suited to the cooling of air and therefore requires the provision of an air handling system. In this case, heat is used to initiate a physical process which exploits the property of air, which is to cool down when moisture is introduced and heat up when moisture is removed (see Ventilation, pp. 106ff.).

Adiabatic cooling
Adiabatic cooling also requires the availability of an air handling unit. In this process, the temperature is reduced by spraying water into the waste air and subsequently guiding it through a heat exchanger where heat is transferred between the warm outside air and the cool waste air. It is through this process that the supply air is cooled (see Ventilation, pp. 106ff.).

Regenerative cooling

If cold is to be generated without the use of finite resources, i.e. by using renewable energies, due to the higher temperature levels, only small heat loads can be removed. The reduction of heat gains through construction-related measures is therefore essential if buildings are to be cooled effectively using renewable energy sources.

Ground coupling
Thanks to its constant, moderate temperature throughout the year in surface-near zones down to depths of 100 m, the ground is an outstanding cold source. Underground heat exchangers for cooling the incoming air have proven successful

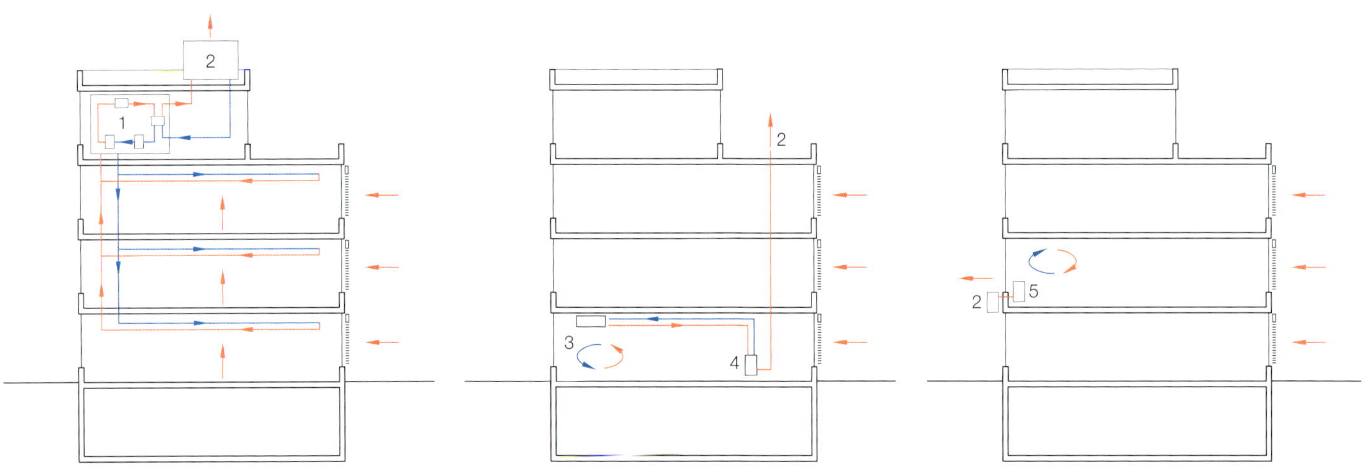

a b c 3.23

for many years. Depending on the capacity and the depth, horizontal collectors, i.e. pipes laid parallel to the surface, or energy piles can be used to reduce the temperature of water or a water/glycol mixture to 12-16°C in summer. This cold source is best combined with a cooling system that activates storage mass, e.g. thermo-active building components (TABS). Owing to the large areas the system requires, the application is often not suited to refurbishment projects (fig. 3.28 d, p. 102).

Groundwater

Groundwater can be used even more efficiently for cooling than the ground. Due to the constant flow, the heat absorption capacity of groundwater is almost infinite. In technical terms, it is also possible to install groundwater cooling in a refurbishment project since the system requires little more than two wells – a production and re-injection well (fig. 3.28 b, p. 102).

Nocturnal cold

Recooling units are normally used to cool down the refrigerants in conventional refrigeration machines. In the recooling circuit, the heat from the refrigerant is either released directly into the outside air or transferred to the environment using water as a medium. Recooling units are therefore always located on the roof of a building as a separate structure. They can also be used to cool thermo-active building components at night (see Cooling ceiling and thermo-active building components, p. 103).

A rarely used principle of regenerative cooling is the possibility of directly using the nighttime skyward radiation in summer. A very simple system includes roof-mounted swimming pool absorber mats - ideally on a flat roof. The building is then either cooled directly by circulating water through the storage mass at night or the cold is stored in a water tank. The average capacity of such a system in our latitudes is 50–80 W/m² of absorber surface.

Solar-powered cooling

The absorption and adsorption refrigeration machines described above can be operated with solar energy by generating the necessary process heat with solar thermal collectors. In order to remove a monthly heat load of, for example, 5000 kWh, the collectors need to generate approximately 7500 kWh of heat. A collector area measuring 40 m² is able to provide this amount of heat during the months June to August (fig. 3.28, p. 102).

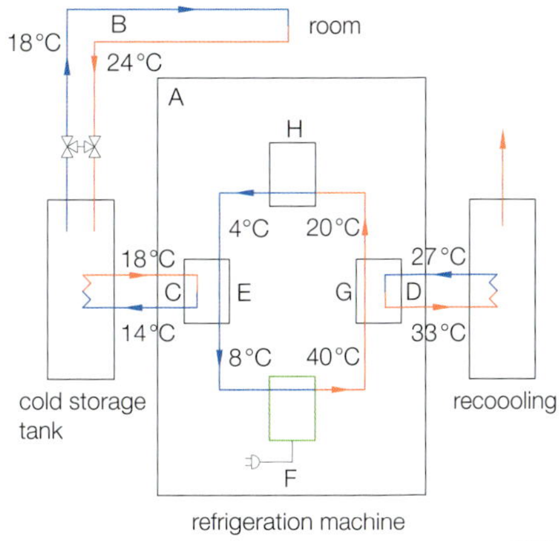

3.24

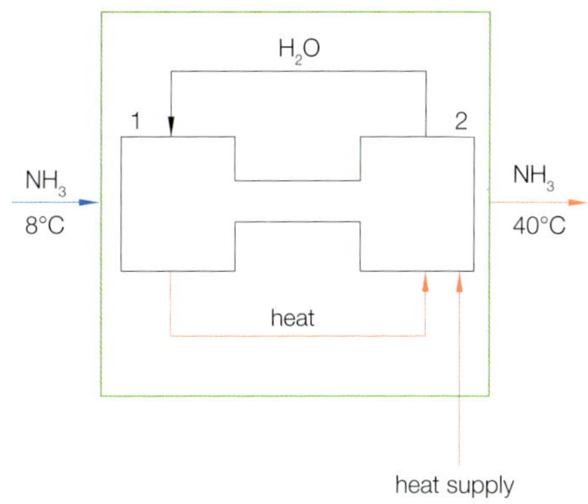

3.25

Energy efficiency assessment of cooling systems

	Final energy [kWh]	Resource/ aux. energy	Primary energy [kWh]	CO_2 emissions [kg]	Costs [%]	Comments
Compression refrigeration machine (elec.)	5000	electricity	13,000	3235	120	flexible in use
Compression refrigeration machine (gas)	10,000	gas	7700	1743	140	flexible in use
Absorption refrigeration machine	14,000	gas	8800	1992	180	flexible in use
Adsorption refrigeration machine	14,000	gas	8800	1992	220	only with air-conditioning
Ground [1]	700	electricity	3250	809	100	only for small heat loads
Groundwater [2]	500	electricity	3250	809	100	only for small heat loads
Night-time ventilation [3]	1200	electricity	3120	776	100	limited use

[1] heat exchanger surfaces in the ground under the base slab, usable for heating purposes in winter
[2] wells down to a depth of approximately 5 m
[3] Additional costs for larger plant in the case of an already existing air handling system

3.26

Refrigerants for heat pumps/compression refrigeration machines and their environmental impact [1]

Refrigerant	Description
HCFC (sale of new goods was prohibited after 2010) [2]	R 123, R 124, R 22, R 401 A, R 401 B, R 402 A, R 402 B, R 403 B
HFC [3]	R 125, R 134 a, R 227ea, R 23, R 236 fa, R 245 fa, R 404 A, R 407 C, R 407 D, R 410 A, R 413 A, R 417 A, R 422 A, R 422 D, R 437 A, R 507, R 508 B
natural refrigerants [4]	R 717 (ammonia), isobutane 1.5, isobutane 1.8, R 600 a (isobutane 2.5), R 600 (n-butane 2.5), R 290 (propane 2.5), R 1270 (propene 2.6), R 744 (carbon dioxide)
Special gas	1-butene, dimethyl ether, R 152 a, R 32

[1] The climate-damaging effect is measured according to the CO_2 equivalent. In Germany, alone the uncontrolled disposal of refrigerants is responsible for releasing 5 million tonnes of CO_2 equivalent emissions into the atmosphere per year.
[2] 1 kg HCFC (hydrochlorofluorocarbon) contributes 10,000 times more towards global warming than 1 kg of CO_2. Furthermore, CFC destroys the ozone layer in the atmosphere.
[3] 1 kg of HFC (hydrofluorocarbon), which is still used today, contributes 3500 times more towards global warming than 1 kg of CO_2.
[4] no climate damaging effect

3.27

3.23 Principles of cooling
a centralised cooling
b single storey cooling
c room cooling
1 refrigeration plant
2 recooling system
3 cold output (circulated air)
4 cold generation
5 compact unit

3.24 Cooling with electricity:
Compression refrigeration machine
A refrigerant cycle
B building's cooling cycle
C cold water cycle
D cooling water cycle
E heat exchanger water/refrigerant (refrigerant evaporates)
F refrigerant is heated through compression (power supply)
G heat exchange between refrigerant/water
H refrigerant expands (liquidisation and cooling)

3.25 Cooling with heat:
Absorption refrigeration machine
In order to increase the temperature of the refrigerant, a thermal compressor with an absorber (1) and a generator (2) is used instead of a conventional compressor (fig. 3.24, F).

3.26 Energy efficiency assessment of different cooling systems for an assumed cooling load of 10,000 kWh per month.

3.27 Refrigerants for heat pumps/compression refrigeration machines according to their environmental impact

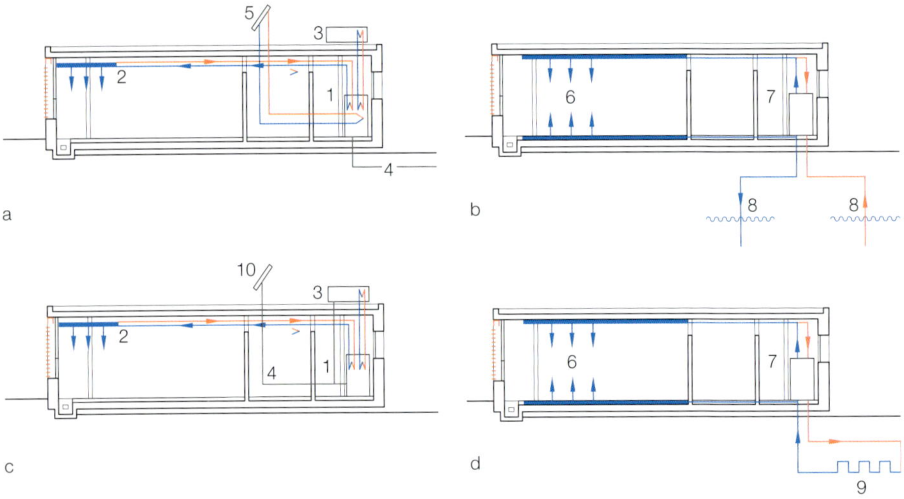

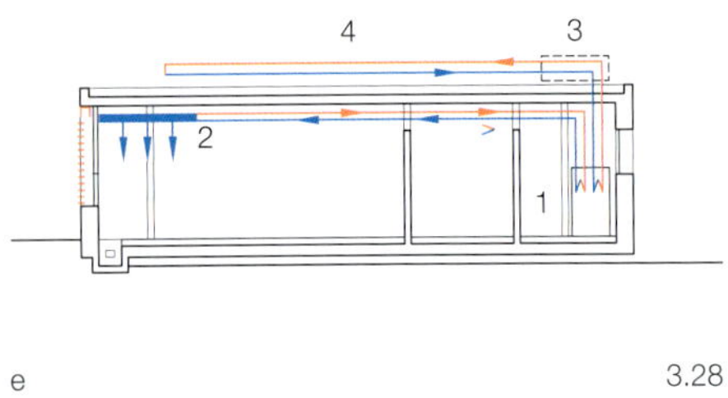

3.28

1 refrigeration machine (absorption or adsorption)
2 cold output using either cooling panels or supply air
3 recooling system
4 power supply
5 collector for heat generation
6 cold output using thermo-active building components (floor/ceiling)
7 heat exchanger
8 production and re-injection well
9 ground heat exchanger
10 photovoltaic panels

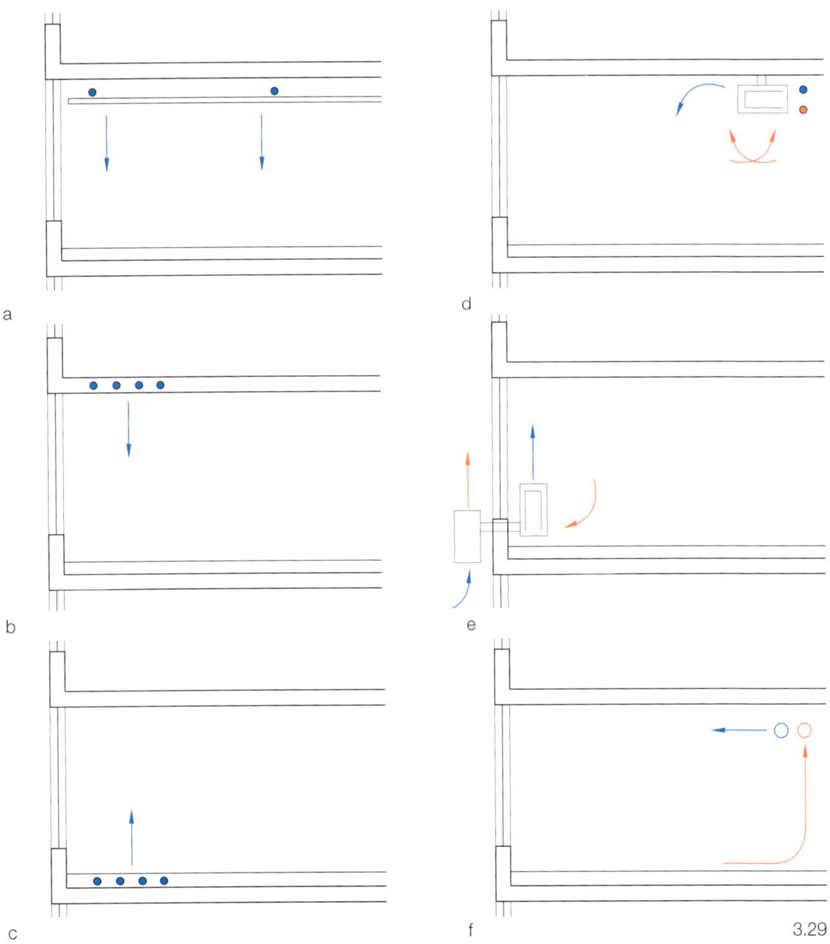

3.29

The electricity required to operate compression refrigeration machines can also be generated on site by using the solar energy of photovoltaics. This immediate use of electricity is particularly efficient since it eliminates the need to feed the electricity into the grid or draw electricity off the grid. It is for this reason that the EnEV considers the direct use of electricity positively in the energy accounting process by performing deductions to the primary energy demand.

Cooling distribution system

Depending on the method chosen to emit the cold in the rooms, the distribution can use either water or air as the transfer medium. In contrast to water, air with its low heat capacity (0.34 Wh/m³K in comparison to 1160 Wh/m³K in the case of water) has the disadvantage that large volumes must be moved (duct dimensions, fan capacity). However, if there are plans to install an air handling unit or if one is already in situ, the air can be used to transport the cold and there is no need to invest in a second distribution system. It is also possible to use a combination of both distribution systems – air and water.

Water

Similar to the way in which heat is distributed, the distribution of cold is particularly efficient when using a water circuit. The cooling capacity depends on the forward flow temperature of the water. The colder the water being supplied to the rooms or room zones, the greater the quantity of heat that can be removed from the areas requiring cooling. However, there is a danger of dew condensation on very cold surfaces. On this account, if the air is not dehumidified, the temperature of the cooling surfaces should be no lower than 3 K below the room temperature. The slightly warmer return flow water is then transported to the refrigeration machine where it is once again cooled. The principles

3.28 Methods used to cool buildings with renewable energy sources
a solar cooling: thermal collectors
b cooling with groundwater
c solar cooling: photovoltaics
d cooling with ground coils
e cooling with nighttime air

3.29 Principles of cold output
water as cooling medium:
a cooling panels
b thermo-active building components (ceiling)
c underfloor cooling
air as cooling medium:
d decentralised cooling with recirculated air and central refrigeration unit
e decentralised air cooling with compact unit
f central air cooling system

3.30 Performance data for different cold output systems (room height is 3 m)

governing the distribution of heat also apply to the distribution of cold:

- the larger the cold emitting surface, the higher the setting of the forward flow temperature
- the lower the heat quantity entering the room due to construction, user-related or technical factors, the warmer the water can be in the cooling circuit

Systems which make use of available storage mass and enable the cold to be emitted across large areas achieve, from an energy efficiency point of view, the best cooling performance.

Air

If the supply air for the destined areas is cooled in the cold generator before being transported to the rooms, air is used as the distribution medium for the cold. However, if the volume of air required for cooling exceeds the volume of air required to provide hygienic air conditions, then the option of adding a water-based system for cooling purposes should be considered. This solution would reduce the amount of energy required to provide cooling solely by using an air-based system.

Cold output

In order to transfer the cold to the interior space, there are, in a similar way to the output of heat, several different possibilities. These naturally differ in terms of their capacity, temperature range, thermal comfort and costs (fig. 3.29 and 3.30).

Cooling ceiling and thermo-active building components

Cooling elements can either be attached directly to the soffit in the form of plastered capillary tube mats or suspended below in the form of panels. When the cooling elements form part of a ceiling system, the system is referred to as a cooling ceiling. If the cooling surfaces are suspended below the soffit as individual components, they are often referred to as cooling sails. The cooling effect is essentially achieved by way of radiation and hence represents a very pleasant way of cooling. As is the case for all cooling systems, it is important to make sure that no condensation forms, i.e. that the surface temperature of the cooling surfaces does not drop below the dew point temperature of the room air.
In the case of thermo-active building components, the water or air pipes are laid in the concrete floor slab. Due to the direct contact between the cooling pipes/ducts and the storage mass, the storage capacity is improved. Despite relatively high forward flow temperatures, the large surface areas in conjunction with the storage mass enable the removal of considerable heat loads. The 24-hour operation of the system is also beneficial in this respect. Owing to the use of high forward flow temperatures, renewable energy sources, such as geothermal heat, groundwater or a recooling unit, can be incorporated very effectively.

Underfloor cooling

An underfloor heating system can be used for cooling purposes in summer. The cost for this solution is considerably less than that for installation of separate heat and cold output systems. The cooling effect is achieved by direct contact and radiation. The drawback of underfloor cooling is the adverse effect on thermal comfort conditions caused by excessive cold. Occupants feel uncomfortable when the temperature difference between head and feet is too large. Underfloor cooling can therefore only be operated at a temperature that is marginally different to that of the room. Its efficiency and application are therefore limited.

Air cooling

In the case of a cold-air cooling system, the cooling effect is achieved by supplying cooled fresh air, cooled air extracted

Performance data for different cold output systems

	Max. temperature difference [K] [1]	Maximum output [W/m²]	Costs [2] [€/m²]	Comment
Cooling elements				
cooling ceiling	8	100	200–600	cooling surfaces combined with ceiling system
cooling sails	8	120	200–800	individual suspended panels
Active storage systems				
underfloor cooling	8	15–30	0–40	thermal storage activated by
concrete core activation	8	20–40	20–40	technical installations
Air cooling systems				
single air exchange	8	8	250–350	heating, cooling, ventilation
double air exchange	8	16	300–400	with a single system
4-fold air exchange	8	24	350–450	

[1] temperature difference between the room air and the cooling medium (danger of condensation, uncomfortable room conditions)
[2] costs do not include cold generation or metering and control mechanisms

3.30

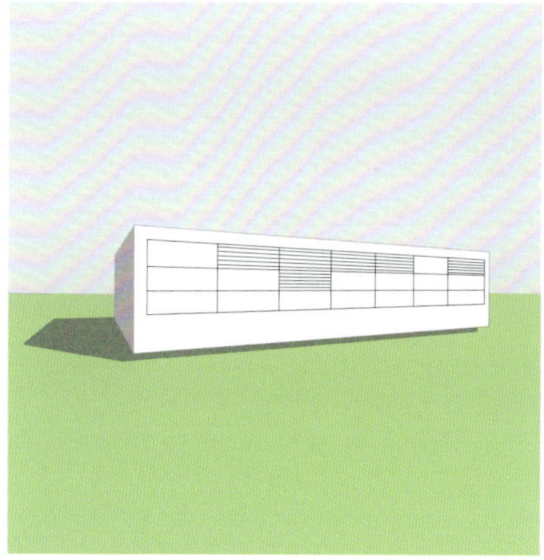

Example: cooling
Simplified accounting method to determine the monthly energy demand for the cooling of a south-facing office (open plan office, 121 m²) in July.

Data

window area (A)	38 m²
air exchange	6 m³/h per m²
air volume (V/h)	726 m³/h
operation period July (t)	20.8 days à 11 h
sun protection	exterior blinds
inside temperature	22 °C (setting for cooling period)
outside temperature in July	24.6 °C (assumption)

Heat sources (Q_{source})
solar heat input through transparent surface areas (Q_s)

insolation (I_s)	104 kWh/m²
g-value DIN 410 (g⊥)	0.65
reduction for frame (W_f)	0.8
reduction for pollution (W_p)	0.9
reduction for overshadowing (W_o)	0.9
reduction for sun protection (W_c)	0.25

$Q_s = F_f \cdot A \cdot I_s \cdot F_p \cdot F_s \cdot F_c \cdot g =$ 416.15 kWh

Heat input through ventilation (Q_v)
$Q_v = V/h \cdot 0.34\ Wh/m^3K \cdot \Delta T \cdot t =$ 146.84 kWh

Heat input through transmission (Q_t)
can be neglected due to insignificance

Heat input through use ($Q_{i,source}$)
$Q_{i,source} = 162\ Wh/m^2d \cdot 121\ m^2 \cdot 20.8\ d =$ 407.72 kWh
(value 162 Wh/m²d from figure 1.27, p. 24)

Total of heat sources (Q_{source})
$Q_{source} = Q_s + Q_v + Q_t + Q_{i,source} =$ 970.70 kWh

Heat sinks (Q_{sink})
(see calculation method for heating period, pp. 45ff.)
600.00 kWh

Heat load to be removed through cooling ($Q_{c,outg}$)
$Q_{c,outg} = 970.70\ kWh - 600\ kWh =$ 370.70 kWh

Power consumption to remove heat load
Compression refrigeration machine

refrigerant	R 134 a
recooling	water-based
temperature of outgoing cold water	14 °C
temperature of incoming cooling water	27 °C
energy efficiency ratio (EER)	4.6
part load value (PLV)	1.34

$$\text{Power supply } Q_{c,f,electr} = \frac{Q_{c,outg}}{(EER \cdot PLV)} = 156.31\ kWh$$

Renewable power supply

insolation (I_s)	104 kWh/m²
power yield approx. 10 %	10.4 kWh
required size of photovoltaic array	15 m²

Comment
When using the renewable cold source, solar energy, with photovoltaics, cooling can be performed directly since the cooling demand and insolation occur simultaneously.

from the rooms (recirculated air) or a mixture of both (mixed air) to a particular space. The cool supply air can either be fed into the room at ceiling level, from where it distributes evenly throughout the room due to the natural thermal properties of air, or it is introduced at a higher speed at floor level.
An especially hygienic and efficient form of supplying fresh air to interior space is through displacement ventilation. The cool fresh air gushes slowly through floor-level openings and forms a pool of cold air since it is heavier than the warm interior air. Internal heat sources, such as persons and office equipment, generate a thermal current and the air rises without the need for any mechanical assistance (fans) to the air outlets at ceiling level.

Energy accounting according to DIN V 18599

The standard DIN V 18599 provides methods to perform the accounting of the most common processes in cold generation, compression and absorption refrigeration machines. All other processes can only be dealt with by carrying out individual technical calculations. Approximately 90% of all installed systems are compression refrigeration machines. It is for this reason that the following description refers to the accounting of their energy balance in accordance with DIN V 18599.
The accounting is performed using the energy efficiency ratio (EER) for the nominal cooling capacity and the part load value (PLV) to express the part-load efficiency. The energy efficiency ratio (EER) indicates the maximum amount of electric energy required to cool the building on the hottest day of the year when all heat loads occur simultaneously. An EER of 3 states that 1 kW of electric energy is required to remove a 3 kW heat load. Higher values indicate higher levels of energy efficiency. The temperature rise the refrigerant has to undergo in the cooling cycle and the type of recooling method chosen (water or air) are fundamental aspects influencing the EER. Subject to the type and the applied refrigerant, DIN V 18599 includes a table of energy efficiency ratios for existing systems without available performance data. These can be adapted to the year of manufacture by applying the appropriate ageing factor (fig. 3.31).
The partial load value (PLV) is used to express the fact that heat loads vary in volume, and the refrigeration machine must therefore not always run at full capacity to remove them. Depending on the possibilities to control the refrigeration machine in regard of the water cycle's temperature and the type of recooling unit, an average performance capacity is determined for the system. This then enables the assessment of the electric energy consumption which is required per month to remove the excessive heat.
The efficiency of an absorption refrigeration system is determined by using the nominal heat ratio ζ, a value that describes the relation between supplied and removed heat. This value is at around 0.7 and expresses that a heat supply of 0.7 kW is required to remove a heat load of 1 kW. Absorption refrigeration machines can also be operated with solar thermal heat, district heat or the exhaust heat from combined heat and power systems. However, the temperature of the medium must be at least 80 °C.

A rough estimate of the refrigeration system's energy demand can be made according to the following four steps:

- Step1: Heat load ($Q_{c,outg}$)
  Based on the example calculation on page 45 (heat sources), it is possible to establish the amount of heat that accumulates during the course of one month and must then be removed.
- Step 2: Energy efficiency ratio (EER)
  This value can be determined for the selected refrigeration system by using the data in figure 3.31.
- Step 3: Part load value (PLV)
  This value depends on the cooling system selected. Appropriate values are listed in DIN V 18599-7, enclosure A.
- Step 4 a: Electric power demand ($Q_{c,f,electr}$) for compression refrigeration machines
  The electric power demand required per month to remove the heat load is determined according to the following calculation:
  $Q_{c,f,electr} = Q_{c,outg} \div (EER \cdot PLV)$
- Step 4 b: Heat demand ($Q_{c,f,therm}$) of the absorption refrigeration machine
  The heat demand per month for the absorption heat pump is calculated as follows:
  $Q_{c,f,therm} = Q_{c,outg} \div \zeta \cdot PLV$

The removal of heat can be achieved very efficiently if:

- a water-based compression refrigeration machine is used,
- the temperature of the outgoing water is high (e.g. 14 °C instead of 6 °C) and
- the temperature of the incoming water is low (e.g. 27 °C instead of 40 °C).

Ageing factors for existing cold generators

Year of manufacture	Ageing factor $f_{c,exist}$
after 1990	1
after 1985 before 1990	0.92
after 1980 before 1985	0.87
before 1980	0.82

a

Energy efficiency ratio (EER) of air-based compression refrigeration machine

Refrigerant	Temperature of outgoing cold water [°C]	Average condensing temperature [°C]	Energy efficiency ratio for normal capacities	
			Piston and scroll compressor 10 to 1500 kW	Screw compressor 200 to 2000 kW
R134a	6	0	2.8	3.0
	14	8	3.5	3.7

b

Energy efficiency ratio (EER) of water-based compression refrigeration machine

Refrigerant	Temperature of incoming/outgoing cooling water [°C]	Temperature of outgoing cold water [°C]	Average condensing temperature [°C]	Energy efficiency ratio for normal capacities		
				Piston and scroll compressor 10 to 1500 kW	Screw compressor 200 to 2000 kW	Turbo compressor 200–2000 kW
R134a	27/33	6	0	4.0	4.5	5.2
		14	8	4.6	5.3	5.9
	40/45	6	0	3.1	2.9	4.1
		14	8	3.7	3.7	4.8

c

3.31

3.31 Energy efficiency ratios (standard values) according to DIN V 18599-7, table 20 and 22
a correction factors for the energy efficiency ratios according to DIN V 18599-7, enclosure F
b air-based compression refrigeration machine
c water-based compression refrigeration machine

3.32 Typical schematic diagram for the heating and cooling of an office building using renewable energies. The heating is provided by ground probes which are connected to a brine/water heat pump. The heat output is achieved using underfloor heating. The low temperatures of the ground are used for the passive cooling of the floors in summer. In addition, the air is preheated/cooled by drawing it in through the ground.

The temperature of the outgoing water can be raised if the heat loads are reduced and the area made available for the cold output is increased.
Similarly, if the temperature of the outside air for the recooling process is low, for example by positioning the equipment in an underground car park, the temperature of the incoming water can also be kept at a very low level.

Energy efficiency refurbishments

The aim of an energy efficiency refurbishment, particularly in the case of non-residential buildings, is to minimise the heat gains as much as possible in order to reduce the output capacity of existing refrigeration systems or even remove them entirely. On the other hand, the installation of a refrigeration system into an existing building is frequently one of the key elements in maintaining the use of a property in the light of higher standards of thermal comfort. A problem often encountered in these circumstances is finding the space to install the equipment, which usually includes the refrigeration machine itself, a recooling unit and facilities to distribute and emit the cold. When retrofitting a plant, the distribution of cold using air as the transfer medium is not recommended or at least only in combination with other systems, due to the amount of space required for the air ducts. If only individual rooms need cooling, decentralised refrigeration units, that can be fitted into cupboards or suspended ceilings, are most suitable. The only difficulty arising from this flexible solution is the recooling process, i.e. the transfer of heat to the outside air. If an exhaust air system is available, the exhaust air is able to cool the refrigerant; the alternative is to integrate the recooler into the building in such a way that it connects to the exterior space.

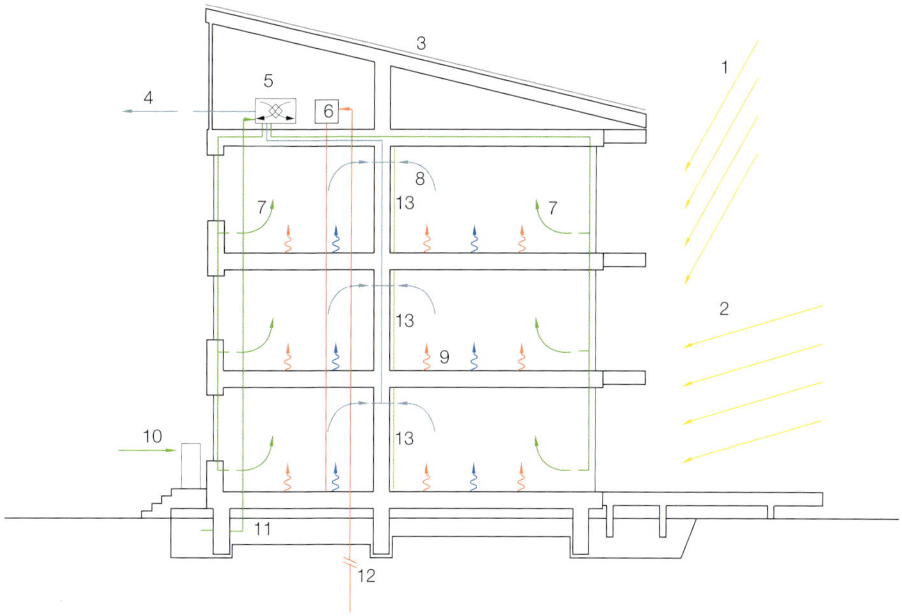

1 summer sun
2 winter sun
3 photovoltaics
4 exhaust air
5 air handling unit
6 heat pump
7 supply air
8 waste air
9 underfloor heating/cooling
10 outside air
11 ground heat exchanger
12 ground probe
13 plant wall for air humidification

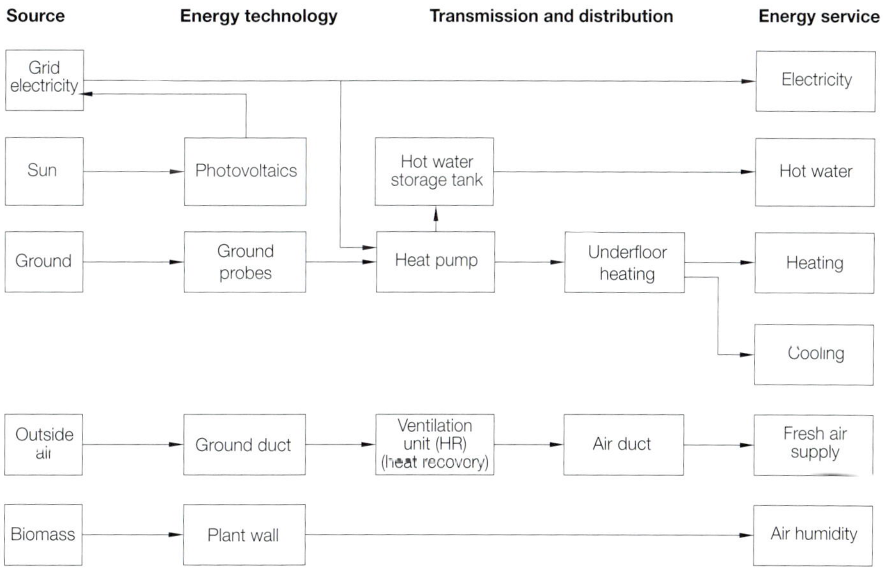

3.32

A further way of cooling individual rooms or even larger zones is to retrofit so-called cooling sails (metal panels with coiled pipe that is either welded or screwed onto a base). The elements, which are flexible in the way they can be added, are suspended from the ceiling and connected to a water-based distribution network.
If large areas or even the entire building require cooling, the best solution for distributing the cold is to install a central refrigeration machine with a water-based network. The cold is then introduced into the rooms individually by using induction units, which, with the help of a fan, cool the room air by circulating it over a water coil. If a hot water system is incorporated alongside the cold water network, the induction units can also be used to supply warm air for heating purposes.
In terms of reducing the primary energy demand in refurbishment projects, the use of renewable cooling systems, such as ground probes or groundwater wells, should be considered. However, the requirements for such systems are a small temperature difference between the forward and return flow water. This, for example, applies to large-area thermo-active building component systems with their fairly high forward flow temperatures. Because it is not possible to integrate cooling pipes into the floor slabs of existing buildings, the only alternative is to use the still rather expensive latent heat storage panels, which can be fixed to the underside of existing ceilings. The fairly thin elements contain coiled pipe and are either fixed directly to the soffit or suspended. With the help of these elements, it is also possible to install an active thermal storage system into old buildings, which then allows the use of a renewable cold source. The heat removed from the building in summer is stored in the ground by means of ground probes (temperature rise) and reused in winter with the help of a heat pump. The higher temperature of the heat source (ground) increases the heat pump's coefficient of performance (fig. 3.32).

Ventilation

In order to maintain hygienic air conditions, every room must be supplied with the right amount of fresh air for the particular space usage. The objective of controlling the quantity of air has resulted in the use of various mechanical systems. Depending on the scope of the air-treatment measures, a distinction is made between complete air conditioning, partial air conditioning and ventilation systems. In the case of full air conditioning systems, the air is heated, cooled, humidified and dehumidified to produce the desired room climate totally independent from the outside conditions. The air-treatment measures in partial air conditioning systems are reduced and the technical plant provides one, two or three of the parameters mentioned above (heating, cooling, humidification and dehumidification). Units purely for ventilation purposes are exclusively used to provide fresh air and protection against moisture (fig. 3.34 and 3.35).

Ventilation systems

The use of a ventilation system is often only discussed in conjunction with the potential energy savings that can be achieved by reducing ventilation heat loss. Especially in the case of non-residential buildings, facilities to control the amount of outside air entering the building help to reduce the risk of rooms becoming overheated in summer. However, more often than not, the problem is less one of too much ventilation, but more one of too little, which means that the provision of oxygen and the removal of pollutants and moisture is no longer ensured. Due to the increasing airtightness of building envelopes, it is fundamental to ensure an adequate exchange of air to avoid the negative effects of unhygienic air conditions (see Ventilation, pp. 76ff.).
When noise pollution is an issue, ventilation units are often used to provide the necessary exchange of air. Appropriate filters can also prevent certain pollutants from entering the interior space (fig. 3.36).

Demand controlled ventilation system (DCV)
The simplest form of a demand controlled ventilation system, which is frequently employed in residential buildings, ensures that fresh air enters the room through controllable openings in the window surround or in the exterior wall. A central fan is provided to remove the waste air. It continuously extracts the air from the various storeys in the building and automatically draws in fresh air through the vents. The exhaust-air duct is best located in the centre of the dwelling unit, preferably within the sanitary area. This has the effect that the extractor fans, which would otherwise be necessary in internal bathrooms, are no longer required. The extraction of waste air within the dwelling unit is simply solved by drawing air through the gaps beneath doors, i.e. without installing an air duct system.
A simple air extraction system can easily be retrofitted in old buildings by making use of existing ventilation shafts (buoyancy driven ventilation). Due to the continuous suction at the top of the shaft, the ventilation is able to take place independent of weather conditions (fig. 3.33 a).

Controlled ventilation with heat recovery (HR)
It is only possible to perform a direct exchange of heat between warm waste air and cold outside air if the supply air is guided through a system of pipes into a heat exchanger. This is where the supply air absorbs heat from the waste air before distribution into the rooms takes place. It is advisable when using such a concept to adapt the positioning of the ventilation units to the usage of the building. In residential buildings this means one ventilation unit per apartment so that the yield from the heat recovery process can be allocated to the benefit of the respective heat consumer. In the case of a central air suction line, these systems can also be combined with an earth tube or ground collector (brine) for preheating the air in winter or precooling it in summer. To a limited extent, it is also possible to humidify the air by exchanging the moisture between the fresh air supply and the waste air in a similar way to the exchange of heat (humidity recovery).
A controlled ventilation system with a heat recovery unit can usually be retrofitted fairly easily in refurbishment projects even though the installation of supply air ducts demands a headroom of approximately 15 cm. The answer in existing buildings is to use the corridors with suspended ceilings to distribute the air and then introduce the supply air into the rooms via sound insulated vents above the doors. The waste air extraction is carried out in the sanitary areas without the need for an independent network of air ducts (fig. 3.33 b).
Compact units with heat recovery are considered a special form of ventilation. They can be fitted into an exterior wall to serve one room only. These units tend to be used when individual rooms are exposed to severe external noise and ventilation via the windows is not always possible (fig. 3.33 c).

Partial air conditioning system

To condition the air in passive houses, systems with a centralised unit are applied. Since the heat demand is reduced in the case of passive house standards, the building's controlled ventilation system is also used for heating purposes. The air supplied to the rooms is heated in the ventilation unit, which means that both the air exchange as well as the heat transmission is performed by using one distribution system. In these circumstances, the heat distribution using water with a forward and return flow system is no longer required. The system is particularly effective if the heat contained in the waste air is recovered and used to preheat the fresh air.
The relative humidity of air decreases when heated, which means that the air should be humidified before being supplied to the rooms. Modern ventilation

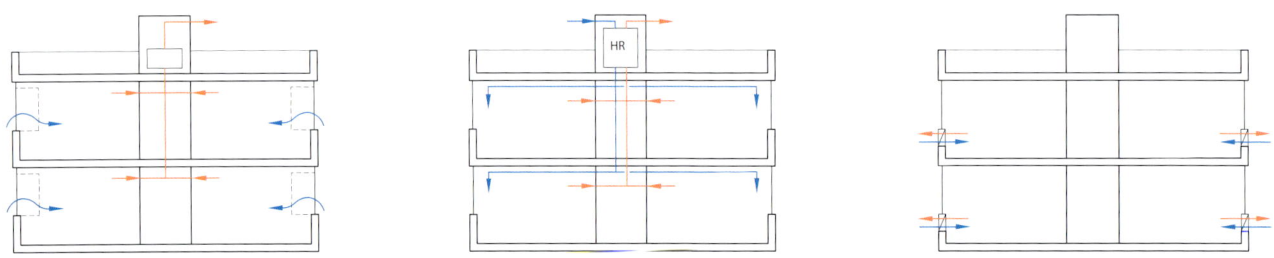

a b c 3.33

units therefore also provide facilities for humidification and moisture recovery. The ventilation unit can also provide cooling, if the building so requires. However, the degree of cooling is limited since the relative humidity increases as soon as the air is cooled, unless it is dehumidified at the same time.

Adiabatic air cooling
This special type of air conditioning uses a humidification technique to cool the air, which requires only a very small amount of energy. In an adiabatic air cooling process, the flow of waste air is saturated with water vapour, which makes the temperature decrease without the heat actually having been exchanged (adiabatic). The cooled exhaust air then flows through a heat exchanger where it cools the incoming outside air (fig. 3.37 and 3.38, p. 108).
The adiabatic cooling method is ideal to remove small heat loads and is therefore especially suited for buildings with little need for cooling.

Complete air conditioning system
When using a complete air conditioning system, the air temperature can be increased by supplying heat, cooled and dehumidified by supplying cold, or humidified by supplying water vapour. The heat, cold and moisture are generated using suitable technical equipment. In contrast to separate technical installations to provide heating, cooling and ventilation, each with their separate distribution system, a complete air conditioning system is able to provide all these services with one system, only using the appropriately conditioned air.
Air conditioning systems have a modular structure and consist of an exhaust air fan, a heat exchanger, a heater and cooler (with dehumidifier), a humidifier, a secondary heater and a supply air fan. Systems with limited air conditioning facilities are made up of fewer modules (fig. 3.40, p. 109).

3.33 Ventilation concepts
a controlled vent. with air extraction system
b controlled ventilation with heat recovery
c controlled ventilation with HR (individual units)
3.34 Overview of various air supply systems in buildings according to DIN 1946-1
3.35 Energy accounting for various ventilation systems based on the following assumptions: 350 m² floor space, 2.50 m clear room height, 185 heating days, degree of efficiency 90 %, gas price €0.70/m³, electricity price €0.24/kWh, energy price increase 6 %, life expectancy 20 years, interest rate 4 %; fan output 0.2 W/m³ in extraction mode, 0.5 W/m³ in supply and extraction mode.
3.36 Efficiency of natural ventilation, extraction systems, and supply and extraction systems

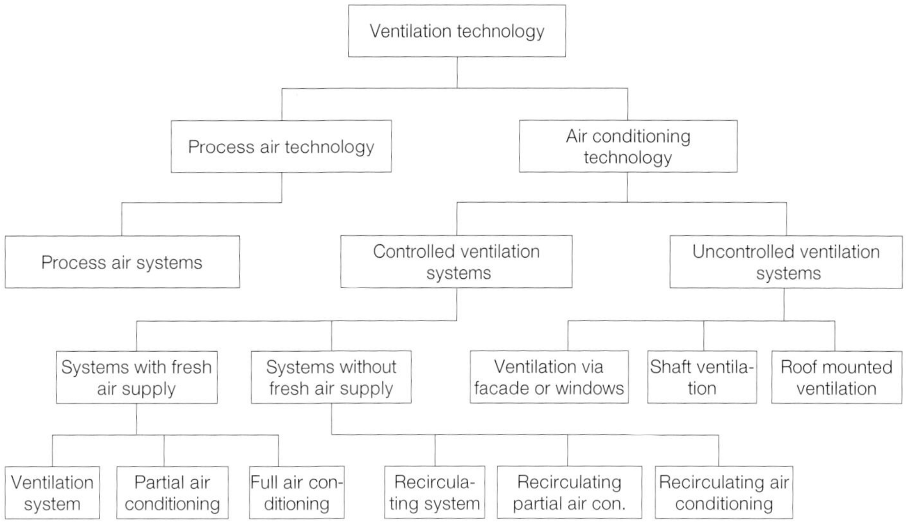

3.34

Energy assessment – ventilation

	natural	controlled [1]	controlled with HR [2]
Heating			
heat demand [kWh]	28,000	24,000	18,000
heating energy demand [kWh]	31,111	26,666	20,000
energy medium	gas	gas	gas
heating energy costs [€]	2178	1866	1400
primary energy [kWh]	34,222	29,333	22,000
Auxiliary energy			
power demand [kWh]	–	311	777
energy medium	–	elec.	elec.
auxiliary energy costs [€]	–	75	186
primary energy [kWh]	–	809	2020
Annual balance			
total costs [€]	2178	1941	1586
total primary energy [kWh]	34,222	30,142	24,020
Annual savings			
(in relation to natural ventilation)			
costs [€]	–	237	592
primary energy [kWh]	–	4080	10,202
Costs/benefits (20 years) [3]			
investment costs in 20 years (annuity) [4] [€]	–	11,773	36,791
energy cost savings in 20 years [€]	–	8718	21,777
extra costs in 20 years [€]	–	3065	15,014

[1] infiltration air exchange rate 0.15 and system air exchange rate 0.4
[2] infiltration air exchange rate 0.2 and system air exchange rate 0.4
[3] cost increase or reduction over a period of 20 years in comparison to natural ventilation
[4] assumption: investment costs for air extraction system €8000, for air supply/extraction system €25,000. The costs for the required ventilation system in bathrooms without windows (DIN 18017) must be deducted.

3.35

Range of ventilation systems

	natural	central air extraction	air supply/extraction system [1] (decentralised/apartment based)
Air supply	too little/too much	adjustable	adjustable
Heat recovery	not possible	possible (central air extraction heat pump)	possible (waste air to supply air)
Moisture control	limited	possible	possible
Air filter	not possible	limited	perfectly possible
Noise protection (outside)	not possible	limited	perfectly possible
Noise protection (inside)	perfectly possible	limited (gaps around doors)	possible (installation of sound attenuators)
Construction	little expense	little expense	high expense (air supply distribution ducts)
Passive preheating	not possible	limited (air intake via buffer storage tank)	perfectly possible (ground duct, air collector)
Operation	user dependent	not user dependent (basic ventilation)	adjustable according to user
Night-time ventilation	limited (protection again burglary/weather)	no problems	no problems
Costs	€0/m²	€20–30/m²	€60–80/m²

[1] in the case of bathrooms without windows, the ventilation according to DIN 18017 is not required

3.36

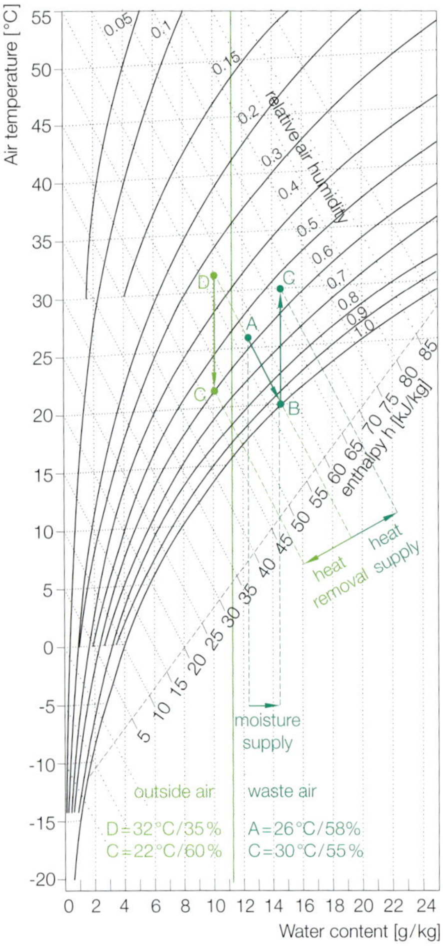

3.37

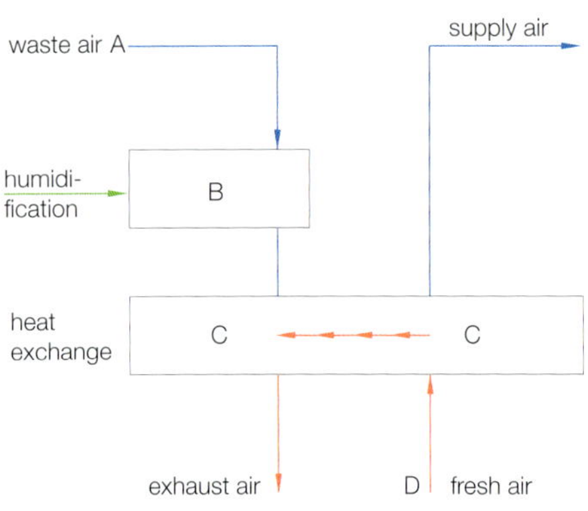

3.38

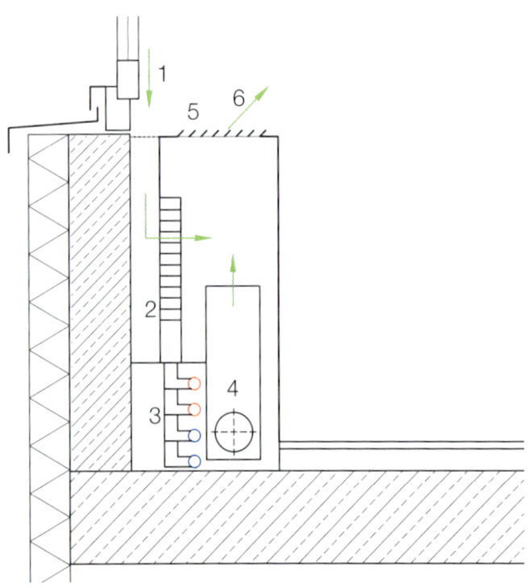

3.39

In non-residential buildings, individual rooms or zones with large room depths or specific user requirements must sometimes be ventilated mechanically. In these cases, for example conference facilities, meeting rooms and areas for medical care, it can make more sense to condition the room using air as the medium, in particular if there are specific requirements in terms of humidity. This is when so-called multi-split appliances (single storey cooling) or individual units (split appliances) are applied.
The greater the requirements in terms of air quality, the more complex the conditioning process of the air becomes. The alteration of one air conditioning feature always has a knock-on effect on others. The relationship between the different features is presented in the Mollier diagram (fig. 3.37).
Especially in buildings which are to be sufficiently flexible to cater for changes in use, the type and structure of building services must be planned very accurately since air-based systems require large amounts of space (ducts, equipment) and all changes involve considerable expense.

Air distribution

Depending on the degree of control, the distribution of air is carried out using a single or multi-duct system with a constant or variable air volume flow.

Air volume flow

When the aim is to simply supply the rooms with fresh air, a constant volume flow is adequate. In the case of more intensive room usage, the windows can be opened to increase the fresh air supply. If the constant volume flow is also to be used to provide heating or cooling, the air is conditioned accordingly in the central unit.
A variable volume flow has the advantage that the amount of supply air conditioned in the central unit can be controlled per room through volume flow regulators and therefore set according to the different user demands in terms of temperature. Plants with a variable volume flow adapt quickly to the changing heating and cooling loads in rooms and are therefore particularly suitable for environments with varying occupancy rates.

Duct systems

In the case of single duct systems, the air is supplied to the rooms using a single air duct. Differentiated requirements concerning the room temperature can be met for various building zones by incorporating post heaters into the duct system (so-called terminal heaters). If however some parts of the building, such as the north-facing rooms, require heating and others, for example rooms facing south, require cooling, the areas must be supplied by separate air ducts.
A double duct system consists of two parallel ducts, one supplying cold air and the other warm air, which are joined prior to the air outlet (single room or group of rooms) to mix the air to the desired target temperature. Due to the great expense in terms of material and installation work, double duct systems are rarely used.

Air velocity

The air can either be distributed through the duct system at high pressure (high pressure system with an air velocity of 10–14 m/sec) or at low pressure (low pressure system with an air velocity of 2–8 m/sec). High pressure systems have the advantage of smaller duct sizes and the corresponding small amount of space required to install the system. On the other hand, greater attention must be paid to the reduction of noise in the ducts. Furthermore, the electricity required to operate the fans is much greater than that of a low pressure system.

Air output

The air is introduced into the room via diffusers or grilles. The number of outlets, their size and positioning must be determined by a building services engineer. Apart from planning the duct system effectively, the main aim should be to prevent draughts (maximum output velocity of 0.3 m/sec).
In regard of energy efficiency, the conditioning of a room should not solely be achieved through air, but together with a water-based system. This combination reduces the air volume and thus the amount of energy required to operate the fans. Induction units, which can be installed below windows, in the floor or in suspended ceilings, provide the opportunity to individually heat or cool the pre-conditioned supply air arriving from the central unit. The primary air is blown into the induction unit, guided over water-based heating or cooling coils and then discharged into the room. During this process room air is drawn in (induction), which is then likewise heated or cooled. The space required for the air ducts and the technical equipment is reduced significantly due to the fact that a smaller amount of air is handled and conditioned in the central unit. An induction unit does

not enable the air to be humidified or dehumidified.
The heating or cooling coils can either be installed as a two or a four-pipe system:

- two-pipe induction system
  Depending on the time of year, the air is either cooled or heated. To achieve this, either hot or cold water is transported through two pipes (forward flow and return flow).
- four-pipe induction system
  In this case, heating and cooling water pipes run parallel so that it is possible to heat parts of the building at the same time as others are cooled (according to zones or choice). The elaborate and expensive system is extremely flexible and ideally suited to buildings with mixed use or very high demands concerning thermal comfort (fig. 3.39).

Fan power
Compared with separate systems for cooling, heating and ventilation, a major disadvantage, when conditioning a building exclusively through air, is the higher energy demand to operate the fans, which results from the larger volumes of air and the greater distances. In contrast, the air exchange and thus the electricity consumption of fans in buildings equipped with separate compact units, depends solely on the required volume of fresh air. In a central system, the air volume to transport the necessary heat or cold to individual rooms has to be increased to meet additional demands for heating in winter and cooling in summer. The low heat capacity of air (0.34 Wh/m³K in comparison to 1160 Wh/m³K in the case of water) highlights the need to improve the building's construction-related features in terms of summer and winter energy demand, when using this type of plant. This is the prerequisite for making use of the fascinatingly simple principle of central air conditioning while consuming an acceptable amount of energy.
The electric energy consumption and therefore the fan power is dependent on three factors:

- air volume
  If only as much air is moved as is required for the fresh air supply, then the output can be designed as a basic air change. Possible higher demands for short periods of time can be covered by opening the windows. If only the medium air is used to provide heating or cooling, the air volume must be adjusted according to the prevailing thermal requirements. This generally leads to a higher air exchange than is actually required.

Separate air supply and air extraction systems:

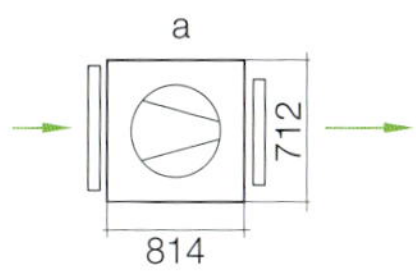

Air extraction unit

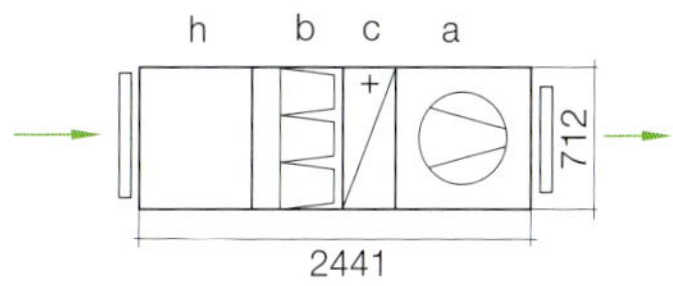

Air supply unit with heater

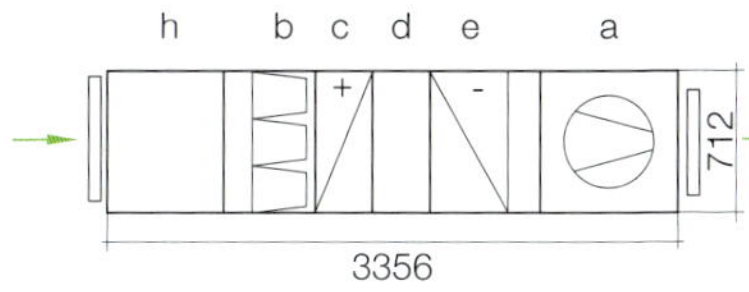

Air supply unit with heater and cooler

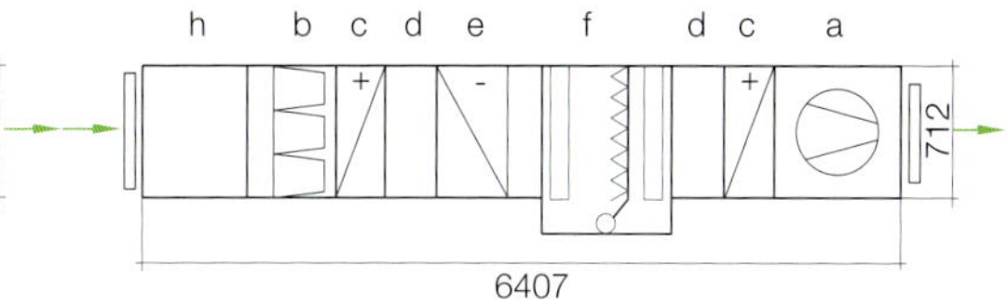

Air supply unit with heater, cooler and humidifier

Combined air supply and air extraction systems:

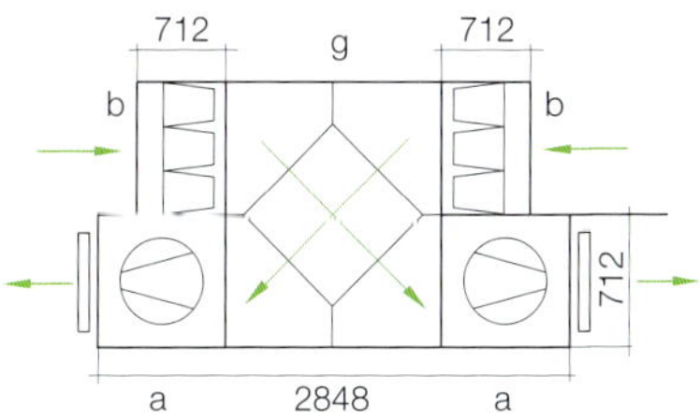

Combined air supply and air extraction unit with heat recovery

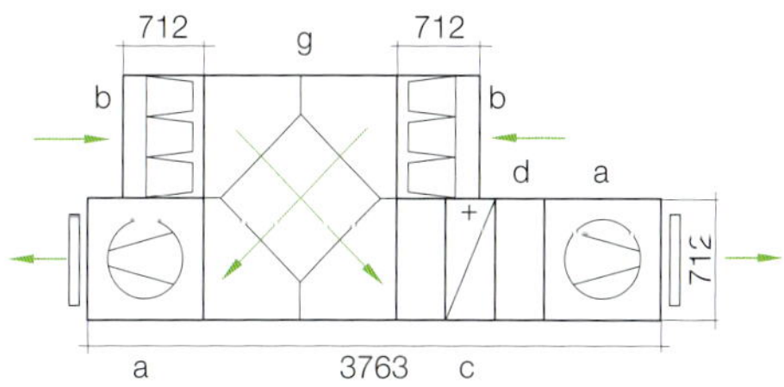

Combined air supply and air extraction unit with heat recovery and heater

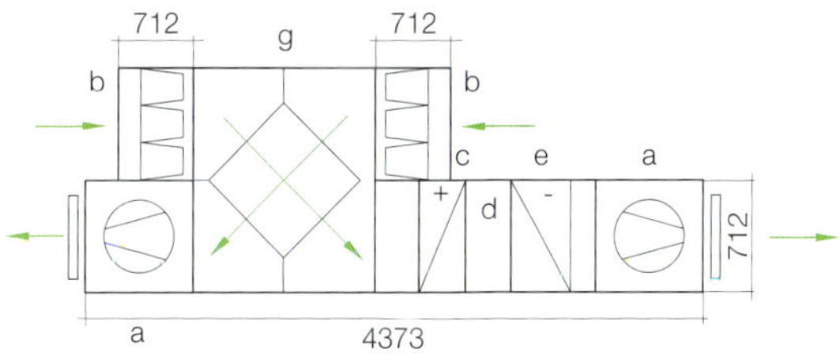

Combined air supply and air extraction unit with heat recovery, heater and cooler

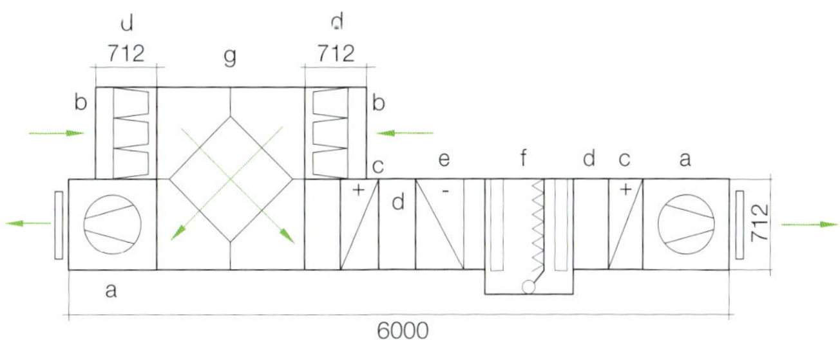

Combined air supply and air extraction unit with heat recovery, heater, cooler and humidifier (full air conditioning unit)

3.40

3.37 Mollier diagram: adiabatic cooling
3.38 Adiabatic cooling (if cold is removed through air)
A waste air (26 °C/58 %)
B humidification: temperature decreases
C heat exchange: supply air cools down, waste air absorbs heat
D outside air (32 °C/35 %)
3.39 Four-pipe induction unit for installation below windows
1 room air is drawn in
2 heating/cooling surface
3 cold and hot water (4 pipes)
4 primary air
5 air discharge grille
6 cooled/heated supply air
3.40 Air supply and extraction units for different kinds of air treatment (approximate cm dimensions)
a fan
b filter
c heater
d inspection
e cooler
f humidifier
g heat recovery
h mixing box (room air can, if required, be mixed in)

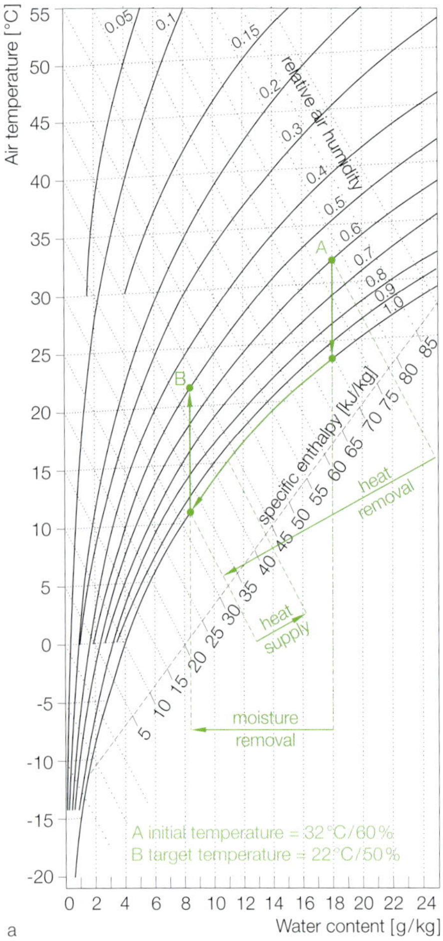

a

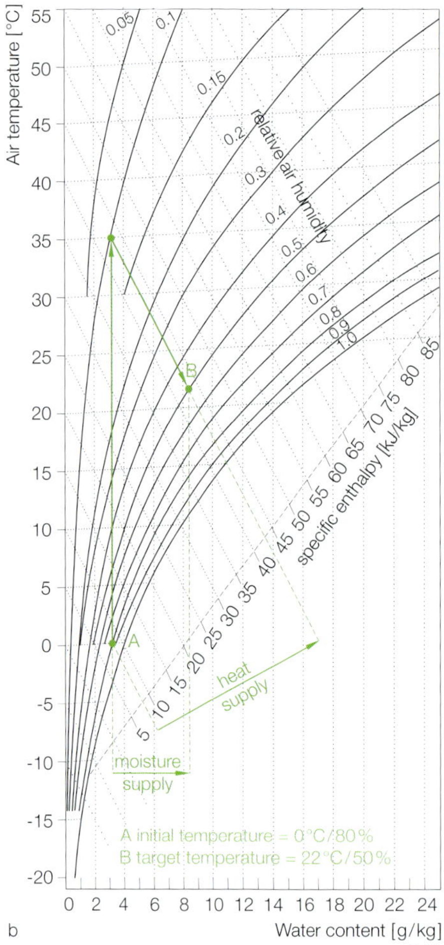

b

3.41

- air flow resistance
  Round air ducts with a low proportion of elbows offer the least air flow resistance. They are superior to all other duct shapes with the same cross sectional area.
- cross section of the air duct
  The bigger the cross sectional area of the duct, the lower the air velocity and therefore also the energy demand of the fan.

A lack of space is usually the reason for installing small, flat ducts with a rectangular cross section. However, a high air velocity is needed to transport the required air volume to the respective rooms. The necessary pressure increase, expressed in pascal, is determined by summing up the partial pressure increases (apparatus, duct cross section, pipeways) in accordance with the air velocity. Figure 3.42 explains the facts and circumstances for larger air conditioning plants. The pressure increase for systems with shorter pipeways is much lower, which also has a positive effect on the necessary fan capacity. The detailed planning of the duct system and the ensuing pressure loss is part of the building services engineer's scope of work.

Standard values for fans

Fans	Specific fan power [W/(m³/s)]	Total pressure increase for a fan efficiency of 60 % [Pa]
Extraction fans	1250	750
Air intake fan with heater	1600	960
Air intake fans in (partial) air conditioning units	2000	1200

If better values are to be assumed for the energy assessment than those listed in this table, it is necessary to provide proof according to DIN EN 13779.

3.42

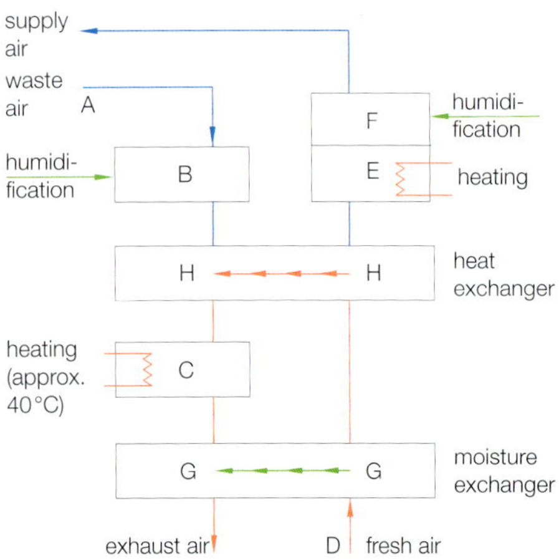

a

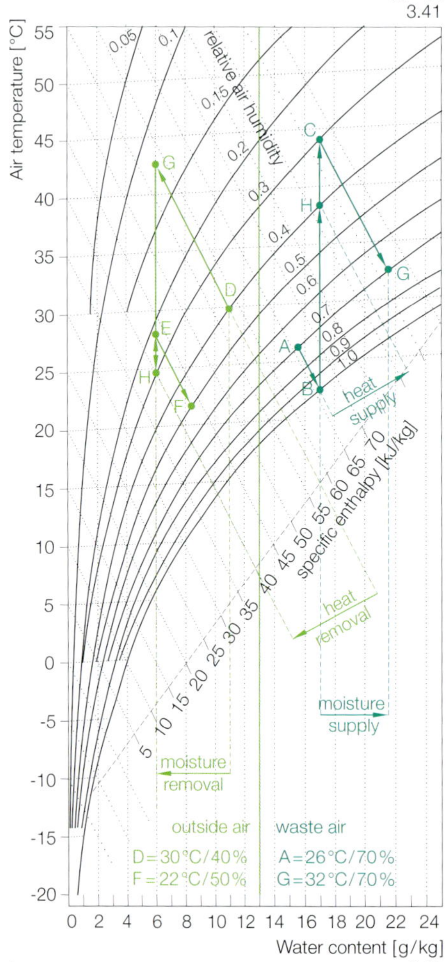

b

3.43

Energy demand for air conditioning

The conditioning of air can be divided into four sub areas - heating, cooling, humidification and dehumidification - each with their specific energy demand. These are explained in greater detail in the following sections.
The use of heat from renewable sources and additional humidification constitutes a special type of air conditioning (fig. 3.43 a and b). It is an especially interesting alternative if heat is available from a district heating network or waste heat from a block-type thermal power station is used.

3.41 Mollier diagram: air conditioning with
a cooling and dehumidification (summer)
b heating and humidification (winter)
3.42 Standard values for the power output of fans according to DIN V 18599-7
3.43 Cooling with heat: adsorption refrigeration machine (heating and dehumidification)
a schematic diagram of technology
b Mollier diagram
A waste air (26 °C/70 %)
B humidification: temperature decreases
C heating: relative humidity decreases
D outside air (30 °C/40 %)
E heating
F humidification to 22 °C/50 % (supply air)
G moisture exchange in sorption wheel:
waste air absorbs moisture and cools down
waste air releases moisture and heats up
H heat exchange:
outside air cools down
waste air absorbs heat

Air heating
A heat input of 0.34 Wh is required to raise the temperature of 1 m³ of air by 1 Kelvin. The heating process can be performed by guiding air over a heating coil, which is provided with the necessary amount of heat by the heat generator. The Mollier diagram highlights the fact that the relative humidity always decreases when the temperature of the air increases. This phenomenon can be the cause for uncomfortable room conditions in winter especially if the humidity value drops below 30 % (fig. 3.41 b).
To save heating energy, the cold outside air can be preheated by drawing it in through a ground tube or through ducts in an underground garage before entering the building. From an energy perspective, it also makes sense to transfer the heat contained in the waste air to the incoming outside air. The following methods can be used to achieve this:

- recuperative heat exchanger
  Supply and waste air are kept separate in their respective closed circuits without direct contact between the air flows. Merely the heat energy contained is exchanged by increasing the contact area between the air ducts (e.g. plate heat exchanger). In large plants, this method ensures that up to 75 % of the heat contained in the waste air is recovered; in the case of smaller plants, heat recovery is up to 90 %.
- regenerative heat exchanger
  This heat exchanger does not only exchange the sensible (measurable) heat, but also the latent heat contained in the moisture. The heat transfer is performed using a rotating sorption wheel, which absorbs the heat and moisture from the waste air (medium exchange) and transfers it to the much colder supply air. Due to the recovery of heat and moisture, regenerative heat exchangers are more efficient than recuperative heat exchangers.
- decentralised heat exchanger
  Regenerative and recuperative heat exchangers work only if the waste air and fresh air ducts are brought into direct contact. The specific duct systems these methods require are often difficult to realise or not feasible due to structural circumstances. In the case of a decentralised heat exchanger, the heat transfer is performed in a decentralised fashion using a transfer fluid that circulates between the fresh air and waste air ducts.

Air cooling
Air coolers are constructed in a similar way to air heaters. The air flows past cooling coils, which a refrigeration unit supplies with cold water. When the air temperature is reduced, the relative humidity of the air increases. This phenomenon leads to uncomfortable room conditions and possibly also damage caused by moisture if the relative humidity in the room rises above 60 %. To avoid this situation, the air must be dehumidified (fig. 3.41 c).
To reduce the energy demand for cooling, the outside air can be precooled (drawn in through the ground or an underground garage) and the thermal energy contained can be recovered by using one of the aforementioned procedures.

Dehumidification
Air can either be dehumidified by reducing the temperature or through contact with a hygroscopic material, which absorbs water like a sponge. If the aim is to ensure that the room air is set to a precise temperature and humidity level in summer, as is the case for sophisticated room functions, such as laboratories, museums, medical treatment areas, etc., the first choice should always be to achieve dehumidification through cooling. This is where the physical phenomenon that cold air absorbs less water vapour than warm air comes into effect. So in order to dehumidify the air, the temperature is reduced below dew point and condensation starts. The cooling process is continued until the air meets the temperature at which the target temperature can be reached purely through heating. The dependencies between initial and target temperature, relative and absolute humidity as well as the energy content are presented in the Mollier diagram (fig. 3.41 a). With the help of the diagram, it is possible to calculate the process precisely and determine the necessary energy input for both steps (see example opposite). The high energy demand can be attributed to the large difference between initial, intermediate and final temperature.

Humidification
If the air is too dry and must be humidified, for example in winter when cold outside air is drawn in and heated, the process must be performed in two steps, similar to the dehumidification process. This explains the high energy demand required in these circumstances. In a first step, the air is heated to a temperature

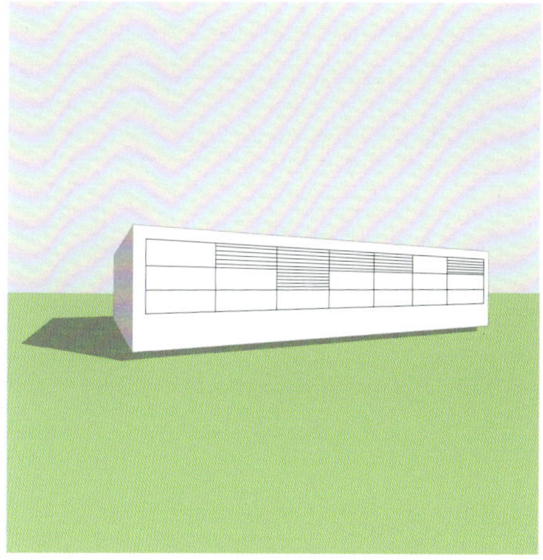

Example: ventilation and air conditioning
The open-plan office (floor area 121 m², operation period 11 h) is to be conditioned. The energy demand for air movement and air treatment provided by an air conditioning unit is compared with a reduced solution with individual units for heating, cooling & ventilation.

Air movement
Air exchange rate (zone 3 DIN V 18599) 726 m³/h
Electricity demand
- waste air 0.40 Wh/m³ (at 750 Pa)
- supply air 0.45 Wh/m³ (at 800 Pa)
- supply air for air con. unit 0.55 Wh/m³ (at 1200 Pa) (due to pressure increase)

Individual units (only for supply of fresh air)
Energy demand for air movement per day:
- supply air 0.45 Wh/m³ · 11 h · 726 m³/h = 3.6 kWh
- waste air 0.40 Wh/m³ · 11 h · 726 m³/h = 3.2 kWh

Air conditioning system
Energy demand for air movement per day:
- supply air 0.55 Wh/m³ · 11 h · 726 m³/h = 4.4 kWh
- waste air 0.40 Wh/m³ · 11 h · 726 m³/h = 3.2 kWh

Comment
If, as has been assumed here, the volume of fresh air suffices to move the required amount of heat/cold, there is only a small extra expense for air movement in the case of a central air conditioning system.

Air humidification (winter)
In order to provide exactly 22 °C / 50 % in the interior space at an outside temperature of 0 °C and a relative air humidity of 80 %, the following energy input is required to humidify the air per day (fig. 3.41 b):
- additional air heating (7900 m³) from 21 to 39 °C in order to finally achieve the target temperature through humidification
  23 kJ/kg · 0.28 Wh/kJ · 1.2 kg/m³ · 7986 m³ = 62 kWh
- water input
  6.5 g/m³ · 7986 m³ = 52,000 g = 52 l
- energy input for the vaporisation of the water
  52 l · 1.16 Wh/l · K · 90 K = 5.4 kWh

Dehumidification (summer)
In order to provide the desired condition (22 °C / 50 %) in the interior space at an outside air of 32 °C / 60 %, the following energy input is required to dehumidify the air per day (fig. 3.41 a):
- air is cooled from 26 to 11 °C in order to finally achieve the target temperature through heating:
  42 kJ/kg · 0.28 Wh/kJ · 1.2 kg/m³ · 7986 m³ = 114 kWh
- heating of the air from 11 to 22 °C:
  12 kJ/kg · 0.28 Wh/kJ · 1.2 kg/m³ · 7986 m³ = 3.2 kWh

Additional energy input for air conditioning system

winter day:	electricity (fan)	0.8 kWh
	heat	67.4 kWh
summer day:	electricity (fan)	0.8 kWh
	heat	3.2 kWh
	cold	114.0 kWh

Comparison:
The energy input for air conditioning with individual systems (without humidification and dehumidification) amounts to 35 kWh in winter and 16 kWh in summer per day.

Specific energy parameters for July

Type number	Humidity requirements			Type of humidification			Type of HR			Energy parameters for supply air 18 °C, operation period 12 h, 31 days		
	none	with tolerance	without tolerance	uncontrolled humidification	controlled humidification	evaporative humidifier	none	heat 75 %	heat and humidity	heat q_h [Wh/(m³h)]	steam q_{st} [Wh/(m³h)]	cold q_c [Wh/(m³h)]
1	■						■			99	–	474
4	■							■		–	–	471
5		■		■			■			112	–	496
8		■		■				■		13	–	489
12		■			■		■			103	–	488
15		■			■			■		4	–	481
19		■				■	■			103	–	484
22		■				■		■		44	–	481
26			■	■			■			697	–	1394
29			■	■				■		682	–	1385
33			■		■		■			591	–	1273
36			■		■			■		463	–	1269
40			■			■	■			538	76	1307
43			■			■		■		464	76	1292
Specific energy parameters for December												
1	■						■			1938	–	–
4	■							■		6	–	–
5		■		■			■			2668	–	–
8		■		■				■		1214	–	–
12		■			■		■			2670	–	–
15		■			■			■		601	–	–
19		■				■	■			1893	–	–
22		■				■		■		4	–	–
26			■	■			■			3292	–	–
29			■	■				■		1239	–	–
33			■		■		■			3295	–	–
36			■		■			■		1207	–	–
40			■			■	■			1855	1462	–
43			■			■		■		2	1461	–

3.44

Data plate of an air conditioning unit (sample)

Type/year of manufacture: NN/1982	Work order number: NN
Air capacity: 5000 m³/h	Motor capacity: 2.2 kW
Δp ext.: 420 Pa	Nominal current: 8.5/4.9 A
Heating capacity: 72 kW	Voltage: 220/380 V
Cooling capacity: – kW	Rotation speed: 1500 min⁻¹

3.45

3.44 Monthly specific energy parameters for cooling, heating, humidification and dehumidification consumed by a variety of plants to condition the air according to DIN V 18599-3 (selection)
3.45 Data plate of an air conditioning system (sample)
3.46 Example of an office building: air-based heating and cooling system without humidification and dehumidification, use of renewable energy sources
- reduction of mechanical ventilation due to natural extraction in summer through atrium
- preheating/precooling by drawing air in through ground tube
- air heating and cooling with a heat pump (groundwater is the heat/cold source)

a section with energy and ventilation concept
b technical concept
c building component activation by installing air ducts into the concrete floor slab

clearly above that of the room; in a second step the air is humidified, which has the effect that the air is cooled, meeting - if the settings are accurate - the desired target temperature. The Mollier diagram can be used to identify the quantities of energy and water needed to meet the initial and target values (fig. 3.41 b, p. 110). The heat demand is especially high because the air is not only heated to room temperature, but significantly higher in order for the humidification process to take place. The energy demand is increased further because the water must be heated to boiling temperature.

Energy accounting according to DIN V 18599
Since the outside air conditions vary constantly, the energy demand to provide cooling for a year can only be determined by performing simulation calculations. It is for this reason that DIN V 18599-3 has determined some average values for the accounting of different systems. The energy demand for air conditioning has been categorised for different installations by providing monthly parameters for heating, cooling/dehumidification and humidification (fig. 3.44). The energy consumption, which is required to heat or cool 1 m³ of air to 18 °C and possibly also humidify and dehumidify the air, forms the basis of the observation. Other air temperatures are considered in the calculations by using the appropriate factors. The elaborate assessment procedure can only be performed in full by specialist engineers.

Energy efficiency refurbishments
Existing air conditioning plant frequently demonstrates the following deficiencies:
- there is no form of heat recovery
- the air volumes are too large
- the fan motors are inefficient
- most central units lack the flexibility to respond to diverse or changing boundary conditions

It is therefore generally the rule that the entire plant is removed and a new concept is developed. Repair and retrofitting work, for example the installation of a heat exchanger, can only be realised at great expense and difficulty, since the equipment must be dismantled, changed and finally reassembled.
The data plate, which is attached to every appliance, provides a first impression of the air conditioning unit's condition. Figure 3.45 lists the information included on a data plate.
When performing an energy efficiency

refurbishment of a complete air conditioning, partial air conditioning or ventilation system, it is necessary, after carrying out a survey and documentation of the existing plant, to assess and possibly improve the construction-related features and thus the respective heating or cooling loads. The analysis should consider:

- the quality of the insulation, including the windows
- the quality of summer heat protection facilities (e.g. windows, sun protection, storage mass)
- possibilities to ventilate the building

In existing building stock, it is never possible to improve all of these parameters in equal measure. However, it is usually feasible to upgrade the glazed areas, including the sun protection devices, which already leads to a significant reduction in the cooling load. The benefit of these measures, when cooling with air, is a clearly reduced air volume, which in turn has a positive impact on the duct cross sections and the amount of electricity consumed by the fans.
Having conducted a structural analysis and eliminated the most severe flaws, a new plant concept is developed based on a new load calculation and by including energy generated from renewable sources. This holistic analysis often leads to the removal of air conditioning plant in existing buildings or its replacement using separate systems for heating, cooling and ventilation. In this way the energy demand for conditioning a building can be reduced by 70 %. However, without air humidification and dehumidification processes, which can only be carried out thoroughly and reliably with a central air conditioning system, there is a risk of not being able to provide perfectly comfortable thermal conditions during peak load periods.
Only buildings with the highest requirements in terms of temperature and humidity levels (e.g. museums, manufacturing buildings in the high-tech industry) need constant, precise climate conditions that can only be achieved by the provision of a complete air conditioning plant. In these circumstances, the existing system has to be replaced by efficient technical installations, which can also operate using renewable energy.
In all other cases, the air conditioning plant can be replaced by alternative systems, providing the construction-related parameters can be adapted to the prevailing climate conditions (see Heat sources, pp. 38ff. and Heat sinks, pp. 68ff.).

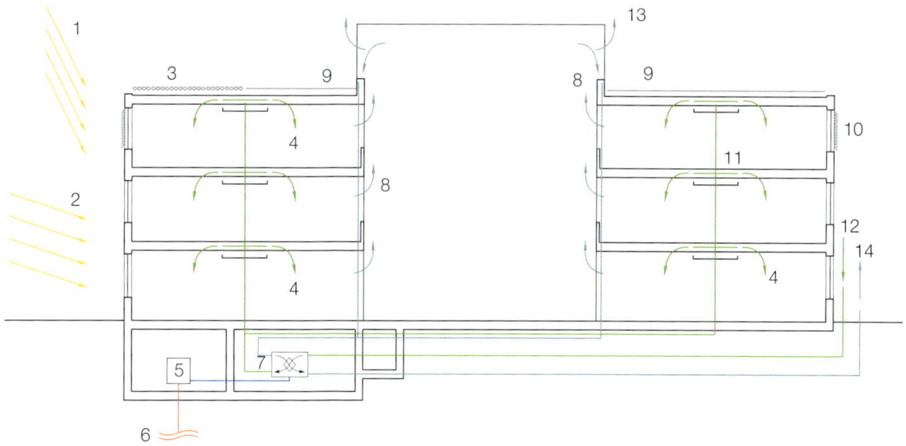

1 summer sun
2 winter sun
3 thermal collectors
4 supply air
5 heat pump
6 ground well
7 ventilation unit with heat recovery
8 waste air (removal through atrium)
9 photovoltaics
10 exterior sun protection
11 thermal building component activation with supply air
12 outside air
13 exhaust air (summer)
14 exhaust air (winter)

a

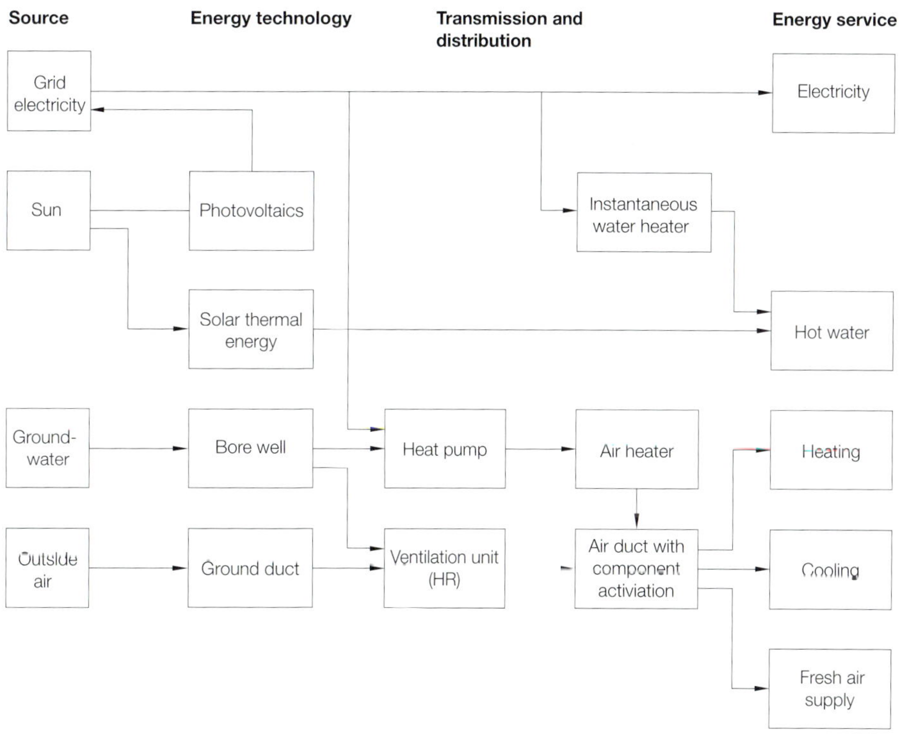

b

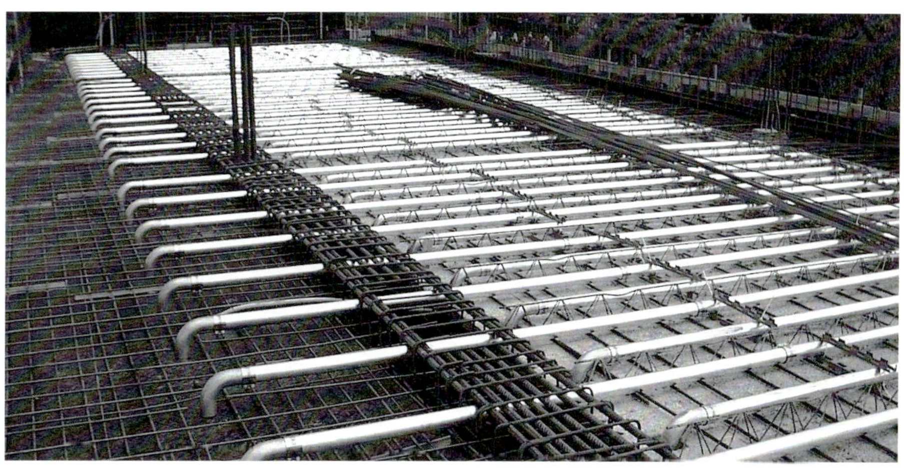

c

3.46

Artificial light

The function of artificial light is, when there is not sufficient natural daylight, to provide a room with the right lighting conditions for the activities assigned to the space. Alongside issues concerning energy consumption, aspects related to ergonomics (comfort, glare) and aesthetics (illumination levels) are also important factors in this respect. In this chapter, the use of artificial light sources is considered solely in terms of energy related issues.

Basics

The operating time of artificial light is dependent on the nature and length of available daylight and the activities being performed in the respective room. The quality or efficiency of the lighting system determines how much electricity is consumed during its operating period. The number of light fixtures required to provide a particular quality of illumination is dependent on the following parameters:

Quality of light source and light fixture
In order to receive a defined illuminance level (lux) in the dark, a specific luminous flux, which is produced by the light source (light bulb), must reach a predetermined surface in the space (fig. 3.47 a). The less energy the lamp requires to provide the luminous flux, the more efficient the lighting. Figure 3.48 lists some typical characteristic values for different light bulbs.
The luminous flux produced by the light source is reduced by the light fixture. This reduction is expressed by the operational efficiency of the light fixture and is usually around 70–80% (fig. 3.47 b). The efficiency of a lighting system is established by measuring the total performance of all installed lamps on a predefined surface (W/m², fig. 3.47 c).

Ballasts in discharge lamps
In the case of discharge lamps (fluorescent lamps), the use of electronic ballasts (EB) can increase the efficiency of a lighting system in comparison to the use of conventional ballasts (CB) or low loss ballasts (LB). The replacement of old lighting plant with inefficient ballasts (CB) by new efficient light fixtures has a savings potential comparable to that achieved through replacing an old boiler with a new one with much lower heat loss.
The extra power demand, which is dependent on the type of ballast used, is presented in DIN V 18599-4, table 2. The electricity consumption of low loss ballasts is 14% higher than that of electronic ballasts; that of conventional ballasts is 24% higher.

Electronic control
The use of an electronic control system with daylight sensors and/or presence monitors in rooms reduces the operating time of lamps significantly. According to DIN V 18599-4, table 22, presence monitors reduce the energy demand for lighting by 20%. Daylight dependent electronic control is especially effective if it turns off the lighting system as soon as there is sufficient daylight, but provides the opportunity for the user to switch it on again manually. This type of control reduces the on-time of lights by approximately 20% and has a corresponding electricity saving effect.

Type of lighting
The type of lighting (direct or indirect) also has a significant effect on the energy demand of a lighting system. From an energy viewpoint, direct lighting is the most efficient; however, it does have an adverse impact on the visual comfort of the persons present due to the severe luminance contrast generated. Lighting concepts which provide a proportion of indirect light to brighten the ceiling improve the situation.
If the room illumination is generated with indirect lighting only, i.e. by light reflection from the ceiling, the energy consumption is considerably in excess of that of a direct system (fig. 3.49).
With the help of DIN V 18599-4, it is fairly easy to compare the installed power capacity (watt) of direct, indirect and mixed lighting systems. According to the standard, the energy demand for a direct/indirect lighting system is around 50% higher than that for a corresponding system with direct light only; the extra demand for electricity for indirect lighting only is as high as 80%.

Energy accounting according to DIN V 18599

The electricity consumption for lighting in a building is dependent on two main parameters. One of these concerns the installed power capacity (p), which makes it possible to produce the required illumination intensity level in the room at dark. The operating time of the lighting system is the second variable influencing the power consumption (see Example: provision of daylight, p. 85).
Concerning the functional description of a lighting system, several aspects must be considered to carry out the accounting. These must be determined in a step-by-step manner by consulting tables or performing calculations:

Step1: required illuminance $\bar{E}_m$
In the case of open-plan offices (user profile 3 of DIN V 18599-10) this value amounts to, for example, 500 lx (fig. 2.54, p. 81).

Step 2: luminous efficacy η_s
The luminous efficacy (lm/W) describes the efficiency of the light source. It corresponds with the quotient of the total luminous flux (e.g. 4600 lm for a fluorescent light tube T8) by the power consumed (e.g. 51 W). In the example presented, the luminous efficacy amounts to 90 lm/W (fig. 3.48).

Step 3: light equipment efficiency η_e
The light equipment efficiency is determined by the design and the light source

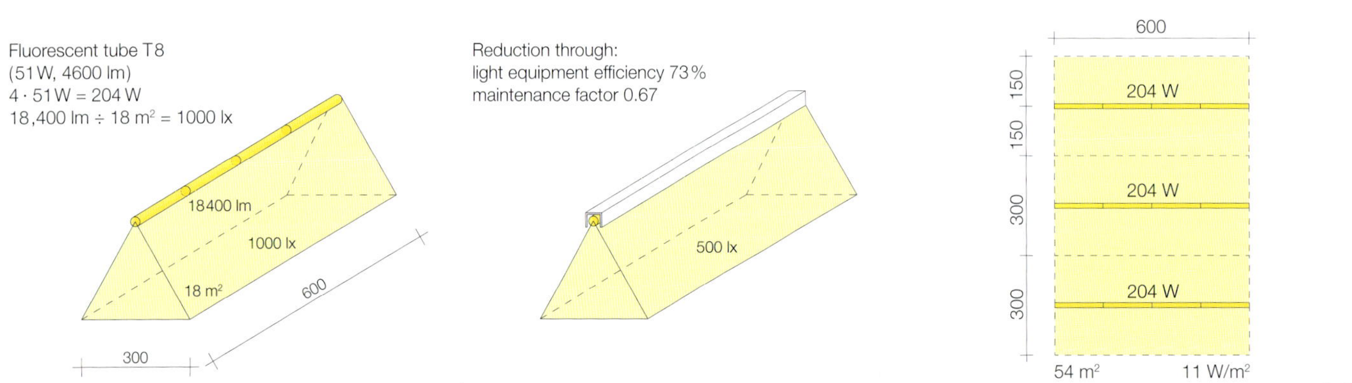

a b c 3.47

used in the selected light fixture. It is therefore dependent on the product and usually accounts for 70–80 % of the luminous flux; in the case of the fluorescent light tube T8, mentioned above, 73 % of the luminous flux can actually be used for lighting purposes (fig. 3.47 b).

Step 4: reduction factor k_a
This blanket value takes account of the fact that, in the case of functional areas with planar illumination, not all surfaces need to receive the predetermined illuminance level. For an open-plan office (zone 3 in the user profile according to fig. 2.54, p. 81), the reduction factor is, for example, 0.92.
There are lighting concepts for offices which combine a basic lighting system generating only 200 lx with additional desk lamps to achieve the required 500 lx on the work surfaces. In these circumstances, the reduction factor can be much smaller; however, it must be calculated for the particular situation.

Step 5: maintenance factor MF
The maintenance factor considers the ageing of the lamp and the dirt and dust accumulation on the luminaire. The required illuminance level must also be achieved when, due to ageing or the collection of dust and dirt on the lamp, the light source emits less light. The maintenance factor MF is a standard value and is set at 0.67. The effect of the maintenance factor is that the installed power capacity must be increased from the outset (see Step 7, p. 116).

Step 6: room lighting efficiency η_r
The *room lighting efficiency* incorporates, on the one hand, the geometry of the space in the form of the room index k, and, on the other hand, the reflective properties of the room enclosing surfaces. The room index k is calculated according to the following equation with the room depth d_r, the room width w_r and the height h'_r of the light fixture above the surface being illuminated:

$$k = \frac{d_r \cdot w_r}{h'_r \cdot (w_r + d_r)}$$

The greater the distance between the light fixture and the surface being illuminated, the lower the room index k. A small room index leads to an increase of the necessary light output.
Once the room index has been established, the *room lighting efficiency* η_r can be determined according to the reflective

Light bulb types: connected load [W] and emitted luminous flux [lm]

Luminous flux [lm]	Incandescent [1] [W]	LV halogen [2] [W]	Energy saving light bulb [W]	Fluorescent T8 [3] [W]	Fluroescent T5 [4] [W]	LED [5] [W]
400	40		7	8		
700	60		11			
800		50				
900	75		15	13	14	14 (7)
1200			20	16	14	
1300				18		
1400	100					
1500		75	23			
2000		100				
2500				30	28	
3000				36	35	
5000				58	54	

In order to assess the efficiency (luminous efficacy η_s), the luminous flux is divided by the connected load [lm/W]; approximate values, dependent on product
[1] incandescent light bulbs below 80 W may no longer be sold
[2] low voltage halogen bulb (with transformer)
[3] fluorescent light tube, 26 mm diameter
[4] fluorescent light tube, 16 mm diameter
[5] light emitting diode (46.6 cm x 10.5 cm x 10.5 cm)

3.48

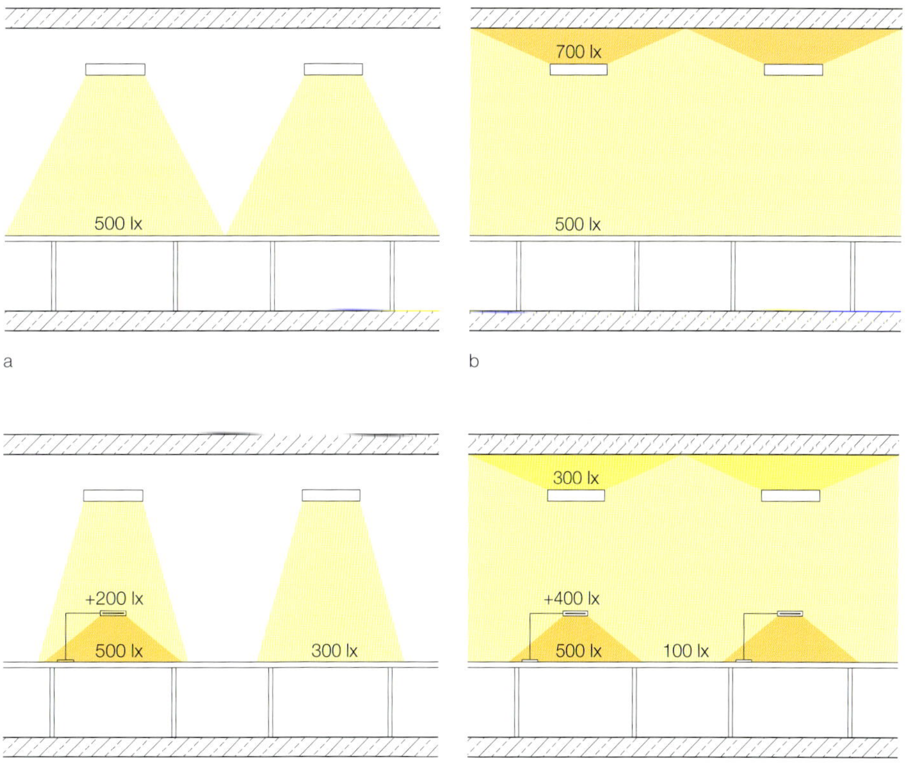

3.49

3.47 Lighting system for office use (concept)
a light source (bulb)
b light fixture (lamp)
c lighting system
3.48 Different light sources
3.49 Various concepts in order to provide the required illuminance of 500 lx on the work surface:
a direct lighting
b indirect lighting
c direct lighting with additional desk lights
d indirect lighting with additional desk lights
When using indirect light only, the electricity consumption to produce the same illuminance level is very high.

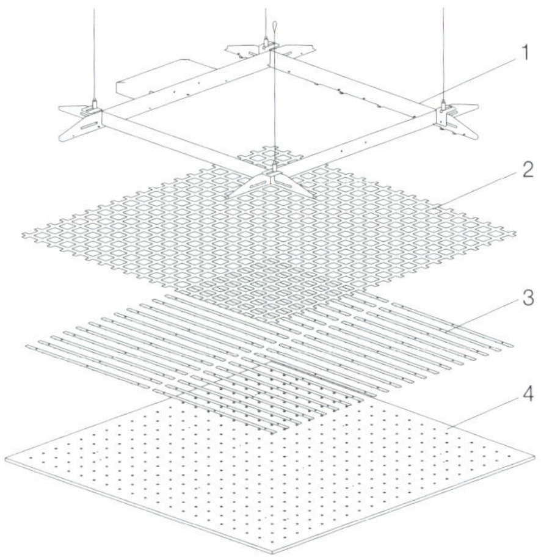

a

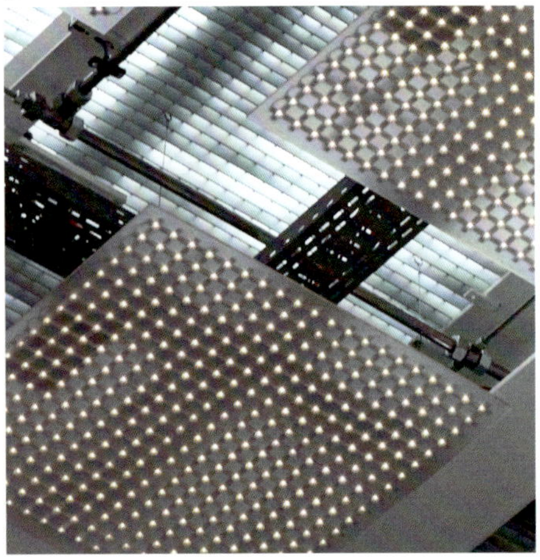

b

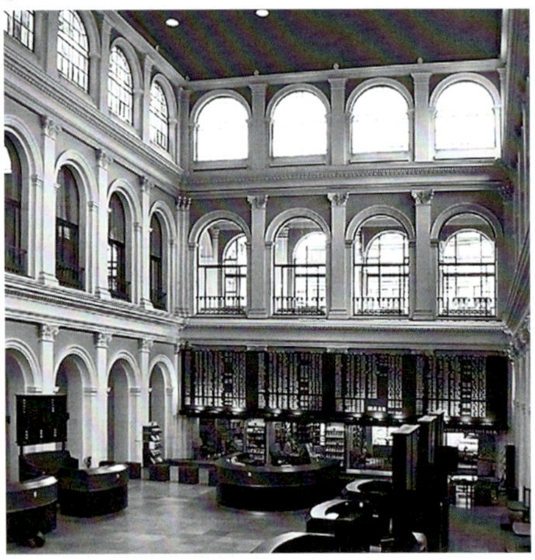

c

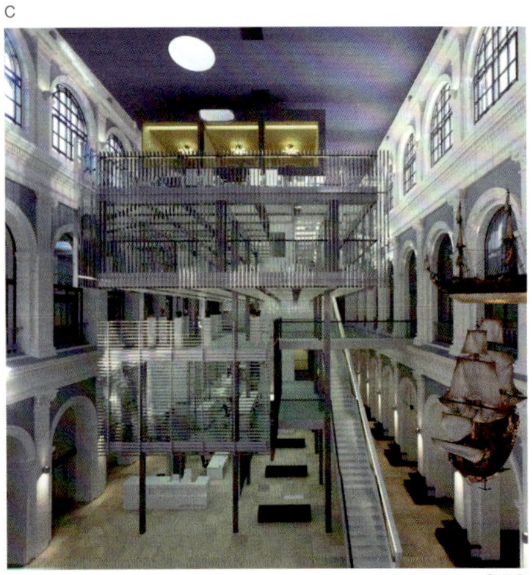

d 3.50

capacity of the room-enclosing surfaces, which are listed in a table (fig. 3.51). The reflection factor is material related and must be calculated by the light designer. On average, the factors 0.7–0.8 are used to describe ceilings and walls with light surfaces; the factors 0.4–0.6 for those with medium light surfaces and the factors 0.2–0.3 for those with dark surfaces. On the other hand, DIN V 18599-4 lists standard values for the reflective capacity of surfaces (dark floor 0.2 / wall 0.5 / ceiling 0.7) so that it is actually possible to consult a table to determine the *room lighting efficiency* in accordance with the type of lighting selected (fig. 3.52). However, possibilities to influence the *room lighting efficiency* by changing the reflective properties of the surfaces are no longer applicable.

Step 7: installed power capacity p
The floor area-related installed power capa-city p required to perform visual tasks is calculated as follows:

$$p = \frac{k_a \cdot \bar{E}_m}{WF \cdot \eta_s \cdot \eta_e \cdot \eta_r}$$

Step 8: lighting power consumption
The multiplication of the area-related installed electric power p with the size of the area to be illuminated and the number of hours during which the daylit areas actually require artificial lighting gives the annual power demand for artificial lighting in daylit rooms.
The result is extended by the power demand for operating hours at night. This aspect is considered with a very simple calculation. The size of the area requiring illumination is multiplied by the number of hours (207 hours at night in zone 3) and the installed power capacity.
These calculations can be used to estimate the energy demand for lighting in a building. If the way in which the various parameters influence each other is understood, it is possible to improve and assess the lighting system in terms of energy related aspects.

Energy efficiency refurbishments
In many existing buildings, the lighting system is responsible for a large proportion of power consumption and therefore also the high CO_2 emissions and primary energy demand of the building. The high level of consumption is caused mainly by insufficient use of available daylight and outdated lighting systems.

The following weak points are often identified in existing building stock:

- excessive lighting (connected loads of up to 40 W/m² in contrast to more suitable values of 10 to 15 W/m²)
- outdated lighting systems (lack of or outdated ballasts)
- inefficient light bulbs with poor luminous efficacy (lm/W)
- lack of facilities to control the lighting system

A first simple step to improve existing lighting systems is to reduce the frequently excessive number of light fixtures by simply removing some of the lamps. The second step is to examine whether and how further improvements can be obtained, for example by exchanging the light bulbs.
The energy demand for lighting systems can be reduced by approximately one third by optimising the provision of daylight (see Daylight, pp. 80ff.) and then implementing a more efficient lighting system.
Typical parameters for old lighting systems and systems that have been improved according to a holistic concept are presented in figure 3.53. The effect of the lighting system's upgrade is measured by the improved installed electric load (W/m²), since the new technology provides the same luminous efficacy with a lower consumption of electricity (line 2). With the help of electronic control systems, such as presence detectors and daylight sensors, the operating time of lamps is reduced significantly (line 3). Changing the illumination system from direct lighting to direct/indirect or even indirect lighting leads to an increase of the

3.50 "House in a house", Hamburg's Chamber of Commerce, Behnisch Architekten, conversion in 2007
The architects Behnisch Architekten inserted a five-storey building into the classicist hall of Hamburg's Chamber of Commerce. It incorporates meeting rooms, a library, a restaurant and a business start-up centre. The rooms are exclusively illuminated using an LED lighting system. A total of 160,000 light spots were installed into 380 rectangular light ceiling panels.
a structural diagram of a ceiling panel
1 support structure made of L sections
2 2 mm aluminium grid
3 LED strip lighting, 40 strips
4 8 mm diffuser panel
b LED panel ceiling
c hall before the conversion
d hall with the five-storey new build
3.51 Room lighting efficiency for direct lighting: in the case of a room index of 2.0 and light room-enclosing surfaces, the room lighting efficiency is 1.09; in the case of dark surfaces 0.72.
3.52 Room lighting efficiency according to DIN V 18599-4
3.53 Energy efficiency evaluation of different lighting systems

connected load, since a greater luminous flux must be generated (line 4 and 5). Despite the greater consumption of electricity, there are considerable benefits as regards comfort conditions. The improved use of daylight through construction-related alterations leads to an increase of areas provided with natural daylight and thus a reduction of the lighting system's operating time (line 6).
It is also worthwhile checking whether and in which way it might be possible to use innovative light products, such as LEDs (light-emitting diodes) (fig. 3.50). The advantage of these light products is not so much the energy efficiency (lm/W) as the low surface temperature, which means a significant reduction of heat gains. It might even be possible to reduce the cooling loads sufficiently to decrease the size of the refrigeration system or even remove it entirely. The expense of these high end lighting systems is a clear disadvantage.
In particular in the case of refurbishments, it is therefore necessary to check whether the cost of replacing the whole system is justified in relation to the savings and better comfort conditions achieved through the alterations.

Room lighting efficiency η_r for direct lighting

	Reflective capacity p (ceiling mounted lamps)									
Ceiling	0.8	0.8	0.8	0.5	0.5	0.8	0.8	0.5	0.5	0.3
Wall	0.8	0.5	0.3	0.5	0.3	0.8	0.3	0.5	0.3	0.3
Work surface	0.3	0.3	0.3	0.3	0.3	0.1	0.1	0.1	0.1	0.1
Room Index k	**Room lighting efficiency η_r**									
0.6	0.73	0.46	0.37	0.44	0.36	0.66	0.36	0.42	0.35	0.35
0.8	0.82	0.57	0.47	0.54	0.46	0.74	0.45	0.51	0.44	0.44
1.0	0.91	0.66	0.56	0.62	0.54	0.80	0.53	0.59	0.52	0.51
1.25	0.98	0.75	0.65	0.70	0.62	0.85	0.61	0.66	0.60	0.59
1.5	1.03	0.82	0.73	0.76	0.69	0.89	0.67	0.72	0.66	0.65
2.0	1.09	0.91	0.82	0.84	0.78	0.94	0.75	0.78	0.73	0.72
2.5	1.14	0.98	0.90	0.90	0.84	0.97	0.81	0.83	0.79	0.77
3.0	1.17	1.03	0.96	0.95	0.90	0.99	0.86	0.87	0.83	0.82
4.0	1.20	1.09	1.03	1.00	0.95	1.01	0.91	0.91	0.88	0.86
5.0	1.22	1.13	1.07	1.03	0.98	1.03	0.93	0.93	0.91	0.89

3.51

Room lighting efficiency η_r as a function of the type of lighting and the room index

Type of lighting	**relative lower semi-room luminous flux of the light fixture φ_u**	**Room index *k***									
		0.6	**0.8**	**1**	**1.25**	**1.5**	**2**	**2.5**	**3**	**4**	**5**
direct	≥ 0.7	0.48	0.59	0.67	0.76	0.82	0.89	0.94	0.98	1.02	1.05
direct/indirect	$0.1 \leq \varphi_u < 0.7$	0.23	0.30	0.36	0.43	0.48	0.56	0.62	0.67	0.73	0.77
indirect	< 0.1	0.17	0.23	0.29	0.36	0.41	0.48	0.53	0.57	0.62	0.65

Intermediate values of the room index can be interpolated.

3.52

Energy efficiency assessment of light systems

	Connected load [W/m²]	**Operating hours [h]**	**Final energy [kWh]**	**Primary energy [kWh]**	**CO_2 emissions [kg]**
1 outdated system	25	1600	4000	10400	2572
2 new system with EB [1]	14	1600	2240	5824	1440
3 new system with EB[1], daylight sensor and presence detector	14	1000	1400	3640	900
4 new system as in 3, however indirect	18	1000	1800	4680	1209
5 new system as in 4, however with desk lamp	16	1000	1600	4160	1093
6 new system as in 5, however with improved provision of daylight	10	800	800	2080	514

The values refer to an office room with a net floor area of 100 m², a room depth of 5 m, a 50 % proportion of window area and light room-enclosing surfaces. Primary energy factor: 2.6 (electricity mix Germany)
[1] electronic ballast

3.53

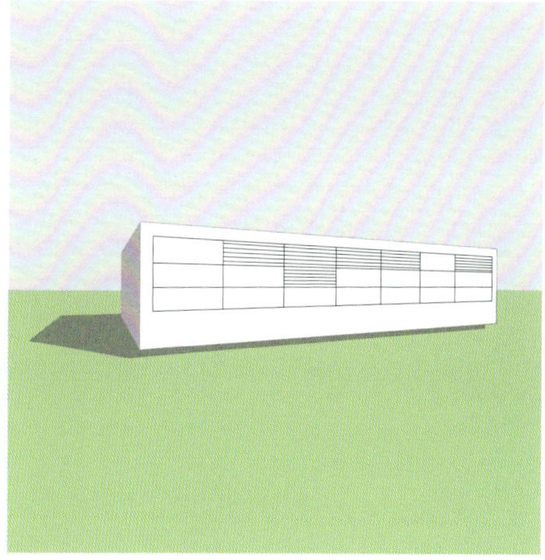

Example: artificial light
In this example, the annual electric power demand for lighting is examined for the open-plan office in the pavilion with a single fully-glazed facade.

Data
see Example: daylight, p. 85)

Step 1–6

illuminance $\bar{E}_m$	500 lx
luminous efficacy η_0	80 lm/W
light equipment efficiency η_e (dependent on product)	0.7
reduction factor k_a	0.92
maintenance factor MF	0.67 (standard value)
room index k (d_r = 16.80 m / w_r = 7,20 m / h'_r = 2.20 m)	2.3
reflective capacity of floor/wall/ceiling	0.2/0.5/0.7
room lighting efficiency η_r	0.95

Step 7: installed power capacity
$p = (0.92 \cdot 500) \div (0.67 \cdot 80 \cdot 0.7 \cdot 0.95) = 12.9\ W/m^2$

Annual electric power demand

total operating time per year	2750 h
daytime operating time per year	2543 h
nighttime operating time per year	207 h
areas provided with natural light	92.4 m²
areas not provided with natural light	28.6 m²

Power demand for nighttime operation per year:
- 207 h · 121 m² · 12.9 W/m² = 323 kWh

Power demand for daytime operation per year:
- areas provided with daylight:
  610 h · 92.4 m² · 12.9 W/m² = 727 kWh
- areas not provided with daylight:
  2543 h · 28.6 m² · 12.9 W/m² = 938 kWh

total power demand per year 1988 kWh
specific value 16.43 kWh/m²
annual primary energy demand 42.7 kWh/m²

Construction-related measures: improved provision of daylight (see Example: daylight, p. 85)
Power demand for nighttime operation per year
- identical 323 kWh

Power demand for daytime operation per year:
- areas provided with daylight:
  410 h · 121 m² · 12.9 W/m² = 640 kWh
- areas not provided with daylight:
  none (total area is provided with daylight)

total power demand per year 963 kWh
specific value 8.0 kWh/m²
annual primary energy demand 19.2 kWh/m²

Measures taken to improve lighting system
The installed system can be improved by providing a differentiated lighting concept (k_a factor), by using more efficient light bulbs (luminous efficacy η_s) and light fixtures (η_e factor); furthermore by installing an electronic control system and choosing lighter colours for the room-enclosing surfaces (room lighting efficiency η_r). These alterations can bring about an installed power capacity of 10 W/m².
total electric power demand per year 597 kWh
specific value 4.9 kWh/m²
annual primary energy demand 12.7 kWh/m²

Photovoltaics

The principle of photovoltaics is the direct conversion of sunlight into electricity. This phenomenon was first discovered by Edmond Becquerel in 1839; it was then further developed after the Second World War for space travel. The aim was to ensure an autonomous energy supply of space stations using solar-generated electricity.
The photovoltaic principle has found greater acceptance since it has become possible to feed the generated electricity into the public grid. Power generation and power consumption are no longer immediately dependent on each other, and the electricity generated by photovoltaics can now contribute towards the public power supply.
In 2010, the proportion of PV power generation in the German electricity mix was 2 %.The aim is to increase this figure to 10 % by 2020 and, at the same time, reduce the costs for the production of the panels by 50 %. In order to achieve this target, the installation of photovoltaic plants is subsidised by offering investors higher feed-in tariffs; however, the number of grants issued is on the decline.

Technology
There are three different technologies which can be used to generate electricity by photovoltaic cells. They differ considerably in their yield and manufacturing costs (fig. 3.57):
- crystalline silicone cells
  (thin films – wafers)
- thin film solar cells
  (coatings produced by vacuum or galvanic deposition)
- organic solar cells
  (printing process with organic nano structures)

The individual cells are added together to make solar modules (panels). These are then assembled in a variety of arrangements to form photovoltaic arrays (fig. 3.54).
To compare the electrical output of different cells, the respective constructions are tested under standard test conditions (STC). These imply a vertical incidence of sunlight of 1000 W/m², a cell temperature of 25 °C and an air mass (AM) of 1.5. The air mass describes the consistency of the air, which in turn has an impact on the solar spectrum of light. In Europe the AM value is 1.5, at the equator it is 1.0. The performance of the solar cell, which is calculated in a laboratory under these test conditions, is referred to as kW_p (p stands for peak). However, the performance data collected under such abstract laboratory conditions does not offer a very precise prediction of the actual yield the various systems can achieve in their practical application.
Innovative solar cells could in future generate electricity in an environmentally-friendly and inexpensive way if they were, for example, mounted in foldable chargers for mobile phones or on car roofs. They are made of organic semiconductor materials, are bend-able and thin like cling-film; they are also very light and can vary in colour. In the building industry, these organic solar cells could be used as thin photovoltaic films covering roofs, facades or even glazed areas.
Whereas photovoltaic elements based on silicon can convert 20 % of solar energy into electricity, technologies based on organic materials can only convert approximately 5 % into electricity (fig. 3.55 and 3.57). The low power yield of organic solar cells is however compensated for by the expected low manufacturing costs.

Power output
The power output of photovoltaic plants is very much dependent on the geographical position and the following plant-related circumstances (performance ratio, PR):
- orientation
- shading issues
- surface temperature of the array (dependent on the situation of the installation)
- performance of the array (kW_p)

Strong sunlight and no overshadowing are the most fundamental prerequisites for the perfect operation of a photovoltaic system. Partial shading interrupts the flow of electricity and leads to reductions of the power output, or even the collapse of the entire solar system. It is for this reason that it is absolutely essential to plan the array carefully using simulations to identify the incidence of sunlight and take into consideration possible buildings and trees overshadowing the panels.
Furthermore, to improve the power output, it is necessary to keep the surface temperature of the cells as low as possible. This is achieved by removing the heat accumulating on the rear side of the cells, for example, by incorporating a rear ventilation of the modules (fig. 3.56, p. 119).
If in addition the heat is removed with

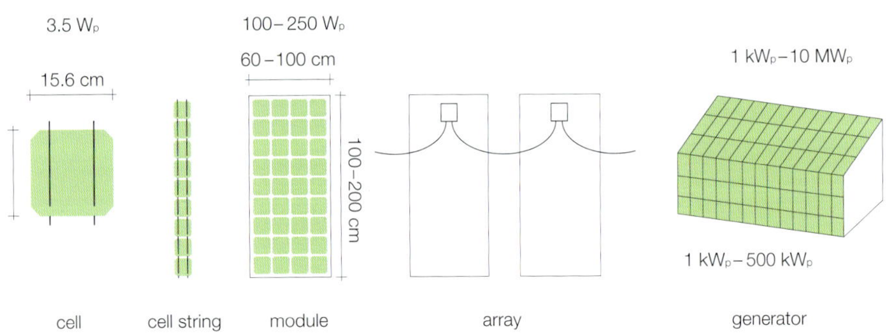

3.54

Performance specifications of different solar cell types

Type of cell	Record cell efficiency (laboratory) [%]	Module efficiency (commercial) [%]	Output per m² of module area [W_p]	Required area for 1 kW_p [m²]	Loss of output due to temperature increase [% / °C]
Silicone monocrystalline	25.0	13–16	130–160	6.5–8	0.4–0.5
back contact solar cells	25.0	18–20	180–200	5–5.5	0.3–0.4
hybrid HIT cells	23.0	16–19	160–190	5.5–6.5	0.3–0.35
Silicone polycrystalline	20.3	12–15.5	120–155	6.5–8.5	0.4–0.5
Silicon, amorphous	12.1	5–9	50–90	11.5–21	0.1–0.2
microcrystalline	10.5	6–8	60–80	13–18	0.5–0.7
micromorph	12.5	8–11	80–110	9–12.5	0.3–0.4
CIS, CIGS	20.3	7–14.5	70–145	7–15	0.3–0.4
CdTe	17.3	9–12.5	90–125	8.5–11	0.2–0.3
Organic solar cells	10.6	(2–3)			
dye cells	11.4	(2–3)			

3.55

3.54 Modular principle of photovoltaics: The solar cell as the basic component can be joined with others to form a solar power system. The values stated refer to monocrystalline silicon cells.
3.55 Performance specifications of different solar cell types

Global radiation horizontal [kWh/m²a]	**Irradiance on generator surface [%]**	**Module area [m²]**	**Module efficiency [%]**	**Performance ratio (PR) [%]**	**Annual yield [kWh/a]**	**Specific yield [kWh/kW$_p$a]**	**Area yield [kWh/m²a]**
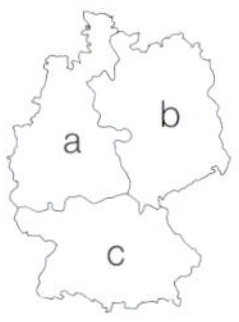 900–1200 kWh/m²a average regional values: a) 975 kWh/m²a b)1025 kWh/m²a c)1100 kWh/m²a	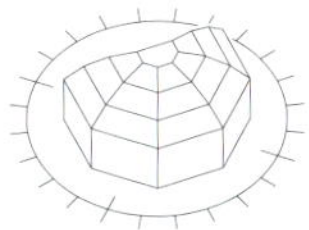 north: 35–100% east/west: 65–116% south: 80–116%		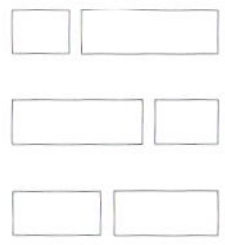 mono/polycrystalline 11.5–20% a-Si, µc-Si 5–9% CIS, CdTe 6–11%	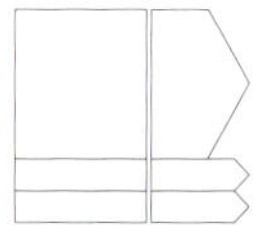 < 70% non-ventilated facade, overshadowed up to 85% in the case of perfect orientation, rear ventilation, detailing, installation, and maintenance	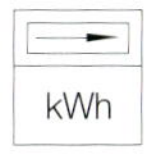 	standard regional values for south-facing or flat roof: a) 850 b) 900 c) 950	mono/poly-crystalline a) 95–170 b) 100–180 c) 105–190 a-Si, µc-Si a) 45–75 b) 47.5–80 c) 50–85 CIS, CdTe a) 50–95 b) 52.5–100 c) 55–105

Examples:

a) 10 kW_p flat roof array in Würzburg with polycrystalline modules (efficiency 14.5%), south-facing, linear arrangement with 30° inclination, module area of 69 m², required roof area approx. 3 · 69 m² ≈ 210 m², mutual overshadowing approx. 2%, estimated performance ratio = 76%

1090 kWh/m²a · 114% · 69 m² · 14.5% · 76% = **9448 kWh/a** ≙ **945 kWh/kW$_p$a** ≙ **137 kWh/m²a**

b) 100 m² non-ventilated facade in Berlin with CIS modules (efficiency 10.5%), south-west facing, no overshadowing, module area of 95 m², PV output = 95 m² · 10.5% = 10 kW_p, estimated performance ratio = 71%

1011 kWh/m²a · 78% · 98 m² · 10.5% · 71% = **5585 kWh/a** ≙ **558 kWh/kW$_p$a** ≙ **56 kWh/m²a**

3.56

Crystalline silicon cells
- monocrystalline
- polycrystalline

Special types:
- high efficiency cells
- hybrid cells

- wafer technology: round to square single elements
- thickness 0.2 mm, side lengths 10–15.6 cm
- market share of approx. 85%, established technology

Thin-film solar cells
- on glass plate
  - amorphous silicon
  - micromorph silicon
  - cadmium telluride (CdTe)
  - copper indium gallium diselenide (CIGS)/ copper indium desulfide (CIS)
- foil or strip cells
  - amorphous silicon
  - copper indium gallium diselenide (CIGS)

- vacuum technology, galvanic: usually large area coating
- thickness 0.5–5 µm, width of cells 0.5–17 mm, width of cell strips 1–36 cm
- market share of approx. 15%, upward trend

Organic solar cells
- foil or glass substrate
  - dye cells
  - polymer solar cells
  - oligomer solar cells

- printing process
- nano structure
- pilot stage

3.57

Installation
- roof
  - flat roof: mounted, integrated
  - inclined roof: mounted, integrated
- facade: non-ventilated, ventilated
- sun protection device
  - fixed system: glass roof, fixed louvres
  - adjustable system: movable shutters, adjustable louvres

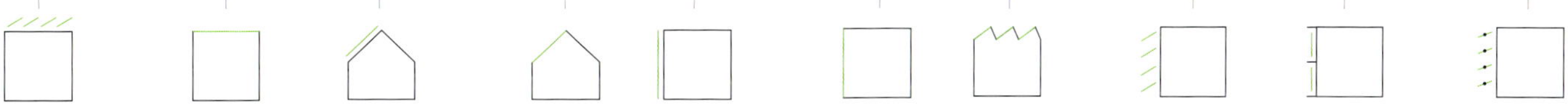

3.58

3.56 Principles and sample calculations for a step-by-step approach to assessing the yield of PV arrays
3.57 Typologies and features of standard solar cells: monocrystalline cells have characteristic rounded corners, polycrystalline cells a distinctive crystalline surface structure and thin-film modules a linear arrangement of cells.
3.58 Installation possibilities for photovoltaics in buildings

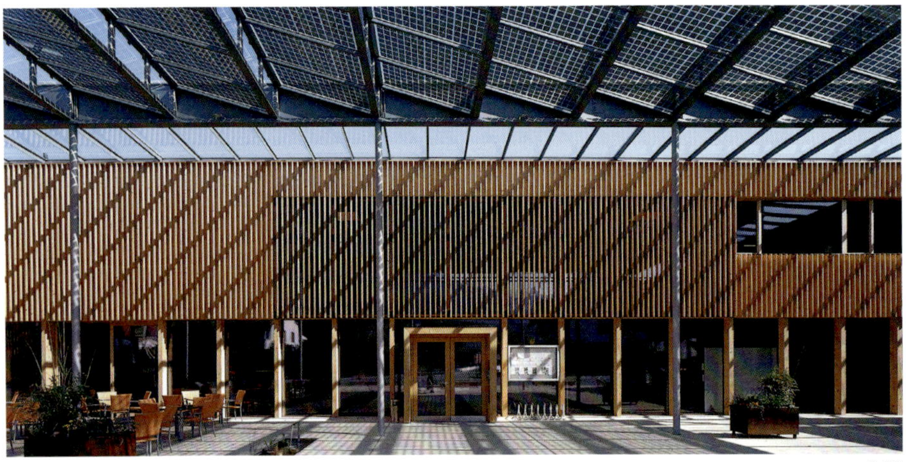

3.59

Temperature increase and reduction of the solar power yield of crystalline modules

Module installation	Temperature increase in comparison to surroundings	Reduction of annual energy yield
non-ventilated facade (warm facade)	55 K	10.5%
non-ventilated roof (warm roof)	43 K	7.5%
badly ventilated facade	39 K	7.0%
well ventilated facade	35 K	6.0%
badly ventilated roof	32 K	5.0%
well ventilated roof	29 K	4.0%
roof mounted at large distance	28 K	3.5%
separate from building	22 K	2.0%

3.60

Feed-in tariffs for electricity from building integrated PV modules [ct/kWh]

	Degression	0–30 kW_p (approx. 0–200 m²)	30–100 kW_p (approx. 200–800 m²)	100–500 kW_p (approx. 800–2000 m²)
1 Jan 2010		39.14	37.23	35.23
1 July 2010	13% additional cutback	34.05	32.39	30.65
1 Oct 2010	3% additional cutback	33.03	31.42	29.73
1 Jan 2011	13% degression (EEG)	28.74	27.33	25.86
1 Jan 2012	15% degression (EEG)	24.43	23.23	21.98
		0–10 kW_p	**10–40 kW_p**	**40–1000 kW_p**
1 April 2012	1% degression [1)]	19.50	18.50	16.50
1 Jan 2013	2.5% degression [2)]	17.02	16.14	14.11

[1] 1 April 2012, general revision of the law for renewable energies (Erneuerbare Energien Gesetz, EEG); from 1 April 2012 regular degression of 1% per month; from 1 Nov. 2012, 2.5% per month

3.61

Remuneration for electricity from building integrated PV modules for personal use [ct/kWh]

Plant size	0–30 kW_p		30–100 kW_p		100–500 kW_p	
Personal use	**30%**	**> 30%**	**30%**	**> 30%**	**30%**	**> 30%**
2010	22.76	27.14	20.85	25.23	18.85	23.23
1 July 2010	17.67	22.05	16.01	20.39	14.27	18.65
1 Oct 2010	16.65	21.03	15.04	19.42	13.35	17.73
1 Jan 2011	12.36	16.74	10.98	15.36	9.49	13.87
1 Jan 2012	8.05	12.43	6.85	11.23	5.60	9.98
1 April 2012 [1)]	—	—	—	—	—	—

[1] Remuneration for personal use of solar power was abolished on 1 April 2012 since the average price of domestic electricity is higher than the feed-in tariff. Personal use of self-generated power is thus also profitable without remuneration.

3.62

3.59 Partially transparent photovoltaic elements as a canopy roof for the central courtyard of the community centre in Ludesch (A), 2006, architect: Hermann Kaufmann, Schwarzach
3.60 Reduction of the solar power yield of crystalline modules according to the situation of their installation
3.61 Feed-in tariffs for electricity from building integrated PV modules in ct/kWh
3.62 Remuneration for electricity from building integrated PV modules in ct/kWh for personal use
3.63 Solar yield and electricity balance for a sample building (see example opposite)

mechanical support, an air-to-water heat pump can help make use of the heat contained in the air. The heat gained can then be used for other purposes, for example hot water production.
This is a principle applied by the so-called hybrid collectors: the waste heat from the photovoltaic elements, which are mounted as the top finish of solar-thermal collectors, is absorbed by the collector and used for heating purposes or the production of domestic hot water.

Energy Saving Ordinance (EnEV)

With the updated version of the German Energy Saving Ordinance (Energieeinsparverordnung, EnEV), issued in 2009, it has become possible to include the electricity generated by photovoltaics into the calculations of the primary energy demand by making a corresponding reduction. However, the requirements according to §5 of the EnEV stipulate that the electricity "must be generated in the immediate vicinity of the building and must be consumed primarily in the building itself. Only excess amounts may be fed into the public grid." Only the power consuming installations included in the calculations performed according to the EnEV can actually be considered as consumers. In the case of central heated residential buildings, in which the energy demand is covered by the combustion of primary resources, the only power demand that may be covered by PV power is the auxiliary energy. If however the production of hot water is performed in a decentralised fashion using instantaneous water heaters, the required electricity can be generated by photovoltaics in summer and deducted from the calculations. The same applies to ventilation units which are operated throughout the year.
The use of electricity generated by renewable sources is particularly interesting when it comes to non-residential buildings, since the power demand for cooling in summer can be satisfied by renewable means. The accounting of electricity is performed on a monthly basis, which means that the electricity yield from the photovoltaic system is offset against the electricity demand in the building each month. If the yield exceeds the demand, the primary energy demand is zero. In the opposite case, the electricity demand which remains after deducting the amount of electricity generated by renewable sources, is multiplied by the standard primary energy factor 2.6. The result is then included in the building's primary energy demand.

Economy
The law on granting priority to renewable energy sources (Renewable Energy Sources Act, Erneuerbare Energien Gesetz, EEG), first issued on 25 October 2008 and last modified on 17 August 2012, regulates the subsidy programmes concerning the use of renewable sources for the generation of energy in Germany. One of the objectives of the legislation is that renewable energy sources shall account for at least 35% of the electricity production by 2020.
The subsidy rates are based on a fixed value established in 2009, which was reduced on an annual basis up until 2012 and on a monthly basis from then on (fig. 3.61). The rates of remuneration established at the time of a system's installation are guaranteed for 20 years.
The remuneration for solar power consumed by the plant owner was abolished with the introduction of the amended law since the average price for domestic electricity exceeded that for the feed-in tariff. The personal consumption of self-generated solar power is therefore clearly worthwhile (fig. 3.62).
Up until 2011, the so-called standard reduction for all solar electricity generated by systems mounted on buildings was 9% per year. The reduction after 2011 was greater if the installed capacity exceeded certain limits. The most recent amendment to the legislation in 2012 was to introduce a monthly degression of 1%, which increased to 2.5 % at the beginning of 2013. However, over the last years, the extreme rate reductions have been counter-balanced by an even greater fall in prices on the photovoltaic market with the result that it is still attractive for house owners to install solar panels. The installation of photovoltaic plant is continuing unabated in Germany where photovolatic power capacity is expected to overtake that of wind power in 2013.

Ecology
Depending on its efficiency, which varies according to location, circumstances on site and the module capacity, a photovoltaic system must be operated for at least 1.5 to 2.5 years in order for the amount of energy generated to compensate for the primary energy demand required for the production of the modules.
For an assumed life expectancy of 20 years, the yield factor is 10–17 (see Appropriateness, pp. 27ff.). The yield factor for photovoltaic systems in Germany is 10. They therefore have an energy payback period of two years.

Energy efficiency refurbishments
There are different possibilities to retrofit photovoltaic modules in existing buildings (fig. 3.58, p. 119). In all situations, it is absolutely fundamental to consider up front the high visibility of the solar panels in the architectural concept of the building. Furthermore, the fact that the yield of an array is reduced by approximately 25% when not perfectly inclined should also be taken into consideration, for example, when fitted in a vertical position on the exterior walls. The drop in efficiency can be compensated for by taking into account the ancillary benefits of so-called building integrated photovoltaics, i.e. including the use of the power-generating components as a weather shield or sun protection system.
Elements that are particularly efficient, due to the added economic, functional, but also urban value included, are ones that act as adjustable louvers or as a canopy roof, as is the case for the courtyard of the community centre in Ludesch (fig. 3.59).
Retrofitted photovoltaic systems always have potential to improve the building's appearance and energy efficiency if they are included in the development of the architectural master plan. Solar panels have, in many respects, the ability to bring about a positive change in existing buildings.

Yield [kWh]	Jan	Feb	Mar	April	May	June	July	Aug	Sep	Oct	Nov	Dec
Solar irradiance	1857	2205	3611	7409	7767	8820	9188	6782	5537	3391	1940	1132
Power yield (11%)	0	243	367	815	854	970	1011	746	609	373	213	0
Power demand												
[kWh]	100	100	100	100	100	100	100	100	100	100	100	100
ventilation	190	160	150	120	90	60	60	90	120	150	160	190
lighting	80	80	80	40	–	–	–	–	40	80	80	80
auxiliary energy	370	340	330	260	190	160	160	190	260	330	340	370
total												
Balance [kWh]												
grid feed-in	0	0	67	555	664	810	851	556	349	43	0	0
personal use	0	243	330	260	190	160	160	190	260	330	213	0
purchased electricity	370	97	0	0	0	0	0	0	0	0	127	370

3.63

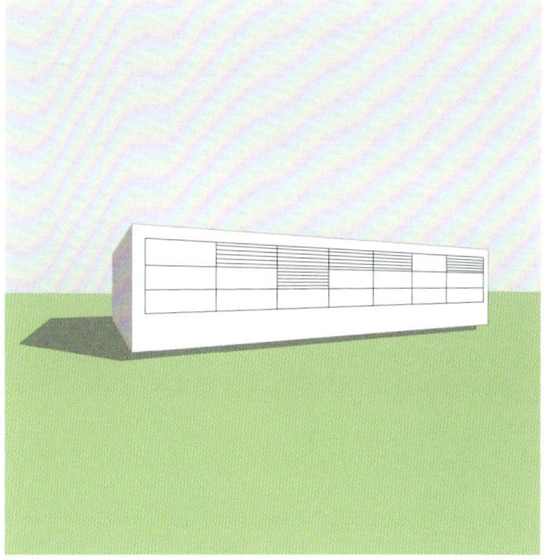

Example: photovoltaics
A photovoltaic array is added to the south-facing and 30°-inclined roof of the pavilion. The distance between the modules has been chosen to ensure that the sun will reach the surfaces without casting any shadows up to an angle of incidence of 28°. The system is switched off in December and January due to overshadowing (fig. 3.63).
This example examines the possible size of the array, its efficiency as well as the crediting of solar power yield according to §5 of the EnEV.

Data
- size of array — 49 m²
- output — 7 kW_p
- costs (300€/m²) — €14,700
- insurance — €36/year
- maintenance — €120/year
- interest on capital — 3.0%
- price increase of maintenance — 3.0%
- life expectancy — 20 years

Costs in 20 years
- annuity of investment and maintenance — €19,923
- insurance — €700
- total — €20,623

Electricity balance
(fig. 3.63)
Solar yield
- solar irradiance — 59,639 kWh/a
- solar power yield (11%) — 6232 kWh/a

Electricity demand
- ventilation — 1200 kWh/a
- lighting — 1540 kWh/a
- auxiliary energy — 560 kWh/a
- total — 3300 kWh/a

Balance (surplus electricity)
- grid feed-in — 3896 kWh/a
- personal use — 2336 kWh/a
- electricity purchased — 964 kWh/a

Income
Percentage of personal use — 37%

Calculation for 1 year
- grid feed-in: 3896 kWh · 0.19 ct/kWh = €720.24
- electricity not purchased due to personal use: 2336 kWh · 0.25 ct/kWh = €584
- total annual savings — €1324.24

Savings in 20 years — €26,484.80

Income in 20 years — €5,861.80

Credit for solar power according to the German Energy Saving Ordinance (Energieeinsparverordnung, EnEV)
By using 2336 kWh of electricity on site, the total primary energy demand is reduced by 6074 kWh.

Residential building

Location:	Munich
Architect:	Clemens Richarz, Christina Schulz, Munich
Structure:	Gustl Lachenmann, Vaihingen
HVS:	Franz Koller, Munich
Building physics:	Clemens Richarz
Building type:	three family dwelling
Year of construction:	1936
Refurbishment:	2001, 2003, 2009

Architecture

A refurbishment to improve energy efficiency not only provides the opportunity to perform necessary structural and technical alterations, but can also be used to unlock the potential which often exists in older residential buildings in terms of functional, spatial, aesthetic and economic features.
Before performing a refurbishment, it is therefore necessary to consider the following issues:

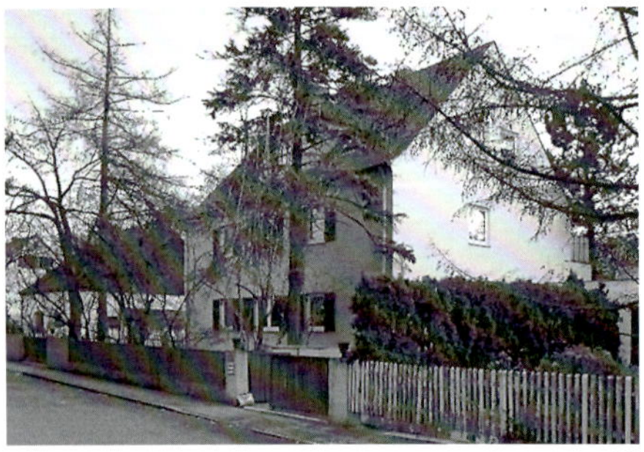

4.1

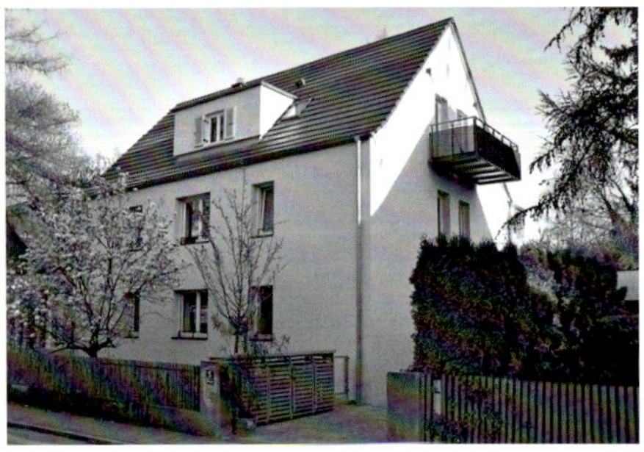

4.2

- Older residential buildings are frequently positioned on relatively large plots and land utilisation is not always the most economic (FAR/PR). It is therefore always necessary to clarify whether the building should be removed or whether a more economic use of the site can be achieved through infill development.
- Opportunities to extend the building are usually fairly straightforward, for example through attic conversions.
- Improvements to the layouts to enhance communication and make better use of the exterior space, or the rearrangement and enlargement of windows are alterations which should be considered and possibly also completed before the refurbishment work actually commences.

Since many residential buildings have to be refurbished whilst occupied – due to existing tenancy agreements – it is sometimes inevitable that the overall concept is realised in several phases. The financial means of the client may also require that not all work is completed in a single phase. In any event, the individual measures should always be an integral component of a greater overall concept.
The three family home analysed here was built in 1936 as a solid structure with one floor below ground, two above and a gable roof.
Since it was vacant, the opportunity was taken in 2001 to strip the interior and fully refurbish the whole building. Since all areas were affected by the alterations, priority was given to a comprehensive renewal of the outdated building services (water supply, heating, electrical installations).
Furthermore, a master plan was developed to implement extensive and durable refurbishment work with the aim of making the most of the property's potential in terms of architectural and energy efficiency features.
From the point of view of building regulations, the property could be extended in the following ways:

- Conversion of the only partially used attic space into a three room apartment
- Extension of the building at the eastern gable with a single-storey addition, which increases the ground floor area by 65 m². This alteration provides the apartment on the first floor with a planted roof terrace measuring 20 m² and therefore valuable exterior space. This alteration adds 5 m² to the apartment's floor area.
- Addition of a storey to the projecting ground floor bay on the south side of the building thereby increasing the size of the room above, on the first floor, by 5 m² and making it appear more spacious.
- Removal of the protruding balcony slabs that were in desperate need of repair. A suspended steel balcony on the opposite side to the previous ones makes up for the loss of exterior space in the roof apartment. It extends the floor area in the roof apartment by 1 m² and helps improve the relationship between the apartments and their separate exterior spaces (terraces, balcony).

These measures, plus the lowering of the parapet height by 20 cm in all windows and the installation of electric roller blinds, which were undertaken during the course of the energy efficiency refurbishment, have considerably improved the living environment within the building. Through the reorganisation of the building, the cubic appearance has been enhanced. The project has also shown that benefits exceeding those of a purely energy-related refurbishment are achieved if architects plan and execute the refurbishment work in a holistic manner.

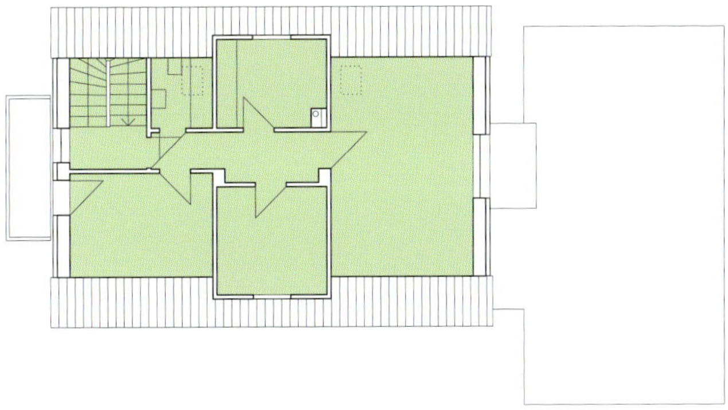

4.3

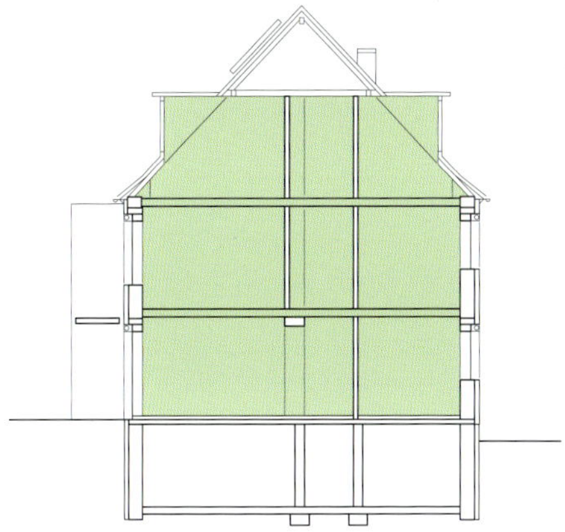

4.4

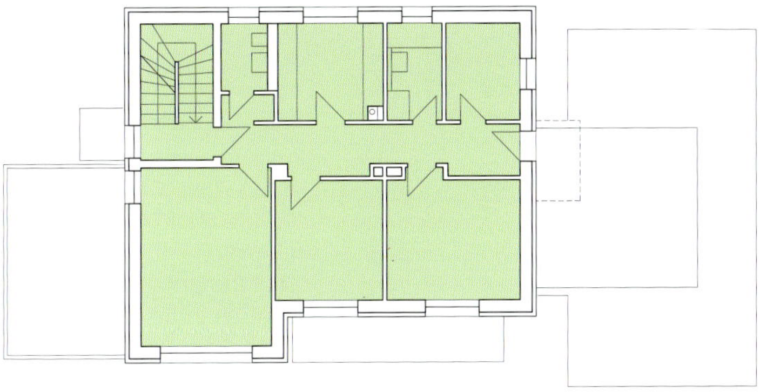

4.5

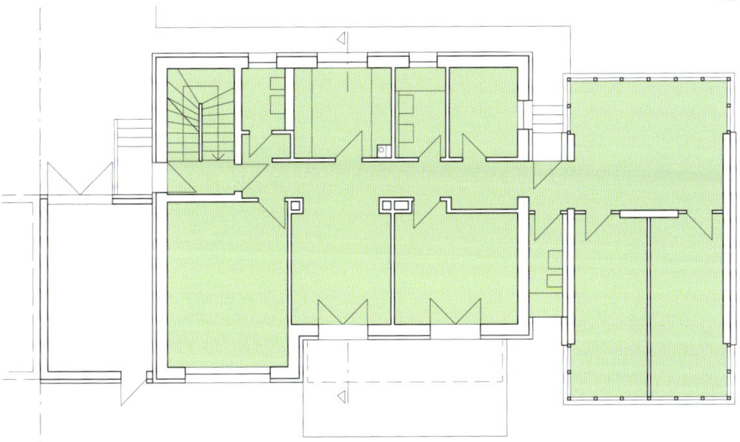

4.6

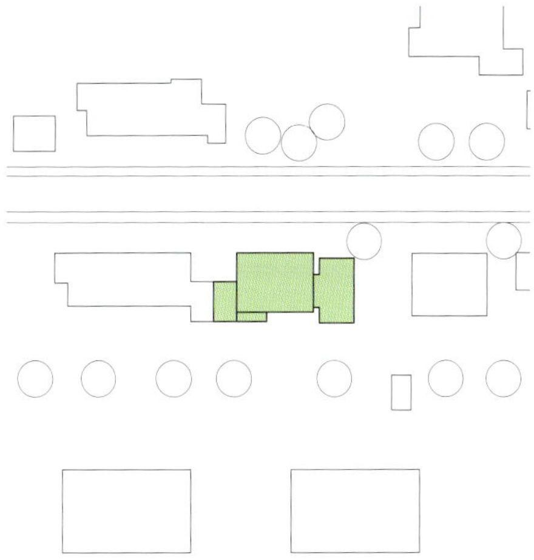

4.7

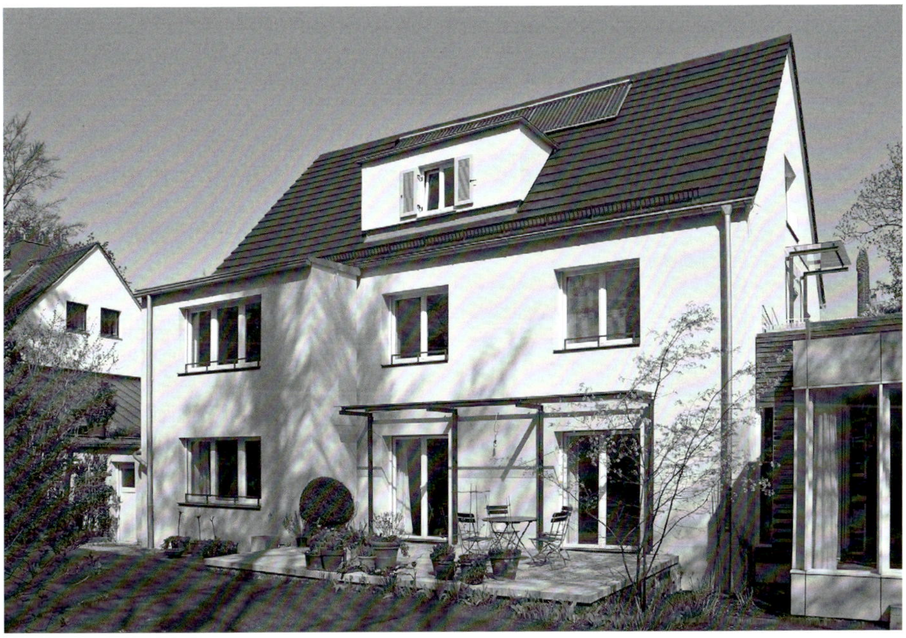

4.8

4.1 North facade (street view) before the refurbishment
4.2 North facade (street view) after the refurbishment
4.3 Floor plan of roof storey
scale 1:200
3-room apartment, 64 m² floor area, (heated area is highlighted in colour)
4.4 Section
scale 1:200
(heated area is highlighted in colour)
4.5 Floor plan of 1st floor
scale 1:200
3½-room apartment, 96 m² floor area, (heated area is highlighted in colour)
4.6 Floor plan of ground floor
scale 1:200
5½-room apartment, 157 m² floor area, (heated area is highlighted in colour)
4.7 Site plan, scale 1:1500
4.8 South facade (garden view) after the refurbishment; the extension is on the right

Passive concept
The passive concept incorporates all construction-related features of the existing building and the sum of refurbishment measures, which have an effect on the heat loss and heat gain of the building via the envelope.

Building stock
Typical of the time it was erected, the building is a solid brick structure with 38-cm-thick exterior walls. Quite remarkably, the walls were already made of vertically perforated brick with a gross density of 1400 kg/m³. The house has a full basement with cellar walls and ceiling made of concrete. The other floors are timber constructions with gravel infill between the joists to improve sound insulation. The attic space, which had only partially been converted, was not insulated. However, the attic knee walls provided a limited degree of insulation.
The condition of the envelope has remained the same except for the installation of PVC windows with insulation glass at the beginning of the 80s. The following values apply to the envelope surface areas:

- roof 1.4 W/m²K
- exterior walls 1.7 W/m²K (calculated)
- windows 2.7 W/m²K (replaced in 1980)
- components adjoining ground 1.2 W/m²K

According to these values, today's minimum requirements for winter thermal protection were not being met – as is the case for most buildings that date from this period. This is true for all components within the building envelope. In contrast, the minimum requirements for summer thermal protection were met thanks to the small proportion of window area and the existence of window shutters.

Refurbishment master plan
The refurbishment concept was developed in such a way that the alterations and improvements could be carried out in three phases; the second and third phase were executed with residents in situ.

Phase 1: basic refurbishment

- roof covering including insulation and conversion of the roof space into a three room apartment
- replacement of all building services

Phase 2: extension
Phase 3: envelope

- removal of balconies and canopy roofs
- addition of extra storey to projecting bay
- cellar insulation
- window replacement
- installation of new balcony and canopy roofs
- facade insulation
- installation of solar energy plant

The standard "KfW Efficiency House 85" according to EnEV 2009 was met on completion of the third phase.

Roof refurbishment
The roof overhang required for installing the thermal insulation composite system was provided for when refurbishing the roof during the first construction phase. The following operations were performed:

- removal of all existing roof tiles and battens
- installation of new roof structure with sheathing fixed to the upper side of the rafters, a roofing membrane, counter battens, battens and roofing tiles
- insulation of the space between the rafters with 12-cm-thick mineral wool featuring a thermal conductivity rating of WLG 035
- fixing of timber members (6 × 6 cm) at a right angle to the rafters and gaps filled with mineral wool insulation
- vapour barrier
- gypsum board panelling

Window replacement
The following operations were performed:

- removal of all windows and window shutters
- removal of the outer part of the existing three-piece window lintels and replacement using brick lintels; these were placed at a slightly higher level to create sufficient space for the new roller shutter boxes without actually having to reduce the size of the window openings
- partial removal of the window parapets down to 70 cm above FFL and rough plastering of the reveals
- installation of new windows:
  laminated frame with air chambers, $U_f = 1.1$ W/m²K
  sealed double glazing 4/16/4 with argon filled cavity, $U_g = 1.1$ W/m²K, $\tau = 80\,\%$, g value = 60 %
  edge configuration using Thermix TN, $\psi = 0.04$ W/mK
  wooden window sill on the inside
- plastering of the inner reveals
- installation of roller blinds with electric drive and remote control (simple assembly, wiring on the outside underneath the thermal insulation composite system)

Exterior wall upgrade
The following operations were performed:

- retouching of the existing render

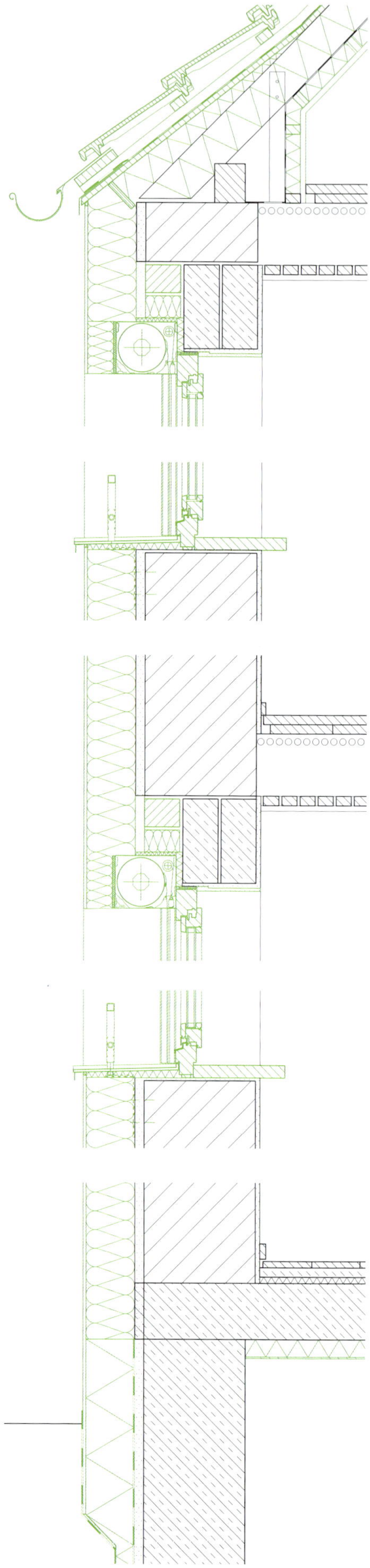
4.9

- installation of a thermal insulation composite system with EPS boards (WLG 032), 16 cm, with system coating
- fixing of the exterior steel window sills into the rendered reveals using premounted brackets. The security railing is fixed to the steel window sills and does therefore not penetrate the thermal insulation system.

Cellar ceiling upgrade
The following steps were performed to protect the habitable space from the cold of the unheated cellar:
- insulation of the soffit in the cellar rooms using sandwich boards with a 5 cm EPS core (WLG 035) and a 5 mm woodwool covering
- suspended ceiling in the corridor to accommodate pipe and cable routing

Refurbishment of exterior cellar walls
Insulation of the exterior cellar walls is not a necessary measure from an energy efficiency viewpoint. Insulation of the plinth would be sufficient to remove the thermal bridge. However, because the damp cellar walls needed attention, it was possible with little extra effort and expense to improve the thermal insulation of the exterior walls. The temperature of the wall surface on the inside has been increased through the exterior insulation, thus reducing the risk of mould growth, frequently caused by condensing moisture in summer.
The following operations were performed:
- excavation down to foundation level
- water jet cleaning of concrete walls, retouching of faulty areas
- application of cement rendering
- two-component bituminous waterproofing membrane
- XPS insulation board (WLG 035), 6 cm, glued
- dimpled membrane to protect insulation
- backfilling of the work area

Removal of thermal bridges
The most severe thermal bridges were identified in the areas of the protruding balcony slabs and canopy roofs. These were eliminated by removing the respective parts. The classic neuralgic issues could be removed during the refurbishment by solving the details without thermal bridges according to DIN 4108, supplement 2. The structural assembly of the new balcony and canopy roofs was achieved almost without thermal bridges by scarcely penetrating the insulation and using very small steel plates. On completion of the refurbishment, there were no thermal bridges worth mentioning.

4.10

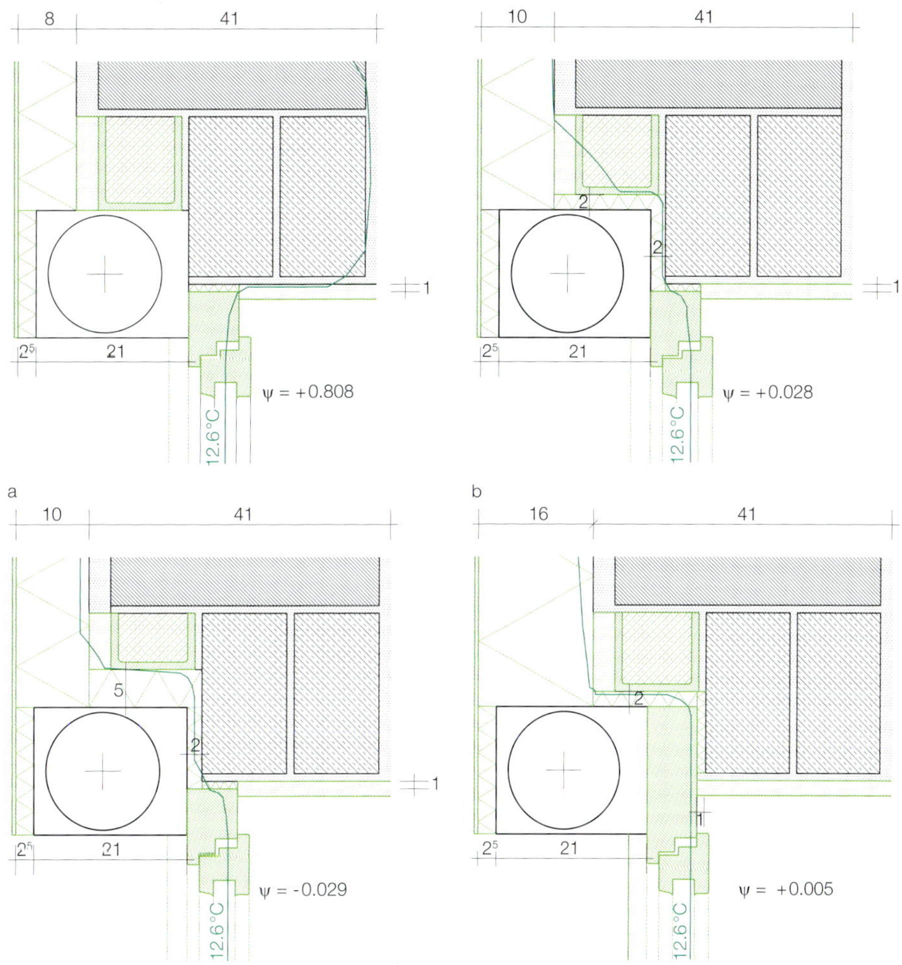

4.11

4.9 Vertical section through building envelope, scale 1:20
4.10 Horizontal section through building envelope, scale 1:20
4.11 Thermal bridges at roller shutter box
Objective: firstly, the detailing is to ensure a uniform appearance of windows with and without roller shutter boxes, and secondly the existing openings are not to be made any smaller (incidence of daylight). Furthermore, the solution is to be without a thermal bridge with $\psi \leq 0.05$ W/mK according to DIN 4108, supplement 2 (see Thermal bridges, pp. 40ff.).
Execution: front-mounted roller shutter box, 21 × 17 cm, electric, standard fixing of roller shutter box on rails, not on building structure. The existing three-piece lintels (standard of 1936) could be replaced easily on the outside by a new lintel in order to create sufficient space for the roller shutter box. With an appropriate thickness of insulation (WLG 032), the lintel and shutter box could be rendered. The green line (isotherm) marks the critical temperature of 12.6 °C. If the line is on the interior surface, there is a danger of mould growth.

Four variations were examined:
a no insulation on lintel
The ψ-value at 0.808 W/mK is much too high, there is a danger of mould growth. The insulation of the exterior wall must be at least 8 cm thick to incorporate the roller shutter box without a seam.
b lintel insulation with 2 cm of PU (WLG 024)
The ψ-value is improved to 0.035 W/mK, which means that the requirements are fulfilled. The thickness of insulation on the exterior wall is 10 cm.
c 5 cm lintel insulation (WLG 032)
This simple measure - in the case of openings up to a width of 150 cm, a lintel height of 7 cm is sufficient - improves the ψ-value significantly to 0.024 W/mK.
d wider window frame
The roller shutter box is set in front of the lintel together with the window, which raises the ψ-value to 0.03 W/mK. This is clearly above the boundary value determined by the DIN standard. However, this from an installation point of view very simple solution requires a wall insulation thickness of 16 cm.

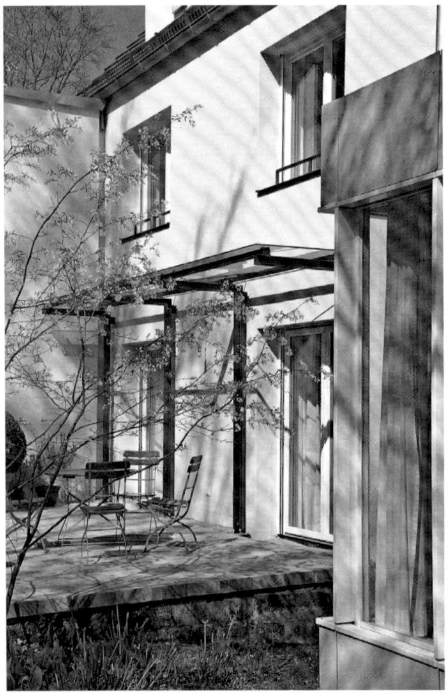

4.12

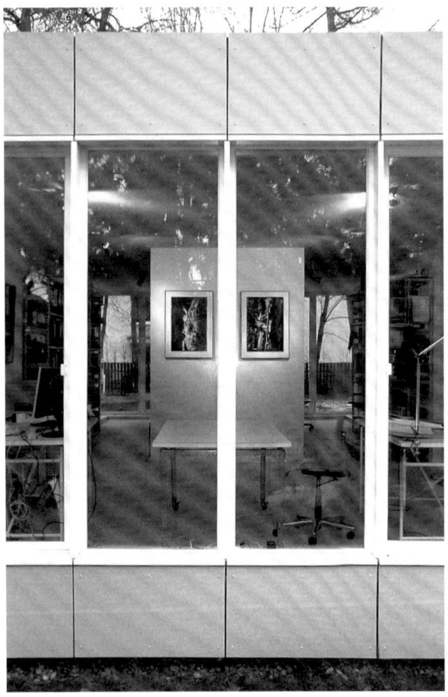

4.13

4.12 Terrace with canopy roof; the extension is on the right
4.13 View into the extension
4.14 Schematic diagram of the heating and hot water production system
1 solar thermal collector
2 solar buffer storage tank
3 domestic hot water tank
4 gas condensing boiler
5 change-over valve
4.15 Heat distribution in apartment
4.16 Cellar ventilation concept
In order to avoid moisture damage caused by incorrect ventilation particularly in summer, the windows in the storage rooms were closed up. A time controlled ventilation system with an air supply duct along the south side and an air extraction duct along the north side provides an adequate supply of fresh air to the cellar rooms.

Technical installations

The building services include all technical installations in a building which are necessary for its operation and to fulfil the designated use. In the case of refurbishments, there is frequently a lack of performance and consumption data of the existing plant without which it is impossible to assess the improved efficiency resulting from new measures. For these cases, the Federal Ministry for Transport, Building and Urban Development (BMVBS) has published average values of existing installations, which can be used instead.

Existing installations

In terms of the distribution pipes, the technical installations in this residential building were still in their original state from the year 1936. Only the heat generator for heating and the supply of hot water had been replaced in 1983.
Data of the existing installations:

- heating
  oil-fired central heating
  thermal output acc. to DIN EN 12831
  60 kW (approx. 240 W/m²)
  distribution heat loss 43.5 kWh/m²
  storage heat loss 0 kWh/m²
  transfer heat loss 3.3 kWh/m²
  efficiency of heat generation 1.21
- hot water production
  decentralised electric storage heating (100 l) in the bathrooms
  distribution heat loss 3.8 kWh/m²
  storage heat loss 1.5 kWh/m²
  transfer heat loss 0
  efficiency of hot water production 1.0

Refurbishment concept

The new building services concept is modular and made up of individual units so that it is possible to replace individual components without affecting the other system parts. The gas condensing boiler installed during the first refurbishment phase could, for example, easily be substituted by a more efficient zeolite heating system at a later date or extended by installing a heat pump.

Heating system

As part of the energy efficiency refurbishment, the heating system was changed from oil to gas. With no need for the oil tank, a further storage room was created in the cellar of the building. On completion of the refurbishment, the heat is generated by a modulating gas condensing boiler with a capacity of 30 kW. The heating is backed up by solar panels.
The entire heat distribution system including the heat output was renewed during the refurbishment. Based on a central ascending pipe, each apartment is connected via a manifold to which the forward and return flow pipes are attached. The heating pipes are installed in front of the wall behind skirting board. The radiators were no longer positioned below the windows, but to their sides, which was fundamental for the enlargement of the windows performed during the second refurbishment phase (fig. 4.15).
Data of the new installations:

- thermal output acc. to DIN EN 12831 30 kW
- distribution heat loss 3.0 kWh/m²
- storage heat loss 0
- transfer heat loss 1.0 kWh/m²
- efficiency of the heating system 0.98

Hot water system

The building's entire hot and cold water supply system was also replaced.
The hot water production is now performed in a centralised fashion using the gas condensing boiler. It heats water in a central domestic hot water storage tank with a capacity of 300 l from which the individual apartments are supplied. The hot water production is backed up by solar collectors.
Data of the new installations:

- thermal output according to DIN EN 12831 included in the heating system
- distribution heat loss 5.7 kWh/m²
- storage heat loss 1.8 kWh/m²
- transfer heat loss 0
- efficiency of hot water production 1.12

When considering this performance data separately, the new central hot water production system does not appear to perform as well as the former decentralised system. What is not expressed here, however, is the actual improvement in terms of energy efficiency brought about by the change from oil to gas (lower primary energy factor of 1.1 in comparison to 2.6 in the total energy balance) and the possibility to make use of solar thermal energy (see. p. 128).

Ventilation

As was the case in the existing building, the ventilation continues to be performed naturally. However, the window replacement made it necessary to develop and execute a ventilation concept in accordance with DIN 1946-6. But because it was decided to dispense with a controlled ventilation system, not least for cost reasons, the necessary air exchange to provide protection against moisture had to be achieved through natural ventilation, inde-

pendent of the users. These "planned leaks" in the building envelope were achieved by installing window fittings which allow for minimum gap ventilation when the window handle is put into a 45° position. According to table 4 in DIN 1946-6, a blower-door test must ensure that an air exchange through infiltration of at least 2.8/h can take place in the building. This value was met with the aforementioned setting of the windows. Because the windows also seal tightly when closed properly, the EnEV requirements concerning airtightness were also met (maximum air exchange through infiltration of 3.0/h). Ventilation sensors prompt the residents to take action regarding air exchange. They send out an optical or acoustic signal as soon as it necessary to open a window due to poor air conditions (moisture, CO_2).
The refurbishment of the damp cellar has been supported by installing a time-controlled automatic ventilation system (fig. 4.16).

Use of solar thermal energy
In the third refurbishment phase, the heating and hot water generation system was extended to include solar thermal energy. This was achieved by mounting vacuum tube collectors on the south-facing roof surface. The load of the 10 m² solar array amounts to 16.6 kg/m², which adds up to a total of 166 kg. The existing roof structure is able to accommodate the extra weight without difficulty.
The heating and hot water system was supplemented by an additional buffer storage tank, which preheats the domestic water. This is achieved by guiding the cold water past a heat exchanger in the solar buffer storage tank before it is supplied to the domestic hot water storage tank. The same principle is used in the heating system to preheat the return water before it is heated by the boiler to meet the desired forward flow temperature (solar heating support system). As soon as the water temperature in the new buffer storage tank is 8 °C above the temperature of the heating's return water, the return water is guided through the buffer storage tank and past the heat exchanger in a closed pipe system.

The performance data of the solar thermal plant is as follows:
- type of collector vacuum tube collectors
- total area 10.44 m²
- aperture area 9.33 m²
- solar buffer storage tank 800 l
- heat yield of collectors 602 kWh/m²

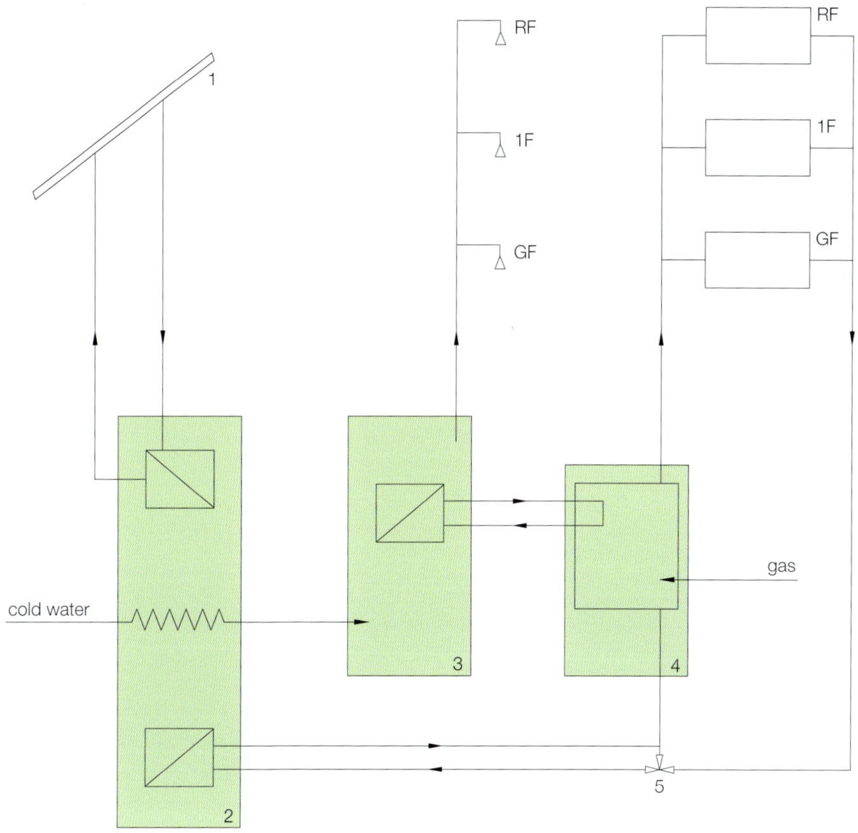

4.14

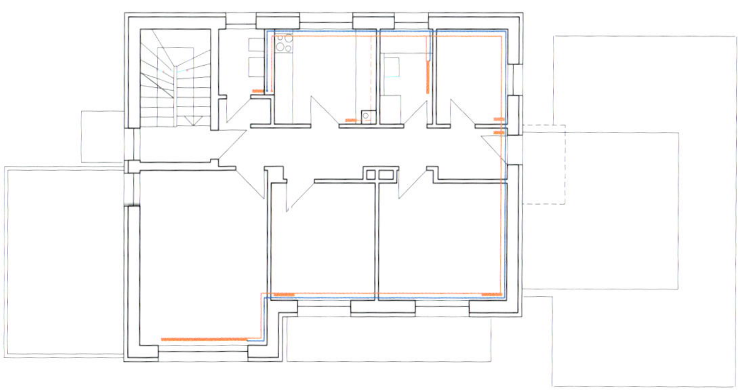

4.15

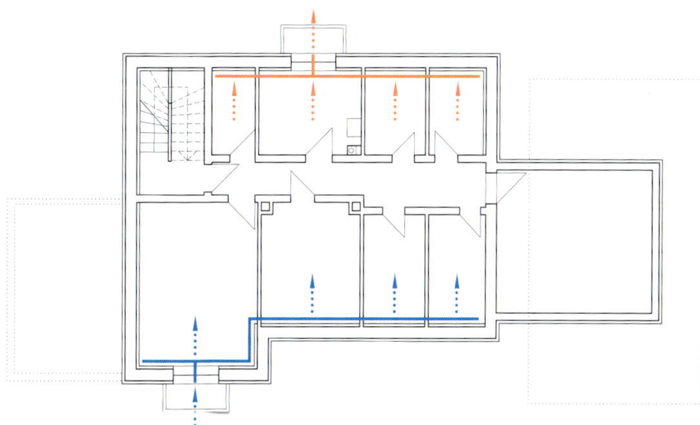

4.16

Accounting

In the case of energy efficiency refurbishments, it is always necessary to distinguish between expenditure for refurbishment measures, which would have anyway been required from a maintenance point of view, enhancement measures, such as increasing the size of the useful floor area, and measures having the sole purpose of improving the condition of the property in terms of energy efficiency. Only the last mentioned category of expenditure, that associated with energy efficiency, may be taken into account when considering the cost effectiveness of a refurbishment, i.e. expenditure (investment) versus income (value of energy savings).

Economy

The total investment for the measures specific to the energy efficiency refurbishment of the analysed three family house amounted to €96,000 (energy-related extra costs only). Funding was available for the following measures:

- funds from the KfW programme "Energetisch Sanieren" with a repayment bonus of 12.5% and an interest rate of 1.4%. In the case of the KfW programme, the total costs of the refurbishment are chargeable and not only the expenditures for the purely energy related work.
- funding through the programme "Energieeinsparung" made available by the city of Munich for exterior walls, windows and solar thermal collectors.

After the refurbishment, the final energy demand for the provision of heat (heating, hot water) was reduced by approximately 80% so that, instead of 7300 l of oil and 4600 kWh of electricity, only a total of 2450 m^3 of gas is now required.
Due to the larger habitable floor area, the demand for gas has increased by 550 m^3 to 3000 m^3. However, all measures concerning the extensions are not included in the economic assessment of the refurbishment. Only the energy demand of the refurbished building stock is used as a standard of comparison for the assessment of the energy efficiency refurbishment's investment.
Based on an applied energy price of 60 ct/m^3 for gas, 60 ct/l for oil and 20 ct/kWh for electricity, the total costs for the provision of heat after the refurbishment no longer amount to €5300, but only €1470 per year. The savings of €3830 must now be used to refinance the additional expenditure for energy-saving measures of approximately €76,000 net, after allowing for grants.
If one considers a period of 20 years and assumes an average price increase of 6% p.a. for the resources oil and gas and approximately 5% p.a. for electricity, the economic situation can be summarised as follows (see calculation procedure, pp. 30ff.):

Expenses:

• investment	€96,000
• KfW repayment bonus	€12,000
• grant from the city of Munich	€8000
• investment less subsidies	€76,000
• annuity of the investment (interest rate 1.4%)	€87,663
• average costs per year	€4383

Savings:

• savings per year	€3830
• price increase per year	6%
• total savings	€140,888
• average savings per year	€7044

Balance:

• average costs per year	€4383
• average savings per year	€7044
• surplus per year	€2661
• amortisation	14.6 years

It is often wrongly believed that the total refurbishment costs, i.e. including the expenses for maintenance work, must be refinanced through energy savings. When this cannot be achieved, as is normally the case, the energy efficiency refurbishment is assumed not to be profitable. This is generally the wrong approach, since the owner will always encounter maintenance costs, irrespective of energy-related measures. Additional work performed to improve the energy efficiency of a building, which is undertaken during the course of a major refurbishment, will generally amortise in 15 to 20 years - even with an assumed annual energy price increase of 6% per year.
Since most of the investments made have a life expectancy well beyond 20 years, it would actually be regarded as detrimental, from an economic point of view, not to perform the energy efficiency refurbishment.

Ecology

The ecological standard of the refurbishment is defined as follows:

• primary energy demand before	
total	92 500 kWh
specific (EnEV)	306 kWh/m^2
• primary energy demand after	
total	33 000 kWh
specific (EnEV)	101 kWh/m^2
• CO_2 emissions before	
total	24 958 kg
specific (EnEV)	82.64 kg/m^2
• CO_2 emissions after	
total	7290 kg
specific (EnEV)	22.2 kg/m^2

Sustainability

The main aim of the refurbishment was to provide a flexible and therefore sustainable use of the residential building. For this reason the separation into three dwelling units was maintained. The extension added to the ground floor apartment can be disconnected and used as a separate autonomous unit, not least because of the sanitary facilities incorporated. Different patterns of use are possible, such as home and work or work and care. The non-load-bearing partition walls enable the interior layout of the extension to be varied as necessary.
Having kept the apartments as single-storey units, it is possible to fit-out the individual levels as barrier-free apartments.

Total construction costs, cost groups 2–7
€570,000 (grants/subsidies deducted)

proportion for construction-related measures €450,000			**proportion for building services measures** €120,000		
proportion for maintenance roof refurbishment window replacement render refurbishment cellar walls concrete refurbishment interior upgrade	**proportion for improved energy efficiency** facade insulation roof insulation improved window frames cellar wall insulation	**proportion for extensions** roof space conversion added storey to projecting bay balcony replacement	**proportion for maintenance** new heating system sanitary fittings electric roller shutters	**proportion for improved energy efficiency** solar thermal collectors with buffer storage tank full electrical installations	**proportion for extensions** cellar ventilation
€180,000	€56,000	€214,000	€95,000	€20,000	€5000

4.17

	Unit	Building stock	Refurbishment	Extension	Total
Data					
heated volume (V_e)	m³	945	1026	208	1234
envelope surface area	m²	650	698	247	945
total floor area	m²	241	250	65	315
heated net floor area (NFA)	m²	271	271	65	355
EnEV floor area (0.32 × V_e)	m²	302	328	67	395
compactness (heated NFA/heated volume)	1/m	0.30	0.29	0.31	0.29
Passive concept					
Winter					
roof	W/m²K	1.28	0.21	0.19	
wall adjoining outside air	W/m²K	1.11	0.17	0.18	
wall adjoining ground	W/m²K	–	–	–	
wall adjoining unheated rooms	W/m²K	1.11	0.17	–	
windows	W/m²K	2.70	1.19–1.22	1.22	
floor adjoining cellar, ground, air	W/m²K	1.44	0.53	0.17	
thermal bridges (individual verification method)	W/m²K	0.20	0.04	0.04	0.04
Summer					
window area	m²	35	47	58	105
window ratio (in relation to the facade surface area)	%	11	15	50	30
g value of windows	%	80	60	50	
F_c sun shading	–	0.3	0.3	0.5 (trees)	
storage mass	–	complicated	complicated	complicated	
nighttime ventilation	–	possible	possible	possible	
Daylight (DIN V 18599)					
area supplied with daylight (in % of the NFA)	%	55	70	85	75
daylight incidence through windows	%	80	80	70	
Ventilation					
air volume (0.8 V_e)	m³	756	820	166	986
floor area requiring ventilation (habitable)	m²	241	250	65	315
proportion of naturally ventilated space	%	100	100	100	100
Active concept (final energy)					
Heating					
energy medium	–	oil	gas/solar	gas/solar	gas/solar
final energy	kWh/a	73,300	22,000	5000	27,000
Domestic hot water					
energy medium	–	electricity	gas/solar	gas/solar	gas/solar
final energy	kWh/a	4600	2500	500	3000
Cooling					
energy medium	–	no cooling	no cooling	no cooling	no cooling
final energy	kWh/a	–	–	–	–
Ventilation					
system	–	natural ventilation	natural ventilation	natural ventilation	natural ventilation
final energy	kWh/a	–	–	–	–
Balance					
Standard of energy efficiency					
EnEV 2007	–	–	New Build – 30 %	New Build – 30 %	New Build – 30 %
EnEV 2009	–	–	New Build	New Build	New Build
EEWärmeG	–	–	fulfilled	fulfilled	fulfilled
KfW	–	–	Efficiency House 85	Efficiency House 85	Efficiency House 85
Economy					
annuity (20 years, 1.4 % interest rate)	€		87,663.22		
savings in 20 years (6 % price increase)	€		140,888		
amortisation	years		14.6		
Ecology					
primary energy	kWh/a	92,590	26,950	6050	33,000
CO_2	kg/a	24,958	5954	1336	7290

4.18

4.17 Allocation of costs for the full refurbishment
4.18 Project data relevant to the refurbishment

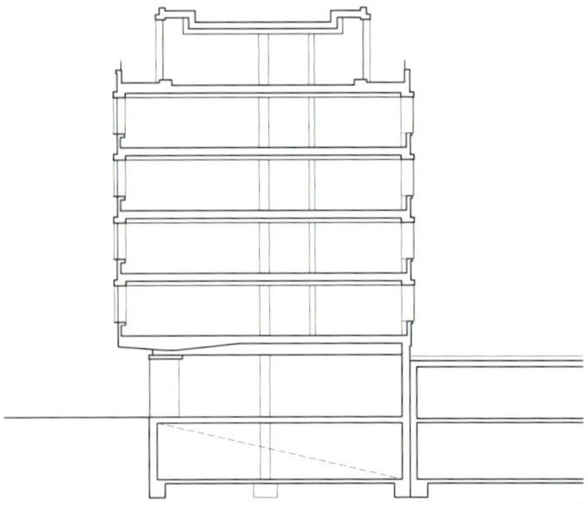

4.19

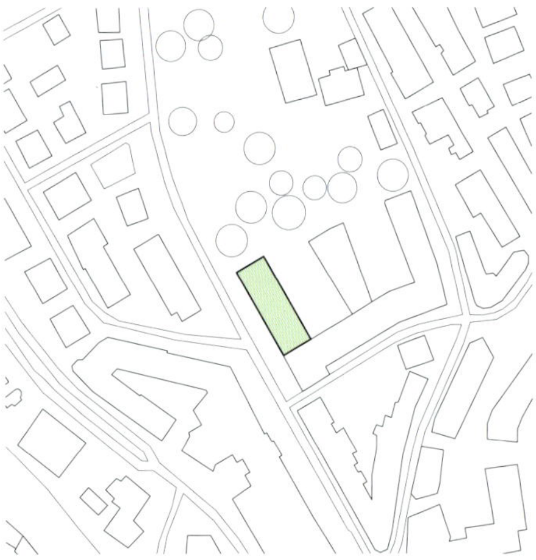

4.20

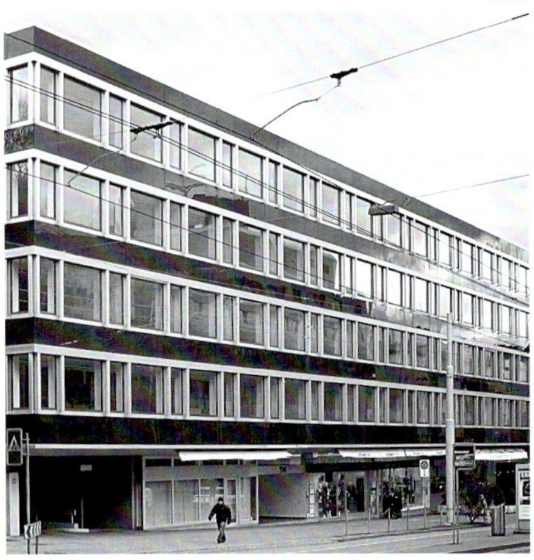

4.21

4.19 Site plan, scale 1:5000
4.20 Section, scale 1:500
4.21 Facade before the refurbishment
4.22 a Floor plan of roof storey, scale 1:500
b zoning according to DIN V 18599-10
4.23 a floor plan, 1st - 4th storey, scale 1:500
b zoning according to DIN V 18599-10
4.24 a ground floor plan, scale 1:500
b zoning according DIN V 18599-10
4.25 Facade after the refurbishment

Zoning according to DIN V 18599-10:

- office
- canteen
- ancillary area
- sanitary

Non-residential building

Building type:	office building
Location:	Zurich
Architect:	Max Dudler Architekten AG, Zurich
Project manager:	Daniel Pescia
Structure:	Pöyry Infra AG, Zurich
HVS:	Pöyry Infra AG, Zurich
Building physics:	Wichser Bauphysik, Zurich
Facade:	Stäger Nägeli, Zurich
Year of construction:	1971
Refurbishment:	2007

Architecture

The refurbishment of this inner city office building from the 60s is in many respects typical of numerous buildings constructed during this period. The reinforced concrete skeleton constructions with non-load bearing facades and minimal insulation, fall well short of today's requirements in terms of thermal insulation and noise protection. Without comprehensive refurbishment work, they are no longer able to meet the needs of changed user conditions (cooling loads through EDP) and the higher standards of thermal comfort.

The office building analysed here is part of a larger complex, which is in the hands of several owners. A question that always arises in the context of such refurbishments is whether and to what extent the renovated part should follow the design concept of the original building, or whether it is possible to depart from the overall appearance of a homogenous building fabric in order to allow individual solutions to the various problems.

In the case of this refurbishment, it was not only necessary to meet high demands in terms of energy efficiency, but also to take into consideration new issues concerning noise and fine dust pollution (the building is positioned on what has become a very busy road).

The economic pressure to make use of the available extension space on the roof of the building posed a further challenge, which had to be considered in the development of the architectural master plan.

These problems were further complicated by the classic issue of how to deal with the refurbishment of the reinforced concrete skeleton construction. Decisions had to be made on whether to retain the solid, non-load bearing window parapets, or whether to remove and replace them with a totally new facade. The latter solution was favoured not least because of the comparatively large amount of floor space gained by removing the parapets.

In the end, the decisive factor for many issues in the refurbishment process was the overriding aim to retain the urban significance of the distinct U-shaped perimeter development from the 60s so that it would actually still be perceived as part of a greater complex. The creative and, subsequently, the structural refurbishment concepts were therefore based on a detailed analysis of the existing building's characteristic elements. Important features from the existing building stock, which were to be adopted in the development of the new facade, included the ribbon windows, linked by exterior mounted sun shading devices of contrasting colour, muntin profiles set in front of the twin columns to provide a vertical rhythm to the facade and enamelled glass panels used in the parapet zones.

The concrete parapets were removed and replaced by insulated panel elements. Nevertheless, the structural arrangement of the facade with parapets and ribbon windows is clear and uses the same combination of materials.

The difference in the ribbon window construction is the fact that an additional pane of glass has been placed on the outside in front of the sun shading blinds, forming a kind of double window. Triple thermal insulation glazing has been used on the inside of the facade. The ribbon windows are clearly structured by vertical construction joints in the outer layer to pick up on the characteristic appearance of the former facade.

The refurbishment work is complemented by making use of the potential to extend the building on the flat roof. A newly erected cubic structure set back from the main facade provides space for a cafeteria with the necessary service areas. The roof of the extension has been planned perfectly to accommodate the new building services (see Building services, p. 134).

The rear part of the ground floor, set into the slope of the site, and the basement were left unchanged in the refurbishment and are still used for parking and other minor functions (heating, oil tank, storage).

The categorisation of uses as presented in the drawings (fig. 4.22b, 4.23b and 4.24b) is the basis for the assessment of non-residential buildings in accordance with DIN V 18599. The so-called zoning of the building is a prerequisite for the calculations concerning the separate energy demands for heating, cooling and ventilation in the individual zones.

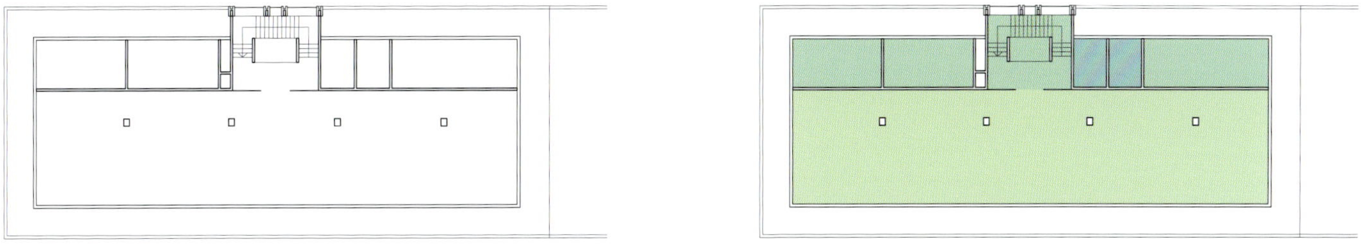

a

b

4.22

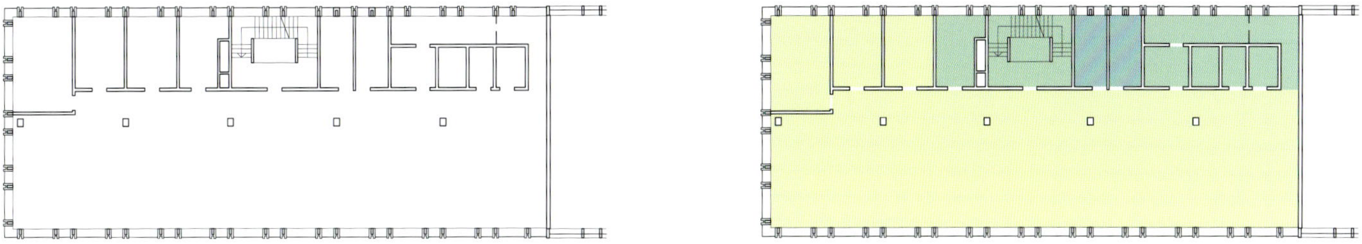

a

b

4.23

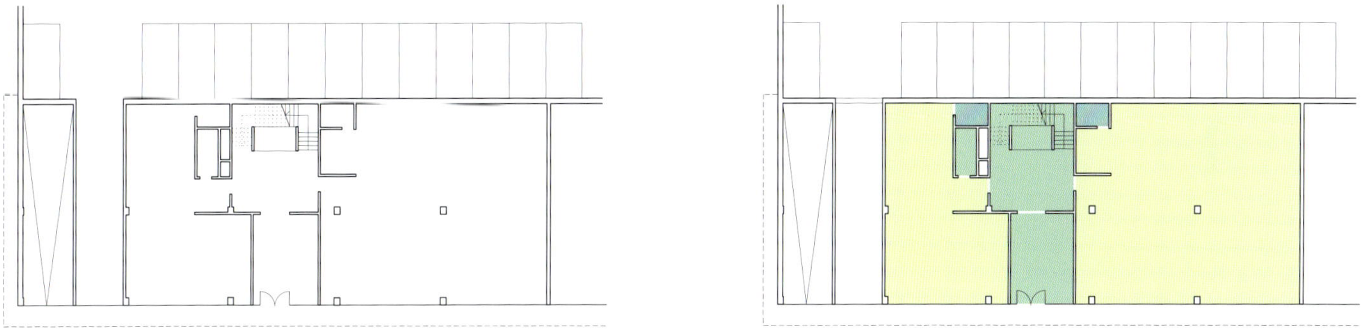

a

b

4.24

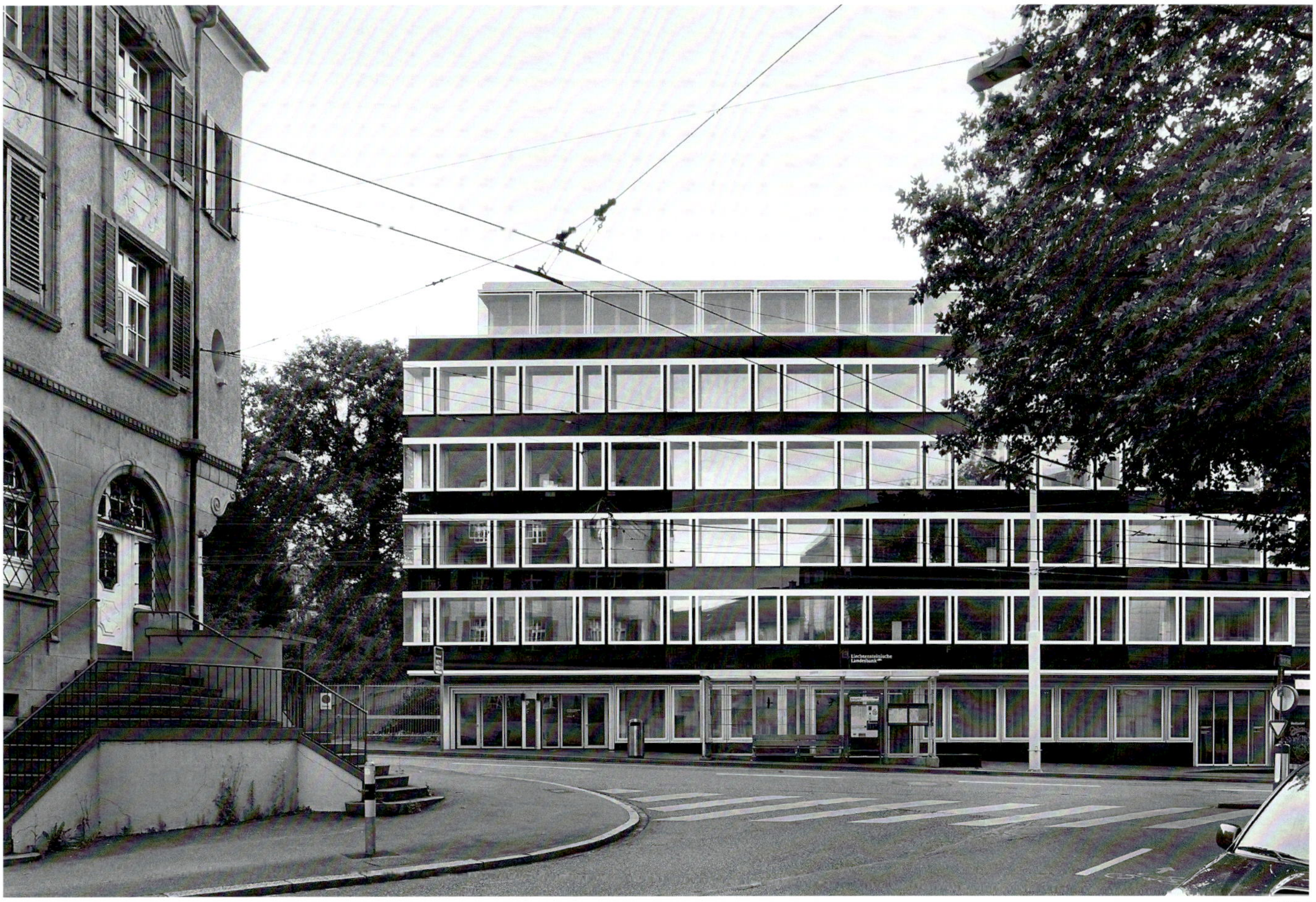

4.25

Passive concept
The passive concept incorporates all construction-related features of the existing building and the total of refurbishment measures, which have an effect on the heat loss and heat gain of the building via the envelope.

Building stock
Typical of buildings from this era, the office block has a reinforced concrete skeleton structure with flat slabs and non-load bearing facades. The floor plans are based on a precise grid of 1.20 m.
The envelope, with the ribbon windows and glass-clad concrete parapets without thermal insulation, was not changed up until the refurbishment in 2007.
According to table 2 of the "publication of rules to adopt and use data for non-residential building stock" issued by the Federal Ministry for Transport, Building and Urban Development (BMVBS) on 30 July 2009, the following values apply to the envelope surface areas:

- roof 0.6 W/m²K
- walls 1.0 W/m²K
- windows 2.7 W/m²K
- components adjoining ground and unheated rooms 1.0 W/m²K

Today's minimum requirements for winter thermal protection were thus not being met – as is the case for most buildings constructed during this period. In contrast, the minimum requirements for summer thermal protection were met due to the externally mounted sun shading blinds and the moderate proportion of window area.

Refurbishment master plan
Because of the differentiated owner structure within the overall office complex, rather than developing a master plan to renovate the entire block, the refurbishment was limited to the part of the building presented here. The obvious decision was therefore to retain the characteristic features of the building envelope in order to not disturb the overall homogenous appearance of the complex.
The energy efficiency refurbishment of the opaque surfaces was based on Swiss regulations (SIA 380/1). However, for the purpose of this documentation, the values of the German Energy Saving Ordinance 2009 (EnEV) have been applied as a basis for the assessment.
No alterations were made to the floor slab above the basement garage, nor to the exterior walls below ground level.
In order to meet today's structural requirements, two cross tie walls were added to the building frame. Furthermore the structure had to be adapted to support the load of the new roof storey (fig. 4.28 and 4.29).

Roof refurbishment
The floor slab reinforcement was strengthened at the column heads to better transfer the additional loads from the extra storey into the load-bearing columns, which were of sufficient capacity. This was executed by exposing the floor slab as a first step and then removing the concrete in an approximately 2 m radius around the columns down to the upper layer of reinforcement. After installing the new plate bonding system, a 5-cm-thick layer of concrete was cast on the existing ceiling to ensure the load-bearing capacity of the floor slab. The remaining roof area of the former building and the new roof area were insulated using 20-cm-thick PU insulation boards (WLS 030), sealed with a PVC membrane and covered with pebbles. The foundations for the machines in the plant room are thermally separated from the ceiling construction by using pressure-resistant insulation board.

Exterior wall refurbishment
The following operations were performed:

- removal of the non-load bearing parapet walls
- assembly of sheet metal panels with 16-cm-thick thermal insulation between the ribbon windows and fixing of exterior glass cladding

Window replacement
The following operations were performed:

- removal of the old windows and blinds
- installation of new aluminium box windows with triple glazing on the inside and a pane of single glazing on the outside; the cavity is ventilated
- fixing of the sun shading blind within the cavity of the box window

The values of the new window construction are as follows:

- thermal transmittance U_w 0.6 W/m²K
- energy penetration coefficient g_{total} 10 %
- light transmission τ 30 %

Removal of thermal bridges
Due to the new facade construction, which is installed in front of the original building structure, there are no thermal bridges worth mentioning. The thermal bridges at the base of the building were not eliminated due to the considerable time and expense this would have involved.

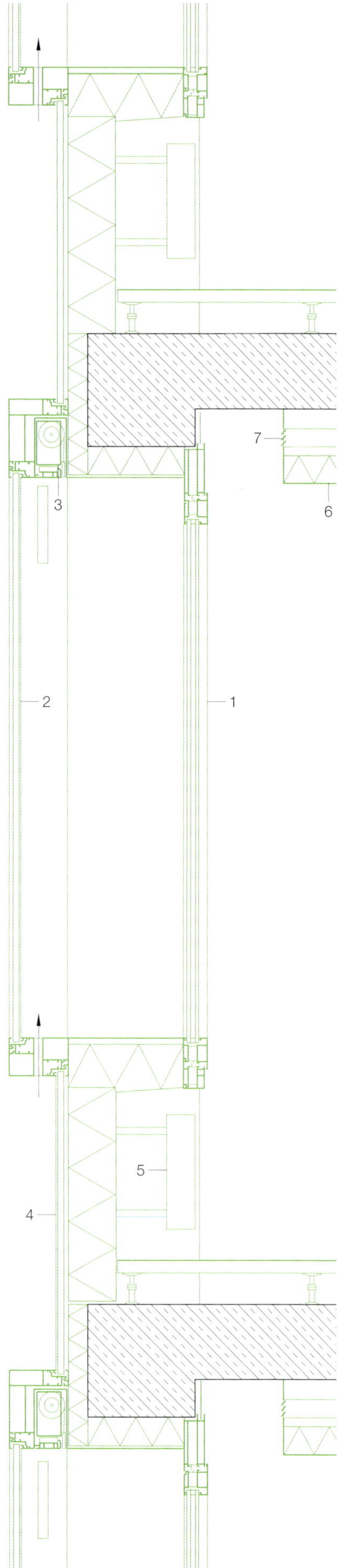

4.26

4.26 Vertical section through building envelope, scale 1:20
1 inner skin:
aluminium frame, U_{frame} = 1.1 W/m²K
triple glazing 4/16/4/16/4 mm with argon fill, U_{glass} = 0.7 W/m²K
τ-value = 71 %; g-value = 48 %
seal ψ = 0.05 W/m²K
2 outer skin:
aluminium frame, U_{frame} = 3.0 W/m²K
air vents on the sides and at the bottom
SSG 8 mm
τ–value = 43 %; g-value = 30 %
3 lamella blind, grey powder coated aluminium
f_c- value = 0.18
4 parapet element: TSG 8 mm
aluminium panel with 16 cm insulation (thermal conductivity level of WLG 030)
5 radiator
6 suspended ceiling:
aluminium perforated sheet, mineral wool
7 waste air extraction vent

4.27 Horizontal section through building envelope, scale 1:20
4.28 Additional tie walls in standard floor level
4.29 Strengthening of floor slab reinforcement on the fourth floor
4.30 Corner of the building with box window (interior)

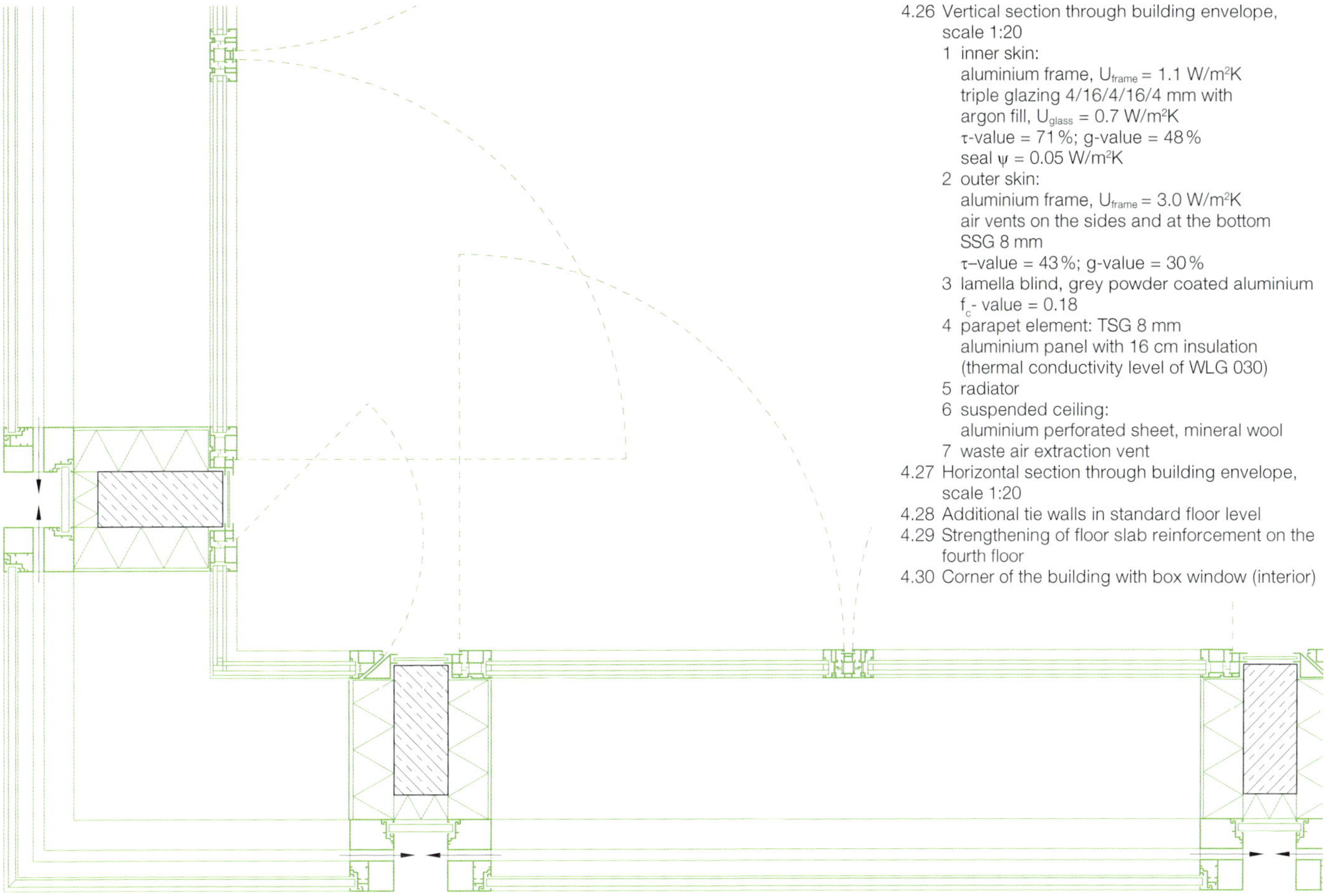

4.27

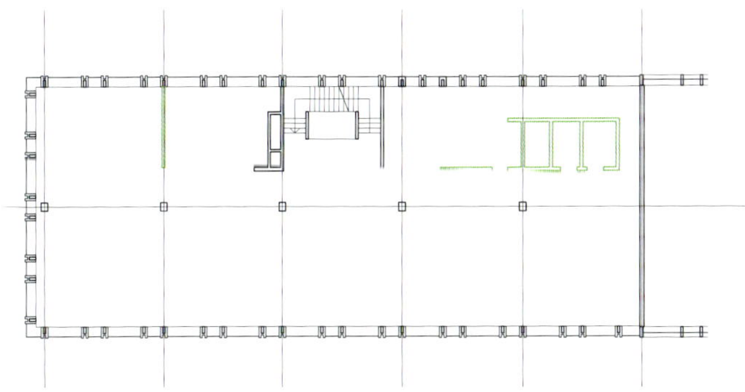

4.28

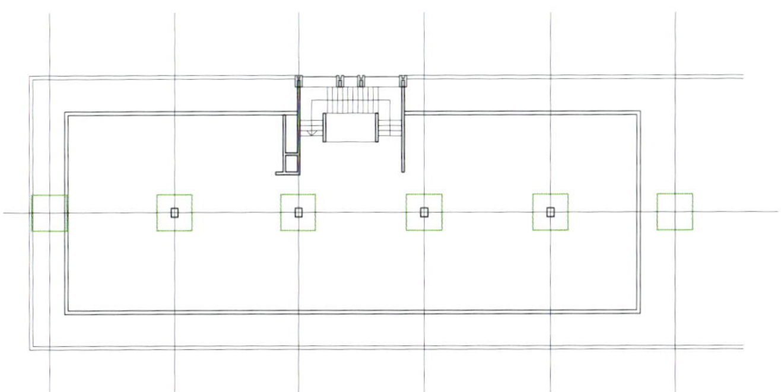

4.29

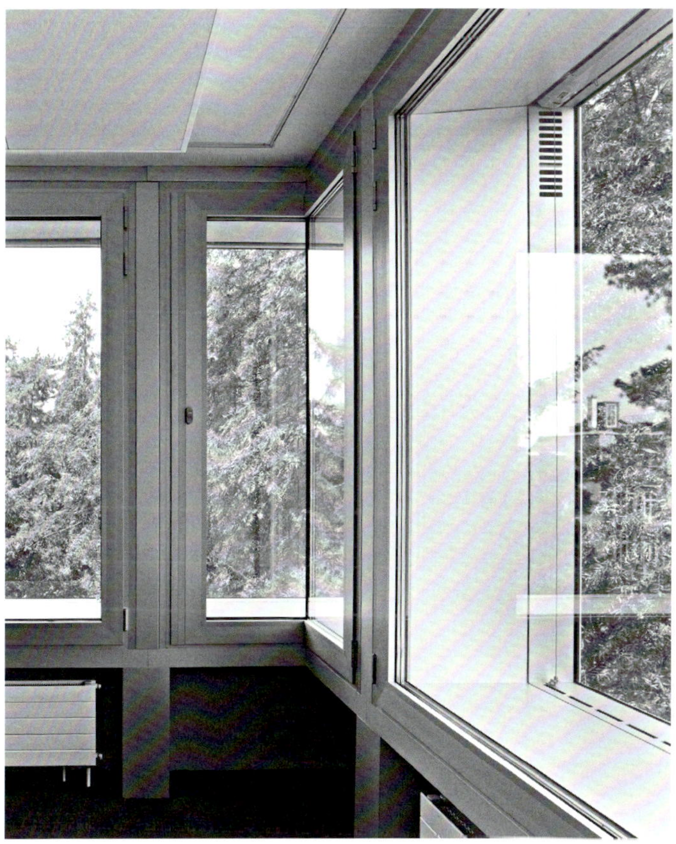

4.30

Building services
The building services include all technical installations in a building which are necessary for its operation and to fulfil the designated use.

Existing installations
In the case of refurbishments, performance and consumption data of the existing plant, without which it is impossible to assess the improved efficiency resulting from new measures, are frequently lacking. The values published by the Federal Ministry for Transport, Building and Urban Development (BMVBS) on 30 July 2009 can be used as a substitute for the missing information in accordance with the "publication of rules to adopt and use data for residential building stock".

Heating
The heat generator, an oil central heating system, and the buffer storage tank for hot water (300 l) were only six years old at the time of the refurbishment and therefore not ready for disposal. The heating plant and the associated oil tank, with a capacity of 30,000 litres, are located in the basement.
Based on table 5 of the above-mentioned publication, the following values can be assigned to the heating plant:
- forward/return flow water 70/50 °C
- thermal output 200 kW (approx. 80 W/m²)
- distribution heat loss approx. 20 kWh/m²
- storage heat loss approx. 5 kWh/m²
- transfer heat loss 3.3 kWh/m²
- efficiency of heat generation approx. 1.2

Hot water supply
The energy demand for hot water is not necessarily taken into account in office buildings. According to table 6 of DIN V 18599-10, the hot water demand must only be included in the calculations of the energy balance if the energy demand for the production of hot water exceeds 0.2 kWh/person per day, which is enough energy to heat about 4.6 litres of water from 10 °C to 60 °C.

Cooling, ventilation
In the existing building, the fresh air was supplied and the waste air was removed naturally through the windows. The building was not cooled mechanically. Owing to the increased noise and fine dust pollution from the traffic and the increased cooling loads caused by the computer networked workplaces, there was considerable need for improvement in this respect.

Electrical installations, lighting
The wiring of the workplaces using parapet ducts was not sufficiently flexible. The over-dimensioned lighting facilities with limited control features, fluorescent tubes and conventional ballasts featured an installed lighting capacity of 25 W/m².

Refurbishment
The conditioning of the building is performed with separate systems for heating, cooling, ventilation and lighting. If individual components have to be renewed, replacements can be made without interfering with other components of the building services (fig. 4.31). It will, for example, be possible to exchange the heating system at a later date or supplement it by installing a second heat generator.
All of the new heavy technical equipment – the refrigeration machine, recooling unit and ventilation system – has been accommodated in the new space assigned for the technical plant, on the roof of the added storey (fig. 4.33).
The planners took great care in arranging the technical superstructures in order to prevent any negative impact on the building's appearance. For this reason, a trough-like zone was incorporated in the centre of the new storey's roof in which the taller equipment is positioned. Hence, it is almost invisible from the outside.

Heating
The generation of heat using the oil cen-

4.31 Schematic diagram of building services
1 room
2 refrigeration machine
3 recooling unit
4 water storage tank
5 heating
6 buffer storage tank for heating
Circuit A:
outside temperature below 15 °C
natural cooling using recooling unit
Circuit B:
outside temperature above 15 °C
classic cooling process, if there is no demand for hot water (4) and heating (5)
Circuit C:
recooling using heat from the domestic hot water (4) and heating (5)
4.32 Ventilation and cooling in a standard storey (ceiling face)
a supply air
b waste air
c cooling ceiling
4.33 Plant room on the roof of the added storey
a exhaust air
b outside air
c waste air
d supply air
e ventilation of underground car park
f cooling/refrigeration machine
g recooling unit
h reserve space for cooling
i reserve space for recooling unit
4.34 Added storey with peripheral roof terrace
4.35 Technical installations on the trough shaped roof

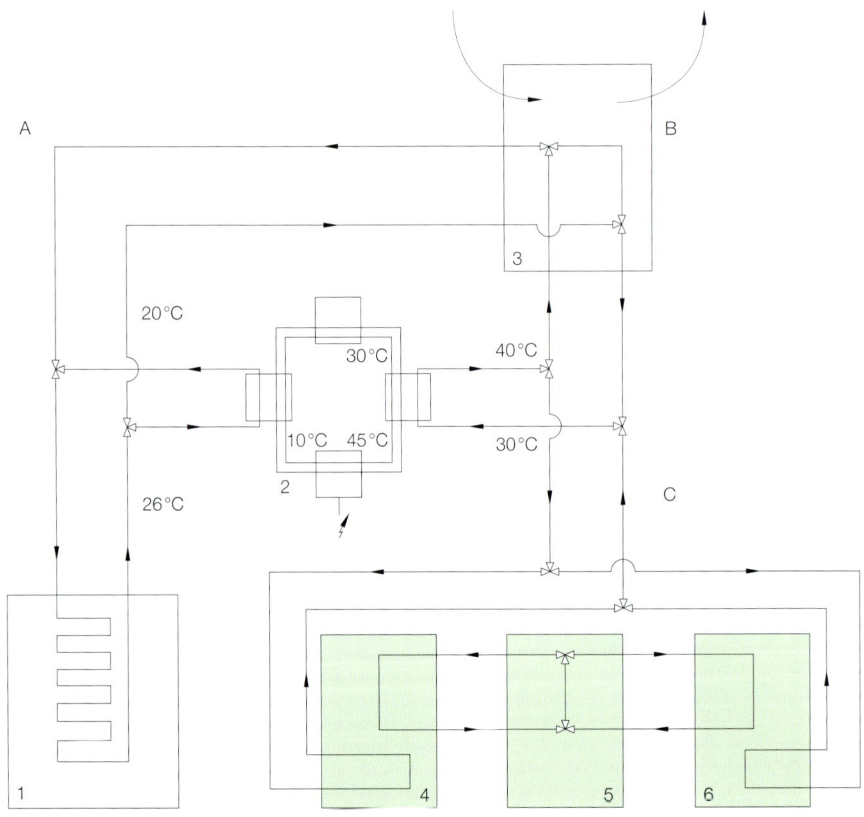

4.31

tral heating system has remained unchanged. However, the output and distribution of heat has been upgraded. New radiators have been installed and positioned, as in the old building, below the windows.
The temperature of the forward and return flow water has been set as low as possible. The aim is to replace the old plant by an air-to-water heat pump at a later date and, in this way, make more efficient use of environmental heat. The waste heat from the refrigeration plant is already being used to generate heat.
After the refurbishment, the key figures for the heating system are as follows:

- forward flow water 50 °C
- return flow water 40 °C
- thermal output 60 kW
- distribution heat loss 10 kWh/m²
- storage heat loss 5 kWh/m²
- transfer heat loss 3 kWh/m²
- production loss 1.2 kWh/m²

Hot water supply
The hot water supply system has been retained. By using the waste heat from the recooling plant, the heat generation system is supported in an energy efficient way.

Ventilation
A mechanical ventilation system has been integrated into the building in order to supply fresh air and remove waste air in an energy efficient manner and without causing noise pollution.
The central ventilation system is positioned on the building's roof. In winter, the required volume of fresh air is preheated in a waste heat recovery unit; in summer, it is precooled by a refrigeration system and supplied to the rooms at a maximum temperature of 26 °C. Vertical shafts and horizontal air ducts in the suspended ceilings supply fresh air to the corridor zones of the individual storeys; waste air is extracted through ceiling vents alongside the facades. This arrangement provides for good distribution of fresh air and effective removal of heat (fig. 4.32).
The key figures for the ventilation system are as follows:

- air volume 14,000 m³/h
- air change rate 1.5–2/h
- electrical load approx. 5 kW (each for the air supply and exhaust fan)

Cooling
Cooling ceilings were fitted in the building to cool the rooms. The necessary cold is generated by a compression refrigeration machine in the plant room. It is only when the outside temperature is below 0 °C that the cold is generated through free cooling methods, which means directly in the recooling unit positioned on the building's roof. The recooling process is preformed using air.
Data for the refrigeration system:

- output 138 kW
- forward/return flow 10/18 °C

Lighting
The only rooms which have been equipped with fixed lamps are the ancillary rooms adjoining the main office space and the access areas. The office zones are illuminated exclusively with standard and desk lamps since it was not possible to incorporate a lighting system into the ceiling structure due to low room height.
The advantage of a decentralised lighting system is the almost unlimited flexibility in terms of position, luminosity and power consumption. The use of LEDs reduces the heat output in the rooms and therefore also the cooling loads significantly.
By improving the lighting system, it was possible to reduce the connected load from 25 W/m² to 10 W/m².

Electrical installations
The electrical installations were replaced throughout the building. A new raised floor system with floor boxes at predetermined distances provides the facilities to link all workplaces and easily connect the desk and standard lamps.

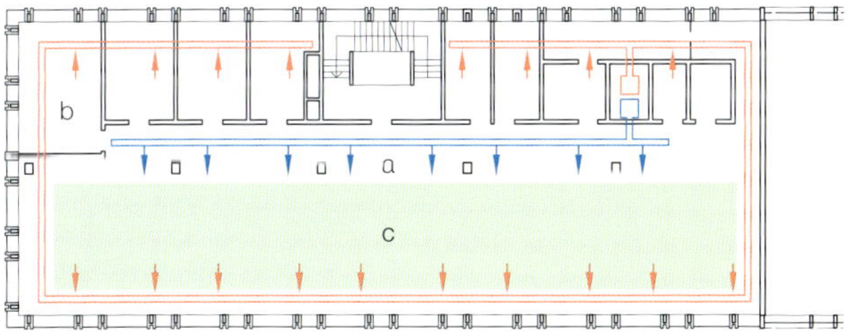

4.32

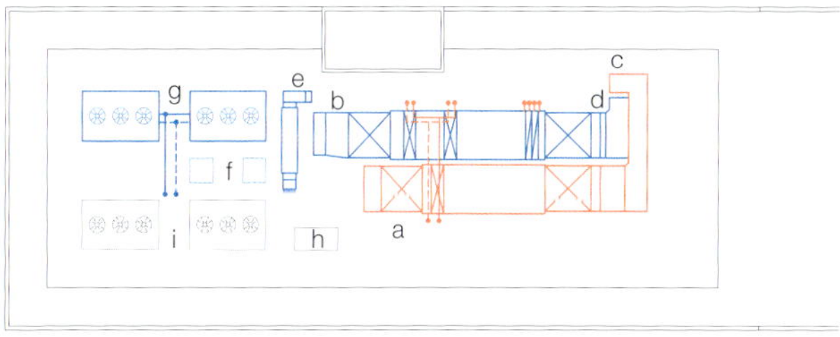

4.33

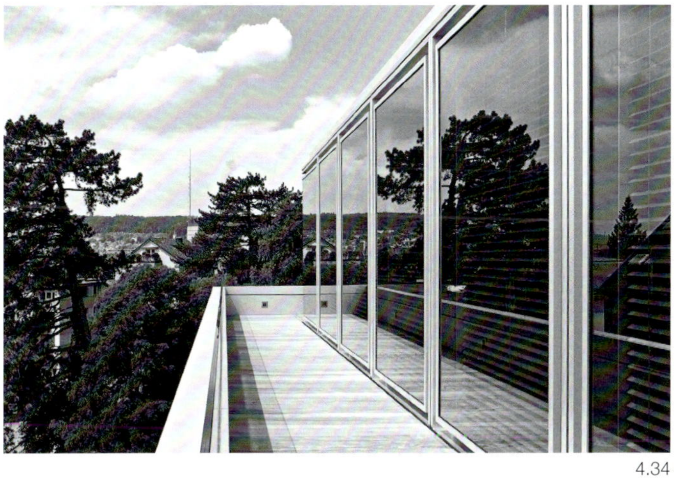
4.34

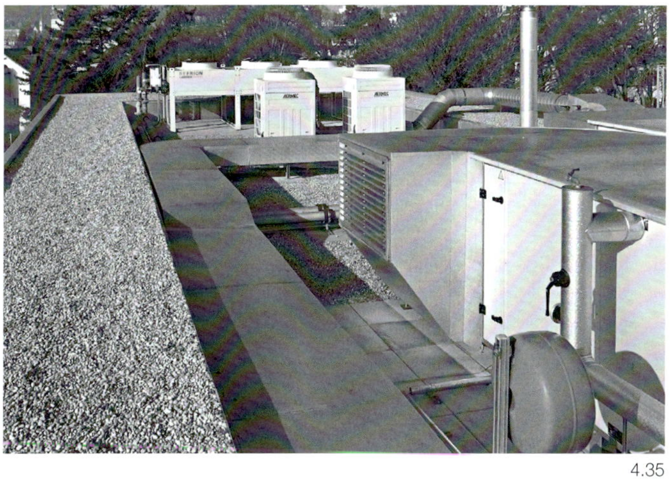
4.35

Accounting

In the case of energy efficiency refurbishments, it is always necessary to distinguish between expenditure for real refurbishment measures, which would have anyway been required from a maintenance point of view, and measures having the sole purpose of improving the condition of the property in terms of energy efficiency. Further expenses are usually incurred through structural and technical upgrades, which are intended to make better use of the site and/or improve comfort conditions in the building. Only the expenditure associated with energy-saving measures can be taken into account when considering the cost effectiveness of a refurbishment in an amortisation calculation, i.e. expenditure (investment) versus income (value of energy savings).

The measures described amounted to the expenditure presented in figure 4.36.

Ecology

The ecological standard of the building on completion of the refurbishment, or more accurately the ecological improvement, is defined as follows:

- primary energy demand before
  total 688,000 kWh
  specific (NFA) 275 kWh/m²
- primary energy demand after
  total 296,000 kWh
  specific (NFA) 118 kWh/m²
- CO_2 emissions before
  total 189,000 kg
  specific (NFA) 76 kg/m²
- CO_2 emissions after
  total 78,000 kg
  specific (NFA) 31 kg/m²

Economy

The final energy demand for the provision of heat (heating, hot water) has been reduced by approximately 90 % through the refurbishment. In order to improve thermal comfort conditions, the refurbishment was used as an opportunity to install new plant for cooling and ventilation. These new installations have increased the energy demand for electricity by 46,000 kWh. The roof extension was not included in the economic assessment of the refurbishment. Only the energy demand of the refurbished building stock is used as a standard for assessing the return resulting from the refurbishment's investment.

Based on an applied energy price of 60 ct/m³ for gas, 60 ct/l for oil and 13 ct/kWh for heat pump electricity, the total costs for the provision of heat after the refurbishment no longer amount to €30,000, but only to €4800 per year. The savings of €25,200 are balanced against the additional expenditure for energy-saving measures amounting to approximately €650,000.

If one considers a period of 20 years and assumes an average price increase of 6 % for the resources oil and electricity, the economic situation can be summarised as follows:

Additional energy efficiency expenses:
- investment €650,000
- grants/subsidies €0
- annuity of the investment (interest rate 3.5 %) €914,694
- average costs per year €47,735

Savings:
- savings per year €25,200
- price increase per year 6 %
- total savings €926,997
- average savings per year €46,349

Balance (rounded):
- average costs per year €48,000
- average savings per year €46,000
- amortisation 20 years

Measures such as the installation of the ventilation system with a heat recovery unit and the replacement of the lighting system should be evaluated as separate investments according to the same criteria. In these cases too, the principal reasons for the investment were maintenance (lighting system) and improvement of comfort conditions (ventilation, noise prevention). If further measures are planned in this context, also in favour of better energy efficiency standards, the additional expenditure and the savings achieved through the investment must be balanced both in terms of economic and ecological performance.

As presented here, the additional energy efficiency measures performed in the context of the refurbishment amortise in 15 to 20 years - even with an assumed annual energy price increase of 6 %. Because most of the investments made concern structural features, which have a life expectancy much greater than 20 years, it would actually be regarded as detrimental, from an economic point of view, not to perform the energy efficiency refurbishment.

Sustainability

Among other things, the refurbishment concept was designed to provide a flexible and therefore sustainable use of the office building. The work areas can be separated to allow a variety of office functions. It is also possible to provide separate access to each storey and thus let each floor to a different tenant. The newly fitted lift means that the building is now barrier-free. The new cafeteria created in the roof storey is directly accessible from the staircase and can therefore be enjoyed by all building users.

Furthermore, the inner-city property is perfectly located to make use of public transportation networks, which means that a further important criterion of sustainable building has been met.

Total construction costs, cost groups 2–7 €2,800,000					
proportion for construction-related measures €1,850,000			**proportion for building services measures** €950,000		
proportion for maintenance roof refurbishment window replacement facade insulation	**proportion for improved energy efficiency** extra cost for box windows extra cost for triple glazing extra cost for insulation extra cost for removal of parapets	**proportion for extensions** added storey to roof	**proportion for maintenance** heating replacement lighting	**proportion for improved energy efficiency** extra cost for air-to-water heat pump	**proportion for roof extension** cooling controlled ventilation with heat recovery system
€750,000	€600,000	€500,000	€150,000	€50,000	€750,000

4.36

	Unit	Building stock (1971)	Refurbishment	Extension (RF)	Total
Data					
heated volume (V_e)	m³	8500	8500	1000	9500
envelope surface area	m²	2500	2500	350	2850
total floor area	m²	–	–	–	–
heated net floor area (NFA)	m²	2500	2500	250	2750
EnEV floor area (0.32 × V_e)	m²	2700	2700	320	3020
compactness (heated NFA/heated volume)	1/m	0.29	0.29	0.25	0.28
Passive concept					
Winter					
roof	W/m²K	0.6	0.20	0.20	
wall adjoining outside air	W/m²K	1.0	0.20	0.20	
wall adjoining ground	W/m²K	0.6	–	–	
wall adjoining unheated rooms	W/m²K	–	0.17	–	
windows	W/m²K	2.70	0.6	0.8	
floor adjoining cellar, ground, air	W/m²K	1.0	1.0	–	
thermal bridges (individual verification method)	W/m²K	0.10	0.03	0.03	0.03
Summer					
window area	m²	790	790	190	980
window ratio (in relation to the facade surface area)	%	11	15	50	30
g value of windows	%	80	40	60	
F_c sun shading		0.25	0.25	0.25	
storage mass		complicated	simple	simple	
nighttime ventilation		not possible	possible (system)	possible (system)	
Daylight (DIN V 18599)					
area supplied with daylight (in % of the NFA)	%	70	70	100	75
daylight incidence through windows	%	80	40	70	
Ventilation					
air volume	m³	7500	6700	800	7500
area requiring ventilation	m²	2500	2500	350	2850
air change rate when occupied	–	1–6	2	2	2
proportion of naturally ventilated area	%	100	0	0	0
Active concept (final energy)					
Heating					
energy medium	–	oil	air-to-water heat pump	air-to-water heat pump	air-to-water heat pump
distribution	–	radiators	radiators	radiators	radiators
final energy	kWh/a	500,000	37,000 (electricity)	3000 (electricity)	33,000 (electricity)
Domestic hot water					
energy medium	–	oil	air-to-water heat pump	air-to-water heat pump	air-to-water heat pump
final energy	kWh/a	ns	ns	ns	ns
Cooling					
energy medium	–	not cooled	electricity	electricity	electricity
distribution	–	–	cooling ceiling/water	cooling ceiling/water	cooling ceiling/water
final energy	kWh/a	–	19,000	4500	23,000
Ventilation					
system	–	natural ventilation	ventilation with HR	ventilation with HR	ventilation with HR
final energy	kWh/a	–	27,000	4000	31,000
Lighting					
system	–	direct	direct/indirect	direct/indirect	direct/indirect
connected load	W/m²	25	10	10	10
final energy	kWh/a	53,000	31,000	2000	33,000
Balance					
Standard of energy efficiency					
EnEV 2009	–	–	New Build – 15 %	New Build – 15 %	New Build – 15 %
EEWärmeG	–	–	fulfilled	fulfilled	fulfilled
KfW	–	–	–	–	–
Economy					
annuity (20 years, 3.5 % interest rate)	€/year		48,000		
average savings in 20 years	€/year		46,000		
amortisation	years		20		
Ecology					
primary energy	kWh/a	688,000	296,000	35,000	331,000
CO_2	t/a	189	78	9	82

4.37

4.36 Allocation of costs for the full refurbishment
4.37 Project data relevant to the refurbishment

Authors

Prof. Clemens Richarz
1954 born in Freiburg/Breisgau
1973–1980 studied architecture at the Eidgenössische Hochschule Zürich
1980–1983 employment as architect
1983–1988 research assistant at the University of Stuttgart in the Institute for Building Construction and Design under Prof. Dr. h. c. Peter C. von Seidlein
1988–1990 participation in the research project "Industriebau der Zukunft" at the University of Stuttgart, Institute for Design and Construction, under Prof. Dr. h. c. mult. Kurt Ackermann
since 1993 professor at Hochschule München, faculty of architecture, for the subjects building construction and climate design
since 1983 self employed
since 1990 cooperation with Christina Schulz

Numerous prizes for competitions and completed projects
Member of the Chamber of Architects since 1983
Energy consultant (BAFA) for residential and non-residential buildings, ZV EnEV surveyor
Author and co-author of various publications (books, professional journals)
Planning and execution of further training programmes at the Bavarian and Baden-Württemberg Chamber of Architects

Christina Schulz
1959 born in Bad Honnef
1978–1981 studied economics at the Berufsakademie Stuttgart
1981–1987 studied architecture at the University of Stuttgart
1987–1990 employment as architect
1990–1995 research assistant at the University of Stuttgart, Institute for Building Construction and Design, under Prof. Dr. h. c. Peter C. von Seidlein
since 1990 freelance work with Clemens Richarz

Numerous prizes for competitions and completed projects
Member of the Chamber of Architects since 1990
Energy consultant (BAFA) for residential buildings
Author and co-author of various publications (books, professional journals)
Editorial assistance for specialist books
Lecturer for building construction at Hochschule München

Literature (selection)

General
Bundesarbeitskreis Altbauerneuerung e.V. (pub.): Bauen im Bestand. Cologne 2009
Daniels, Klaus: Low-Tech, Light-Tech, High-Tech. Munich/Basel 1998
Daniels, Klaus: Technologie des ökologischen Bauens. Munich/Basel 1995
Giebeler, Georg et al.: Atlas Sanierung. Munich/Basel 2008
Glücklich, Detlef (pub.): Ökologisches Bauen. Munich 2005
Gonzalo, Roberto; Habermann, Karl J.: Energieeffiziente Architektur. Munich/Basel 2006
Gore, Al: Eine unbequeme Wahrheit. Munich 2006
Hausladen, Gerhard et al.: ClimaDesign. Munich 2005.
Hegger, Manfred et al.: Energie Atlas. Munich/Basel 2008
Meadows, Dennis; Meadows, Donella; Zahn, Erich: Die Grenzen des Wachstums. Munich 1972
Pehnt, Martin: Energieeffizienz. Heidelberg 2010
Schmidt-Bleek, Friedrich: Das MIPS Konzept – Weniger Naturverbrauch – mehr Lebensqualität durch Faktor 10. Munich 2000
Voss, Karsten et al.: Bürogebäude mit Zukunft. Cologne 2005
Ziegler, Jean: Der Kampf gegen Armut und Unterdrückung. Munich 2008

Principles
Energy demand
Bockhorst, Michael: ABC Energie. Bonn 2002

Thermal comfort
Grandjean, Etienne: Wohnphysiologie – Grundlagen gesunden Wohnens. Zurich 1973
Meyer, Christoph; Oppermann, Jens; Wimmer, Andreas: Behaglichkeitsatlas – Bewertung der Anordnung von Heizflächen und Lüftungselementen hinsichtlich Raumklima und thermischer Behaglichkeit. Stuttgart 2004

Interior climate
Arbeitsgemeinschaft ökologischer Forschungsinstitute (AGÖF): Innenraumschadstoffe, Fogging und Gerüche; Results from the 8th AGÖF conference on 19/20 Septembre 2007 in Fürth
Schulze-Darup, Burkhard (pub.): Passivhaus-Projektbericht: Energie und Raumluftqualität. Fürth 2002

Exterior climate
Hupfer, Peter; Kuttler, Wilhelm (pub.): Witterung und Klima. Wiesbaden 2006

Design parameters
Ministry for Transport, Building and Housing; Federal Authority for Building and Regional Planning (pub.): Leitfaden Nachhaltiges Bauen. Berlin/Bonn 2001
Federal Ministry for Transport, Building and Urban Development (pub.): Leitfaden Nachhaltiges Bauen. Berlin 2011
Ebert, Thilo; Eßig, Natalie; Hauser, Gerd: Zertifizierungssysteme für Gebäude. Munich 2010
König, Holger et al.: Lebenszyklusanalyse in der Gebäudeplanung. Munich 2009

Economy
Keller, Siegbert: Baukostenplanung für Architekten. Wiesbaden 1995
Schulte, Karl-Werner: Immobilienökonomie. Volume 1: Betriebswirtschaftliche Grundlagen, Volume 2: Rechtliche Grundlagen, Volume 3: Stadtplanerische Grundlagen, Volume 4: Volkswirtschaftliche Grundlagen. Munich 2005–2008

Energy accounting
Federal Ministry for Transport, Building and Urban Development: Leitfaden für Energiebedarfsausweise im Nichtwohnungsbau. Berlin 2007
Rainer, Dirk: Energieeinsparverordnung Schritt für Schritt. 5th edition. Neuwied 2010
Schoch, Torsten: EnEV 2009. 2nd edition. Berlin 2009
Wichtermann, Karl-Heinz: Praxis der Gebäudeenergieberatung – EnEV 2009. Bad Nenndorf 2010

Construction-related measures
General
Liersch, Klaus W.; Langner, Normen: Bauphysik kompakt. Berlin 2008
Richter, Ekkehard et al.: Lehrbuch der Bauphysik: Schall–Wärme–Feuchte–Licht–Brand–Klima. Wiesbaden 2008
Willems, Wolfgang M.; Dinter, Simone; Schild, Kai: Vieweg Handbuch Bauphysik Teil 1: Wärme- und Feuchteschutz, Behaglichkeit, Lüftung. Wiesbaden 2006
Zürcher, Christoph; Frank, Thomas: Bauphysik – Bau und Energie, Volume 2. Zurich 2004

Heat sinks
Hauser, Gerd; Stengel, Horst: Wärmebrückenatlas für den Mauerwerksbau. Wiesbaden 2002
Institut für Bauforschung e.V. Hannover (pub.): U-Werte alter Bauteile. Stuttgart 2005
Pfundstein, Margit et al.: Dämmstoffe. Munich 2007
Schittich, Christian (pub.): Solares Bauen. Munich/Basel 2003
Volland, Karlheinz; Volland, Johannes: Wärmeschutz und Energiebedarf nach EnEV 2009. Cologne 2009

Heat sources
Fraunhofer Solar Building Innovation Center (pub.): Sonnenschutz. Stuttgart 2008

Ventilation
Fraunhofer Institute for Building Physics (pub.): IBP Bericht WTB-0-2007: Further development and evaluation of technologies and assessment methods to increase the overall efficiency of buildings (final report EnEff06). Stuttgart 2008

Daylight
Köster, Helmut: Tageslichtdynamische Architektur. Grundlagen, Systeme, Projekte. Munich/Basel 2004

Building services measures
General
Daniels, Klaus: Gebäudetechnik. Ein Leitfaden für Architekten und Ingenieure. Munich 1992
Daniels, Klaus; Hamman, Ralph: Energy Design for Tomorrow. Fellbach 2009
Hayner, Michael; Ruoff, Jo; Thiel, Dieter: Faustformel Gebäudetechnik für Architekten. Stuttgart 2010
Lenz, Bernhard; Schreiber, Jürgen; Stark, Thomas: Nachhaltige Gebäudetechnik. Munich 2010
Pistohl, Wolfram: Handbuch der Gebäudetechnik. Volume 1 – Sanitär/Elektro/Förderanlagen. Neuwied 2009
Volume 2 – Heizung/Lüftung/Beleuchtung/Energiesparen. Neuwied 2009
Quaschning, Volker: Erneuerbare Energien. Munich 2008
Quaschning, Volker: Regenerative Energiesysteme. Technologie – Berechnung – Simulation. Munich 2007
Recknagel, Hermann; Sprenger, Eberhard; Schramek, Ernst-Rudolf: Taschenbuch für Heizung und Klimatechnik 11/12. Munich 2010
Reiß, Johann et al.: Solare Fassadensysteme. Energetische Effizienz – Kosten – Wirtschaftlichkeit. Stuttgart 2005.

Heating, domestic hot water
Haller, Andreas; Humm, Othmar; Voss, Karsten: Renovieren mit der Sonne. Staufen 2000

Ventilation
Horstkotte, Klaus; Fritzsche, Nico: Taschenbuch für Lüftungsmonteure und -meister. Heidelberg 2011
Richter, Wolfgang et al.: IRB – Fraunhofer Institute for Building Physics (pub.): Bewertung von dezentralen raumweisen Lüftungsgeräten für Wohngebäude sowie Bestimmung von Aufwandszahlen für die Wärmeübergabe im Raum infolge Sanierungsmaßnahmen. Stuttgart 2004
Trogisch, Achim: Planungshilfen für Lüftungstechnik. Heidelberg 2009

Cooling
Ranft, Fred; Frohn, Bernhard: Natürliche Klimatisierung. Munich/Basel 2004

Artificial light
Brandi, Ulrike et al.: Tageslicht – Kunstlicht. Munich 2005
Ganslandt, Rüdiger; Hofmann, Harald: Handbuch der Lichtplanung. Wiesbaden 1992

Photovoltaics
Weller, Bernhard et al.: Photovoltaik. Munich 2009

Standards and guidelines

Principles
Thermal comfort
DIN EN 15251: 2007-08 Indoor environmental input parameters for design and assessment of energy performance of buildings addressing indoor air quality, thermal environment, lighting and acoustics
DIN EN ISO 7730: 2006-05 Ergonomics of the thermal environment - analytical determination and interpretation of thermal comfort using calculation of the PMV and PPD indices and local thermal comfort criteria
DIN EN 14500: 2008-08 Blinds and shutters – thermal and visual comfort – test and calculation methods
DIN 14501: 2006-02 Blinds and shutters – thermal and visual comfort – performance characteristics and classification

Interior climate
DIN EN 15251: 2007-08 Indoor environmental input parameters for design and assessment of energy performance of buildings addressing indoor air quality, thermal environment, lighting and acoustics
DIN ISO 16000 Indoor air
Parts 1–23 determine the quality of indoor air. Selection:
Part 8: Determination of local mean ages of air in buildings for characterizing ventilation conditions
Part 16: 2009-12 Detection and enumeration of mould – sampling by filtration

Design parameters, compactness
VDI 2050 Sheet 1: 2006-12 Requirements at technique centres – technical principles for planning and execution

Appropriateness
DIN EN 15643 Sustainability of construction works – sustainability assessment of buildings
Part 1: 2010-12 General framework
Part 2: 2011-05 Framework for the assessment of environmental performance
Part 3: 2010-04 Framework for the assessment of social performance
Part 4: 2010-04 Framework for the assessment of economic performance
DIN EN 15942: 2009-05 Sustainability of construction works - environmental product declarations – communication format from business-to-business

Economy
DIN 276 Building costs
Part 1: 2008-12 Building construction
DIN 18960: 2008-02 User costs of buildings
VDI 2067, sheet 1-22 Economic efficiency of building installations

Energy accounting
DIN V 18599: 2007-02 Energy efficiency of buildings – calculation of the net, final and primary energy demand for heating, cooling, ventilation, domestic hot water and lighting
Part 1: General balancing procedures, terms and definitions, zoning and evaluation of energy sources
Part 2: Net energy demand for heating and cooling of building zones
Part 3: Net energy demand for air conditioning
Part 4: Net and final energy demand for lighting
Part 5: Final energy demand of heating systems
Part 6: Final energy demand of ventilation systems and air heating systems for residential buildings
Part 7: Final energy demand of air-handling and air-conditioning systems for non-residential buildings
Part 8: Net and final energy demand of domestic hot water systems
Part 9: Final and primary energy demand of power generation plants
Part 10: Boundary conditions of use, climatic data
Energy Saving Ordinance (EnEV) (technical regulation concerning the energy-saving insulation and energy-saving technical plant in buildings)
Publications of the Federal Ministry for Transport, Building and Urban Development from 30 July 2009
– Rules for energy consumption indicators in residential buildings
– Rules for energy consumption indicators in non-residential buildings
– Rules for data recording and use in residential building stock
– Rules for data recording and use in non-residential building stock
Laws for the promotion of renewable energies in terms of heat generation 2011 (EEWärmeG)

Construction-related measures
Heat sinks
DIN 4108 Thermal protection and energy economy in buildings
Part 2: 2003-07 Minimum requirements to thermal insulation
Part 3: 2001-07/corrected: 2002-04 Protection against moisture subject to climate conditions; requirements and directions for design and construction
Part 4: 2007-06 Hygrothermal design values
Part 6: 2003-06 Calculation of annual heat and energy use
Part 7: 2001-08 Air tightness of buildings – requirements, recommendations and examples for planning and performance
Supplement 2: 2006-03 Thermal bridges – examples for planning and performance
DIN EN ISO 6946: 2008-04 Building components and building elements – thermal resistance and thermal transmittance – calculation method
DIN EN ISO 10077
Thermal performance of windows, doors and shutters – calculation of thermal transmittance
Part 1: 2010-05 General
Part 2: 2009-07 Numerical method for frames
DIN EN ISO 10211: 2008-04 Thermal bridges in building construction – detailed calculations
DIN EN ISO 13370: 2008-04 Thermal performance of buildings – heat transfer via the ground
DIN EN ISO 13789: 2008-04 Thermal performance of buildings – transmission and ventilation heat transfer coefficients – calculation method

Heat sources
DIN 4108 Part 2: 2003-07
Minimum requirements for thermal insulation
DIN EN 410: 1998-12 Glass in building – determination of luminous and solar characteristics of glazing
DIN EN 12207: 2000-06 Windows and doors – air permeability – classification
DIN EN 14351 Part 1: 2009-05 Windows and doors – product standards, performance characteristics – windows and external pedestrian doorsets without any requirements concerning fire and/or smoke tightness

Ventilation
DIN EN 13829: 2001-02 Thermal performance of buildings – determination of air permeability of buildings – fan pressurization method

Daylight
DIN 5034 Daylight in interiors
Part 1: 2011-07 General requirements
Part 2: 1985-02 Principles
Part 3: 2007-02 Calculation
Part 4: 1994-09 Simplified determination of minimum window sizes for dwellings
Part 5: 2010-11 Measurement
Part 6: 2007-02 Simplified determination of suitable dimensions for rooflights
DIN 5036 Parts 1–4: Radiometric and photometric properties of materials
DIN 5039: 1995-09 Light, lamps, luminaires – definitions, categories
VDI 6011, Sheet 1: 2002-08 Optimisation of daylighting and artificial lighting – Fundamentals

Building services measures
General
DIN EN 15232: 2007-11 Energy performance of buildings – Impact of Building Automation, Controls and Building Management

Heating, domestic hot water
DIN V 4701 Energy efficiency of heating and ventilation systems in buildings
Part 12: 2004-02 Heat generation and domestic hot water generation
DIN EN 832: 2003-06 Thermal performance of buildings – calculation of heating energy requirement, residential buildings

DIN EN 12831: 2003-08 Heating systems in buildings – method for calculating the design heat load Supplement 1: National Annex NA
DIN EN 12975 Thermal solar systems and components - solar collectors
Part 1: 2011-08 General requirements
DIN EN 15377 Part 1–3 Heating systems in buildings – design of embedded water based surface heating and cooling systems
VDI 4650 Sheet 1: 2009-03 Calculation of heat pumps – simplified method for the calculation of the seasonal performance factor of heat pumps – electric heat pumps for space heating and domestic hot water
Sheet 2: 2010-11 Simplified method for the calculation of the annual coefficient of performance and the annual utilisation ratio of sorption heat pumps – gas heat pumps for space heating and domestic hot water – electric heat pumps for space heating and domestic hot water
VDI 6012 Integration of decentralised and regenerative energy systems in buildings
Sheet 1: 2010-09 Fundamentals and energy storage
Sheet 3: Fuel cells

Cooling
VDI 2078: 1996-07 Cooling load calculation of air-conditioned rooms
VDI 2078, Sheet 1: 2003-02 Cooling load calculation of air-conditioned buildings with room-conditioning from cooled walls and ceilings

Ventilation
DIN EN 12792: 2004-05 Ventilation of buildings - symbols, terminology and graphical symbols
DIN EN 13779: 2007-09 Ventilation for non-residential buildings, performance requirements for ventilation and room-conditioning systems
DIN 1946-6:2009-05 Ventilation and air conditioning Part 6: Ventilation of residential buildings – general requirements, requirements for measuring, performance and labelling, delivery/acceptance (certification) and maintenance
VDI 2071: 1997-12 Heat recovery in heating, ventilation and air conditioning plants
VDI 3803: 2010-02, Sheet 1: 2010-02 Air-conditioning – central air-conditioning systems. Structural and technical requirements (VDI ventilation code of practice)
VDI 3804: 2009-03 Air-conditioning – office buildings (VDI ventilation code of practice)
VDI 4706: 2011-04 Criteria for indoor air climate (VDI ventilation code of practice)
VDI 6035: 2009-09 Ventilation and air-conditioning technology – decentralized ventilation systems – wall-mounted air-conditioners (VDI ventilation code of practice)
VDI 6040, Sheet 1: 2010-05 (draft) Air-conditioning – schools – requirements

Artificial light
DIN 5035 Artificial lighting
Part 6: 2006-11 Measurement and evaluation
DIN 6169 Colour rendering
Part 1: 1976-01 General terms
Part 2: 1976-02 Colour rendering properties of light sources in the field of lighting
DIN EN 12464 Light and lighting – lighting of work places
Part 1: 2011-08 Indoor work places
Part 2: 2007-10 Outdoor work places
DIN EN 12665: 2009-08 Light and lighting – basic terms and criteria for specifying lighting requirements

Photovoltaics
DIN EN 50461: 2007-03 Solar cells – data sheet information and product data for crystalline silicon solar cells
DIN EN 60904 Photovoltaic devices
DIN EN 61646: 2009-03 Thin-film terrestrial photovoltaic (PV) modules – design qualification and type approval
DIN VDE 0126-21: 2007-07 (E) Photovoltaics in building

Links (selection)

General information
www.bmvbs.de
homepage of the Federal Ministry for Transport, Building and Urban Affairs provides up-to-date information on changes to legal requirements and standards
www.energie-info.de
database for energy-related matters with an excellent glossary
www.thema-energie.de
database of the German Energy Agency (dena) with a good structure of different topics
www.bine.info
reports on structural and technical building work using innovative methods
www.ea-nrw.de
homepage of the Energy Agency North Rhine-Westphalia with a large range information concerning the subject energy
www.umweltbundesamt.de
homepage of the Federal Environment Agency
www.enbausa.de
newsportal for energy efficiency developments and refurbishments

Information on the subject environment
www.greenpeace.de and www.footprint.at
holistic approach to environmental issues, ecological footprint
www.zae-bayern.de
information on new research concerning the field of low-resource construction work
www.enob.de
research portal for low energy building including numerous examples of completed projects
www.wuppertal-institut.de
general information concerning the consumption of energy and resources. Ecological rucksack of (building) materials.
www.klimabuendnis.org
information concerning climate change
www.gemis.de
determination and assessment of exhaust emissions that are generated by the consumption of resources
www.energiekrise.de
tables and diagrams concerning the finite nature of resources
www.deutschebp.de
background information on the global consumption of energy
www.energie-info.de
information on the subject of energy; keyword search in the glossary is very good
www.atmosfair.de
compensation for the environmental damage of air travel by supporting ecological projects
www.klima-sucht-schutz.de
how-to manual on energy saving and climate protection with many specific examples and suggestions
www.architektur.tu-darmstadt.de/powerhouse
website provided by the Department for Design and Energy Efficient Construction at TU Darmstadt with definitions (approx. 800 terms) and project documentations

Principles
Energy demand
www.ise.fhg.de
Fraunhofer Institute for Solar Energy Systems
www.solarserver.de
Internet portal solar energy, solar thermal energy, photovoltaics and solar building
www.eurosolar.de
European Association for Renewable Energy
www.erneuerbareenergien.de
website of the corresponding journal for renewable energies
www.bsw-solar.de
Federal Association of the Solar Industry
www.unendlich-viel-energie.de
Renewable Energies Agency
www.zsw-bw.de
Centre for Solar Energy and Hydrogen Research Baden-Württemberg
www.erneuerbare-energien.de
information portal of the Federal Ministry for Renewable Energy
www.sonnewindwaerme.de
information portal of the corresponding journal for renewable energies

Thermal comfort
www.ibp.fraunhofer.de
Fraunhofer Institute for Building Physics (IBP)

Interior climate
www.anbus.de
information to analyse and assess environmental pollutants
www.lfu.bayern.de
website of the Bavarian Environmental Protection Agency (LfU) with data and information on climate change, air pollution control, environmental management, etc.

Exterior climate
www.klimadiagramme.de
information on climate
www.dwd.de
link to the homepage of the German weather service
www.volker-quaschning.de
solar altitude diagrams, climate data, information or renewable energies and climate protection

Design parameters
www.enob.info
information on the subsidy programme "Energieoptimiertes Bauen" (EnOB) sponsored by the German Federal Ministry of Economics and Technology
www.eneff-schule.de
information on the research project "Energieeffiziente Schule" (EnEff:Schule), accompanying project of EnOB

Appropriateness
www.dgnb.de
German Sustainable Building Council (DGNB), awards the DGNB certificate
www.usgbc.org
LEED (Leadership in Energy and Environmental Design) from USA with a clear definition of the term sustainability
www.intep.de
engineering practice founded by Prof. Peter Steiger specialised in the determination and assessment of materials cycles and ecological building issues
www.nachhaltigesbauen.de
information portal for sustainable building provided by the Federal Ministry for Transport, Building and Urban Affairs with guidelines (e.g. guideline for sustainable building) and online database of building materials (e.g. Ökobau.dat, WECOBIS, durability of building materials)

Economy
www.destatis.de
information portal of the Federal Statistical Office of Germany

Energy accounting
www.enev-online.de
up-to-date information on the German Energy Savings Act and corresponding topics (Heat Act, Energy Performance of Buildings Directive)

Construction-related measures
Heat sinks
www.ift-rosenheim.de
information on windows
www.ivh.de
information on rigid foam board
www.vip-bau.de
information on vacuum insulation materials
www.fvhf.de
information on non-load bearing, rear-ventilated facades

Heat sources
www.baunetzwissen.de
online encyclopedia among other things for issues related to sun shading, windows and facades
www.sonnenschutz.com
information on sun shading devices

Ventilation
www.wohnungslueftung-ev.de
comprehensive, up-to-date information on the subject of ventilation, Excel tools for DIN 1946-6
www.komfortlueftung.at
information on the subject of ventilation and energy efficiency

Daylight
www.fvlr.de
Association of daylight and smoke protection
www.thedaylightsite.com
literature and research results concerning daylight
www.baunetz.de/infoline/licht
information on artificial and natural light

Building services measures
Heating and cooling
www.bhks.de
Federal Industrial Association for Heating, Air Conditioning and Sanitary Engineering
www.waermepumpe.de
information on heat pumps
www.solarwaerme.at
company-independent information on solar thermal plants

Ventilation
see Structural measures, ventilation

Artificial light
www.licht.de
cooperative organisation Fördergemeinschaft Gutes Licht
www.baunetz.de/infoline/licht
information on artificial and natural light

Photovoltaics
www.solarwirtschaft.de
Federal Solar Economy Association
www.erneuerbare-energien.de
information portal of the Federal Ministry for Renewable Energies
http://re.jrc.ec.europa.eu/pvgis/
geographic information system to simulate photovoltaic plants
www.ise.fraunhofer.de
Fraunhofer Institute for Solar Energy Systems (ISE)

Illustration credits

The authors and editor wish to extend their sincere thanks to all those who helped realise this book by making illustrations available, granting permission to reproduce them and supplying information.
All drawings contained in this volume have been specially prepared in-house. Photos without credits are from the architects' own archives or the archives of DETAIL. Despite intense efforts, it was not possible to identify the copyright owners of all photos and illustrations. However, their rights remain unaffected, and we request them to contact us. The numbers refer to the figures in the text.

Principles
1.1 Deutsche Stiftung Weltbevölkerung: DSW-Datenreport 2009, Hannover
1.2 www.globalwarmingart.com
1.4 according to Jacobson, Mark Z.; DeLucchi, Mark A.: "Emissionsfreie Welt bis 2030" In: Spektrum der Wissenschaft 12/2009, pp. 82 f.
1.5 Internationales Wirtschaftsforum Regenerative Energien (IWR), Münster
1.6 Bundesanstalt für Geowissenschaften und Rohstoffe (BGR), Hannover
1.8 BP Review, www.bp.com
1.9 Institut für Wohnen und Umwelt, Darmstadt
1.11 DIN 4108-2: 2003-07
1.12–1.16 DIN EN ISO 7730
1.17 Institut für Angewandte Umweltforschung e.V., Oberursel
1.18 Umweltbundesamt und Arbeitsgemeinschaft der Obersten Landesgesundheitsbehörden (AOLG)
1.19 Umweltbundesamt (Innenraumlufthygiene-Kommission/IRK), Dessau-Roßlau
1.20 World Health Organization (WHO), Geneva
1.21, 1.22 http://koeppen-geiger.vu-wien.ac.at/
1.23 according to Deutsches Zentrum für Luft- und Raumfahrt e. V. (DLR), Cologne
1.24 Climate data according to: www.wetteronline.de, www.klimadiagramme.de, http://re.jrc.ec.europa.eu/pvgis and www.geo-reise-community.de
1.25 DIN 18599: 2007-02
1.26 Christian Kandzia, Esslingen
1.27 DIN V 18599-10, table A4
1.29a Frank Kaltenbach, Munich
1.29b Bruchhaus/Lachenmann, Munich
1.30 Jean-Luc Valentin, Frankfurt am Main
1.33 Richarz und Strunz, Munich
1.34 Deutsche Gesellschaft für Nachhaltiges Bauen e.V. – DGNB (pub.): Das Deutsche Gütesiegel Nachhaltiges Bauen. Stuttgart 2009, p. 11
1.35, 1.36 Baukosteninformationszentrum Deutscher Architektenkammern GmbH (BKI), Stuttgart
1.38 Freelunch.com/Moody`s
1.39 Federal Ministry for Transport, Building and Urban Development : Leitfaden Nachhaltiges Bauen. Berlin 1/2001
1.40 Datengrundlage auf: Statistisches Bundesamt, Fachserie 17 Reihe 7, Verbraucherpreisindizes für Deutschland. Wiesbaden
1.41 DIN 18960
1.42 DIN 276-1

Construction-related measures
2.1 DIN 4108-2: 2003-07, table 3
2.2 Oberacker, Reiner: Die Nebel lichten sich, In: Glaswelt 3/2004, pp. 28 ff.
2.3 DIN EN ISO 6946
2.4 DIN 4108-2: 2003-07
2.5 DIN 4108-2: 2003-07 and DIN EN ISO 6946
2.6 DIN 4108-2: 2003-07, table 4
2.8 DIN 4108, supplement 2: 2006-03 (selection)
2.9 according to DIN V 18599-2: 2007-02
2.11 German Energy Agency (dena), Berlin
2.33 DIN 4108-2
2.34 DIN 4108-2: 2003-07
2.35 Glas Trösch Beratungs-GmbH, Ulm
2.36 DIN 4108-2, table 7
2.37c Heinrich Helfenstein, Zurich
2.38 DIN 4108-2, table 8
2.39 DIN 4108-2, table 9
2.40 Zürcher, Christoph; Frank, Thomas: Bauphysik. Zurich 2004
2.44 DIN 1946-6, table 10
2.45 DIN 1946-6
2.46b Fritz Haller Bauen und Forschen GmbH, Solothurn
2.46c Reynaers AG, Frauenfeld
2.47 DIN V 18599-10:2007-02, table 4
2.48 German workplace ordinance (ASR) 5
2.49–2.52 Brandi, Ulrike: Tageslicht – Kunstlicht. Munich 2005, pp. 42ff.
2.53 DIN 5034-1:1999-10, appendix A
2.54 DIN V 18599-10: 2007-02, table 4
2.55 VELUX Deutschland GmbH, Hamburg
2.56 see 2.49, p. 43
2.61 see 2.49. p. 25
2.62 DIN V 18599-4, table 6
2.63 DIN V 18599-4, table 9 (extract)
2.64c Markus Weidlich, Weiden

Building services measures
3.1 DIN 18012
3.2 VDI 2050, sheet 1
3.3 VDI 3803
3.8 Pistohl, Wolfram: Handbuch der Gebäudetechnik. Volume 2 – Heizung/Lüftung/Beleuchtung/Energiesparen. Neuwied 2009
3.9 Quaschning, Volker: Erneuerbare Energien und Klimaschutz. Munich 2008
3.12 according to EnEV, appendix 5, table 1
3.14, 3.15 Federal Ministry for Transport, Building and Urban Affairs: Bekanntmachung der Regeln zur Datenaufnahme und Datenverwendung im Wohngebäudebestand. Berlin, 30 July 2009
3.17 Pistohl, Wolfram: Handbuch der Gebäudetechnik, Volume 1. Munich 2009
3.21 SikoSolar Vertriebs GmbH, Jenbach (A)
3.22 Viessmann Werke GmbH, Allendorf (Eder)
3.27 www.thema-energie.de/German Energy Agency (dena), Berlin
3.31 DIN V 18599-7, table 20 and 22
3.32 Lenz, Bernhard; Schreiber, Jürgen; Stark, Thomas: Nachhaltige Gebäudetechnik. Munich 2010, p. 99
3.40 according to Wolf GmbH, Mainburg
3.43 DIN V 18599-7
3.44 DIN V 18599-3 (selection)
3.46a, b see 3.32, p. 97
3.46c Kiefer Luft- und Klimatechnik, Stuttgart
3.50a Nimbus Design GmbH, Stuttgart
3.50b Jürgen Landes
3.50c Behnisch Architekten
3.50d Roland Halbe, Stuttgart
3.51 Osram GmbH, Munich
3.52 DIN V 18599-4
3.54–3.58 Weller, Bernhard et al.: Photovoltaik. Munich 2009, pp. 11, 14, 22, 30 ff., 50
3.59 Bruno Klomfar, Vienna
3.60 see 3.54, p. 30

Project analyses
4.1, 4.2, 4.8, 4.12, 4.13 Richarz und Schulz, Munich
4.21 Max Dudler Architekten, Zurich
4.25, 4.30, 4.34, 4.35 Wehrli Müller Fotografen, Regensdorf

Index

Sample calculations

Project examples